ЧИСЛЕННЫЕ МЕТОДЫ РАСЧЕТА ЯДЕРНЫХ РЕАКТОРОВ

Chislennye Metody Rascheta Yadernykh Reaktorov

NUMERICAL METHODS FOR NUCLEAR REACTOR CALCULATIONS

NUMERICAL METHODS FOR NUCLEAR REACTOR CALCULATIONS

Gurii Ivanovich Marchuk

Supplement Nos. 3-4 of the Soviet Journal of Atomic Energy
ATOMNAYA ENERGIYA • Atomic Press, Moscow, 1958

TRANSLATED FROM RUSSIAN

Springer Science+Business Media, B.V. 1959

First Printing – July 1959
Second Printing – August 1962
Third Printing – February 1966

ISBN 978-1-4899-4892-2 ISBN 978-1-4899-4890-8 (eBook)
DOI 10.1007/978-1-4899-4890-8

Library of Congress Catalog Card Number: 59-9229

Originally published by Consultants Bureau Enterprises, Inc in 1959.
Softcover reprint of the hardcover 1st edition 1959

227 W. 17th St., New York 11, N. Y.

CONTENTS

PAGE

NOTE

This monograph is a discussion of the fundamental problems arising when using numerical methods for nuclear reactor calculations. Particular attention is devoted to various practical applications. A prerequisite to the understanding of this monograph is familiarity with basic reactor theory as presented, for instance, by Glasstone and Edlund [23] or Galanin [20]. The book is intended for physics graduate students, doctoral candidates, engineers, and scientific workers specializing in nuclear reactor calculations.

FOREWORD

This book is an attempt at a more or less systematic exposition of numerical methods for the calculation of thermal, intermediate, and fast neutron reactors. Particular attention is devoted to the problems of critical mass, the space-energy neutron flux distribution, and the neutron importance (iterated fission probability). The book gives effective methods for reducing the basic and adjoint reactor equations to a set of multi-group diffusion equations. These equations are then successfully solved by the method of difference factorization. Perturbation theory is used to calculate small effects. Considerable space is devoted to heterogeneous reactor calculations by effective homogenization methods. In treating fast-neutron reactors, particular attention is paid to numerical methods for solving the kinetic equations.

This book deals only with problems of a mathematical nature. Problems related to the general theory of nuclear reactors and to the physical interpretation of the results obtained are barely touched upon.

Since this is the first time many of these results have been published in detail, the book may contain some errors.

Numerical methods for reactor calculations were developed from 1953 to 1956 by the author together with the group of scientific co-workers he directed, consisting of E.I. Pogudalina, E. S. Maksimova, V.V. Smelov, G.A. Iliasova, I.P. Markelov, I.P. Tiuterev, and others. All of these methods were first tested on a large quantity of theoretical and experimental data and were later used in operation.

Numerical methods for nuclear reactor calculations were developed and perfected under the influence of the group of theoretical and experimental physicists directed by acting member of the Academy of Sciences UkrSSR A. I. Leipunskii.

Also of great influence was the work of L.N. Usachev and A.S. Romanovich on adjoint equations and perturbation theory.

In developing the numerical methods for reactor calculation, the author was constantly in scientific communication with a large number of scientific workers, both mathematicians and experimental and theoretical physicists. During the creative discussions of problems important from a practical point of view, new mathematical problems were uncovered, and methods for solving them were indicated. Particularly active in such discussions were L.N. Usachev, V.Ia. Pupko, V.A. Kuznetsov, B.F. Gromov, G.I. Toshinskii, N.I. Buleev, S.B. Shikhov, V.V. Orlov, T.Kh. Sedel'nikov, and others. The author is grateful to all these co-workers.

The author expresses his deep gratitude to A.I. Leipunskii for many discussions of theoretical problems and methods for nuclear reactor calculations, as well as for many remarks he made throughout the present work.

The author also acknowledges the aid of D.I. Blokhintsev and E.S. Kuznetsov, who discussed various aspects of the theory and methods of calculation at various stages in the development of the numerical methods.

Of great aid to the author in preparing the present monograph were his scientific co-workers Zh.N. Bel'skaia, N.P. Kochubei and A.I. Vaskin. To these friends the author also expresses his deep gratitude.

G. Marchuk

NOTATION

D	diffusion coefficient $= 1/3\Sigma_{tr}$.
(f)	vector function.
$f(\mu_0, u - u')$	the probability density that a neutron whose lethargy is u' and whose direction is $\vec{\Omega}'$ undergoes a collision which changes these to u and $\vec{\Omega}$ respectively.
G	the region of space occupied by the reactor.
J	resonance integral.
J_{eff}	effective resonance integral.
L	operator.
l	optical path of a neutron in a block (also used as an index).
l'	l/d, where d is the block thickness.
M	mass of the nucleus.
n	neutron density.
nv	neutron flux (also written φ).
P_l	Legendre polynomial of order l.
q	slowing-down density.
$Q(\vec{r})$	number of secondary neutrons per act of capture leading to fission.
$Q^*(\vec{r})$	fission neutron importance (iterated fission probability).
r	maximum lethargy loss, $\max(u - u')$.
$S(\vec{r}, \vec{\Omega}, u)$	neutron sources in the reactor.
S	external surface of the reactor.
S_e	extrapolated reactor surface.
T	index designating the thermal group for any quantity.
u	logarithmic energy decrement (lethargy), equal to $\ln E_0/E$, where E is the neutron energy, and E_0 is a constant energy.
$\vec{v}$	neutron velocity vector (also written $\vec{\Omega}v$).
v	neutron velocity $= \lvert\vec{v}\rvert$.
V	the volume of an absorbing block.
V_0	volume of moderator in a cell.
V_{cell}	the volume of the Wigner-Seitz cell.

Y_{lm}	spherical function.
γ	the parameter $\overline{\xi^2}/2\xi$.
Γ^j	resonance width for the energy E_j.
ϵ	the parameter $\overline{\xi^2}/2\xi^2$.
λ	mean free path $= 1/\Sigma$.
μ	the cosine of the angle between the neutron direction and the z axis.
μ_0	the cosine of the scattering angle in the laboratory coordinate system.
$\overline{\mu}_0$	the mean scattering-angle cosine.
νf	effective neutron yield per fission.
ξ	mean logarithmic energy loss in elastic scattering.
$\overline{\xi^2}$	mean square logarithmic energy loss in a collision.
ρ_k	number of nuclei of atomic number k per cm^3 of moderator.
ρ^U	number of uranium-238 nuclei per cm^3.
ρ_0^U	number of nuclei per cm^3 of metallic uranium-238.
ρ_g^U	number of uranium-238 nuclei per cm^3 of a homogeneous mixture of the elements of a cell in a heterogeneous reactor.
Σ_s	macroscopic scattering cross section.
Σ_c	macroscopic capture cross section.
$\Sigma = \Sigma_s + \Sigma_c$	total macroscopic cross section.
Σ_{tr}	transport cross section.
τ	neutron age.
$\langle\varphi\rangle$	the probability of avoiding resonance capture.
φ	neutron flux (also denoted by nv).
φ^*	importance of slowing-down neutrons.
Φ	thermal neutron flux (some special functions are also denoted by this symbol).
Φ^*	thermal neutron importance.
$\chi(u)$	the density with which neutrons of lethargy u are generated.
$\chi(E)$	the density with which neutrons of energy E are generated.

INTRODUCTION

The fundamental physical problems of nuclear reactor calculations are the critical size and space-energy neutron distributions. In most cases these can be obtained in a more or less satisfactory way within the framework of the age theory of slowing down, whose general principles had already been stated by E. Fermi [93, 102].

The theory of nuclear reactor calculations rests on and has developed in terms of the kinetics of neutron moderation, with contributions by many scientists (R. Peierls [64, 128], G. Placzek [129, 130, 131], E. Wigner [10], I.Ia. Pomeranchuk [3], H. Hurwitz [108, 109, 110], G. Wick [139, 142, 143], A. Weinberg [41], and others). This work has established a unified and quite general point of view on neutron interaction with matter in a reactor.

The most general and versatile mathematical tool that can be used to obtain various approximate solutions of the kinetic equations was found to be the method of spherical harmonics (S. Chandrasekhar [87], R. Marchak [118, 121], M. Wang and E. Guth [140], H. Bethe, L. Tonks and H. Hurwitz [94], G. Holt [107], L.N. Usachev [81], and others), as well as the related method of "Gaussian quadratures" (G. Wick [142, 143], S. Chandrasekhar [87], and others). These methods were later developed and used for one-velocity problems (E.S. Kuznetsov [38, 39, 40, 41], R. Ivon and R. Mertens [124, 125, 126], and others [114]).

In the overwhelming majority of critical size calculations, it is sufficient to use the diffusion approximation. Such an approximation is usually valid for reactors whose dimensions are large compared to the mean free path.

If we assume that in the diffusion approximation the neutron collision density hardly changes as a function of the lethargy, we can expand it in a Taylor's series and arrive at the well-known age equation. This equation satisfactorily describes neutron slowing-down by nuclei whose mass is much greater than the neutron mass.

In addition to the age theory of slowing down, other more accurate methods have been proposed. Among these we may consider the method for approximating the kernels of the Boltzmann integro-differential equations ("the synthetic kernel method"), which involves the approximate replacement of the scattering function by another function which is simpler and more convenient to use in practice, yet such that some of its integral characteristic "moments" are the same as those of the exact scattering function (E. Wigner, E. Greuling and D. Goertzel [54], D.V. Shirkov [90, 91], and others).

The thermal neutrons can be united into a single group which is assigned a certain effective cross section averaged over the spectrum.

In the case of strong thermal neutron capture, the one-group model will no longer suffice. It then becomes necessary to include the thermal motion of the nuclei in a more rigorous way (E. Kogen [33], H. Hurwitz et al. [109], V.V. Smelov [73], and others).

An important step forward in reactor calculation methods was the development of the adjoint equations, and in particular, perturbation theory (E. Wigner [54], A.S. Romanovich, L.N. Usachev [82], S. Glasstone and M. Edlund [23], R. Erlich and H. Hurwitz [101], G.I. Marchuk [50, 52, 53] and others). Simultaneously, perturbation theory was applied to the one-velocity kinetic equation (N.A. Dmitriev [82], K. Fuchs [105]) as well as to the general kinetic equation of a reactor (L.N. Usachev [82]).

The use of the adjoint equations made it possible both to calculate the critical mass of a reactor more accurately, and to develop perturbation theory to the point where it became a reliable tool for analyzing experiments performed on nuclear reactors. It should be noted that the adjoint equation theory also allows one to analyze the kinetics of a nuclear reactor in general form [82].

This growth of computational methodology was constantly stimulated by the practical design requirements of ever better equipment. Reactor design developed simultaneously in three directions: thermal, intermediate and fast neutron reactors.

The greatest progress was made in developing methods for thermal neutron reactor calculations (I.V. Kurchatov [44, 45], A.P. Aleksandrov, A.I. Alikhanov [1], D.I. Blokhintsev and A.K. Krasin [6, 7, 8, 36], G.N. Flerov, I.M. Frank and V.S. Fursov [86], I.I. Gurevich and I.Ia. Pomeranchuk [26], S.M. Feinberg [84, 85], A.D. Galanin [16-20], and others). A review of foreign work has been given in two articles by D. Chernik [32, 38].

In the USSR the theory and calculational methods for intermediate neutron reactors have been extensively developed (A. I. Leipunskii, V.A. Kuznetsov, A.S. Romanovich and L.N. Usachev [82], V.Ia. Pupko [67], B.F. Gromer, G.I. Toshinskii, G.I. Marchuk [50-53], and others). In the USA methods for intermediate neutron reactor calculations were developed by R. Erlich and H. Hurwitz [101, 108].

In 1949 A.I. Leipunskii indicated the possibility of using a fast neutron reactor to produce energy and simultaneously to breed atomic fuel (see, for instance, Blokhintsev [9]). The resulting theory and calculational methods for fast neutron reactors has grown constantly (A.I. Leipunskii, D.I. Blokhintsev [9], L.N. Usachev [81], O.D. Kazachkovskii, I.I. Bondarenko, S.B. Shikhov, V.S. Vladimirov [11-13, 15], Iu.A. Romanov, and others). A review of foreign work is given by D. Okrent and others [60].

The existence of the three largely independent directions of reactor research has led to differentiation in the theory and methods of nuclear reactor calculations.

The intermediate and thermal neutron reactor calculations have grown primarily in the direction of multigroup approximations based on the extensive use of effective numerical methods for solving equations of the diffusion type.

The well-known two-group method has been developed to its limit [7, 17, 20, 23, 25, 135], but is clearly inadequate for solving many problems.

As is well known, in the two-group method the thermal neutrons are classified into one group, and the neutrons of higher energy into another. The result is that the reactor equations are reduced to a set of equations of the diffusion type, and these are solved with known boundary conditions on the surfaces of separation between the zones and at the external extrapolated reactor surface.

Because the problem is homogeneous, nontrivial solutions can exist only for certain combinations of values for the parameters, and a nonnegative solution of this problem will describe the neutron spectrum of a critical reactor.

The two-group method gives a satisfactory solution for the critical mass of a reactor when its characteristic dimensions are much greater than the slowing-down length and when most of the fission takes place in conjunction with the diffusion of thermal neutrons.

It is known that in many cases the two-group method can not lead to sufficiently accurate results. This is particularly true when it comes to reactors whose dimensions are either not very large compared to the slowing-down length or when there is strong absorption of slowing-down neutrons. In these cases it is necessary to use the multigroup method. Both methods will give solutions in closed analytic form. When the reactor has many zones, however, or when there is a large number of energy groups, it becomes extremely difficult and hardly convenient in practice to solve the problem in analytic form. It is therefore natural to proceed to more effective methods for approximate reactor calculations.

A. Thompson [138] used a very effective method analogous to Kellog's [113] to find the effective mass. His method, called the "method of source iteration," reduces the eigenvalue problem to the successive solution of Cauchy problems.

There are many successive approximation ways to solve the set of basic reactor equations; the simplest is the already mentioned multigroup method.

The remaining problem is to find an effective solution of the diffusion equations. It is known that analytic methods are not convenient from the computational point of view, so that one must use numerical methods which will guarantee the desired accuracy. One of these is the factorization of the diffusion differential equation originating with E. Ince [2] and developed by A. Thompson [138] and V. S. Vladimirov [14].

A finite-difference equation can be substituted for the differential diffusion equation in many different ways, of which the most convenient and accurate is the "continuous count" method. The concept of this method belongs to A.N. Tikhonov and A.A. Samarskii [79].

The effective solution of the finite-difference equations for one-dimensional regions is obtained using the method of difference factorization proposed by I.M. Gel'fand and O.V. Lokutsievskii [28, 47] and independently by A.S. Kronrod (see [20]) and Stark (see [101]). The two-dimensional finite-difference diffusion equations are solved by the well-developed methods of linear algebra such as those, for instance, of Liebmann, Seidel, relaxation methods, etc. [34, 49, 59, 116, 133, 134, 136].

M.V. Keldysh, I.M. Gel'fand and O.V. Lokutsievskii have developed a new method for solving the two-dimensional finite-difference diffusion equations, a method based on the concept of matrix factorization. Matrix factorization was developed by K.I. Babenko and N.N. Chentsov for the equations of hydrodynamics. They also investigated the numerical stability of the method [47]. This method can also be used successfully for solving a wide class of boundary-value problems in sets of differential equations.

Of great interest is the method for solving the slowing-down equation, using a net in its classical form of a triangle. As is known, in this case the whole region of definition of the solution is divided into cells by a coordinate grid.

The triangle scheme has already been solved for the unknown function of points of the subsequent energy step. Therefore, because of its great simplicity, it becomes very convenient for practical application [22, 31, 43, 66, 70, 78].

As is known, the triangle scheme can be used only under numerical stability of the finite-difference equations, which means that the error in rounding off must not increase from step to step. This condition restricts the energy increment as a function of the space increment [28, 34, 46, 48, 69, 71, 72, 75, 77, 100, 103, 112, 117].

The problem of using finite-difference methods for an approximate solution of partial differential equations requires a knowledge of the computational accuracy and the extent to which the approximate solution reproduces the exact one. These extremely interesting and important problems of the theory have been investigated by many authors (R. Courant, K. Friedrichs and H. Lewy [43], G. O'Brien, M. Hyman and S. Kaplan [95], F. Hildebrand [106], L. Collatz [34], O. A. Ladyzhenskaia [46], and others [55 99, 104, 111, 115, 133]).

The development of computational methods for fast neutron reactors has proceeded both towards increasing the number of energy groups and towards developing more effective methods for solving the kinetic equations.

The present work is an attempt at a more or less systematic description of the basic methods for nuclear reactor calculations, which, on the one hand, will guarantee a given accuracy, and on the other hand, are sufficiently versatile and convenient to be used in a wide class of problems.

Chapter I

BASIC EQUATIONS

1. Statement of the Problem

Consider a nuclear reactor consisting of a core and a reflector.

In the core, fuel nuclei undergo fission and neutrons are produced with a certain energy spectrum. We shall call these neutrons slowing-down sources. Having high kinetic energy, they are scattered by nuclei of the moderator. If, in the scattering process, the momentum and total kinetic energy of the neutron-nucleus system remains constant, we call it elastic, whereas if part of the energy goes to exciting the nucleus, we call it inelastic.

Since every successive scattering event is accompanied by energy loss on the part of the neutron, the neutron velocity constantly decreases.

This slowing-down process is accompanied by scattering of the neutrons in space. Therefore, if the reactor is of finite size the neutron may escape after several collisions.

In order to decrease the probability of such escape, the core is surrounded by a reflector. Some of the neutrons leaving the core are then incident on the reflector. The reflector causes some of these neutrons to be returned to the core, the remainder leaving the reactor irreversibly.

When a neutron collides with a fissile nucleus or a nucleus of the moderator, there is a finite probability that it will be absorbed. In this case the slowing-down cycle is cut off at the neutron absorption energy.

Some of the neutrons, however, may be slowed down to energies below E_C, where E_C is obtained from the condition that the spectrum of slowing-down neutrons is the same as that of the thermal neutrons (more rigorous definitions of E_C are given elsewhere [33, 109]). In this region the thermal motion of the moderator nuclei is an important factor. As is known, the nuclei have a Maxwell velocity distribution.

The scattering of thermal neutrons is accompanied by absorption and leakage. If these are small, the neutrons enter into statistical equilibrium with the medium and take on a Maxwell distribution. In this case all neutrons with energies below E_C can be lumped into a single group with effective constants which are averages over the equilibrium Maxwell spectrum. This is the model usually used in calculations.

The difference between the actual neutron energy spectrum and the equilibrium spectrum may be large if the capture and leakage of thermal neutrons is important. In this case equilibrium is not established between the neutrons and the moderator nuclei, and the neutron spectrum is no longer Maxwellian [33, 54, 73, 92, 109]. If one finds the neutron spectrum, one can again go on to considering the one-group model of neutron scattering.

Some of the neutrons absorbed by the fuel, both in slowing down and in the region of thermal energies, give rise to fission. This process is accompanied by secondary neutron production and the liberation of a large amount of thermal energy when the fission fragments are decelerated in the medium. These secondary neutrons become slowing-down sources, and the chain reaction thus continues.

As is well known, a reactor is said to be operating under critical conditions when it just generates neutrons, that is when the number of neutrons per unit time produced by fission is equal to the number of neutrons leaving the reactor due to absorption or leakage. Because critical operation assumes the number of neutrons to remain constant, it is a stationary state.

A reactor in which the fission takes place primarily at thermal neutron energies ($E \lesssim E_C$) is called a thermal reactor. If most of the fission is caused by neutrons whose energy lies in the interval $E_C \lesssim E \lesssim 100$ kev, the reactor is called intermediate. If most of the fission is caused by capture of fast neutrons (E $\gtrsim$ 100 kev), the reactor is called fast.*

As is known, the energy spectrum of the neutrons averaged over the volume of the core depends, for a given moderator, primarily on the ratio of the number of fuel nuclei to the number of moderator nuclei. The lowest critical charges of fuel are needed for thermal reactors, whereas the highest are needed for fast neutron reactors. At the same time, we note that the dimensions of thermal reactors are the largest, while the fast reactors are the smallest. The choice of one or another type of reactor is determined by the use for which it is intended.

In addition it should be noted that in thermal and intermediate reactors the neutron slowing-down process is caused primarily by elastic scattering. Inelastic scattering usually gives only a small contribution. In fast neutron reactors, on the other hand, it is inelastic scattering which is most important.

In performing calculations for heterogeneous nuclear reactors, it is first necessary to investigate the scattering processes in the individual cells. The spectrum obtained can be used to find the effective physical constants of the reactor and then to perform the calculation on an "equivalent" homogeneous reactor.

2. Elastic Slowing Down of Neutrons in Matter

Elastic scattering is the most important mechanism by which the neutrons are slowed down in intermediate and thermal reactors. As is known [3, 23, 118], elastic slowing down of neutrons can be described satisfactorily by classical mechanics.

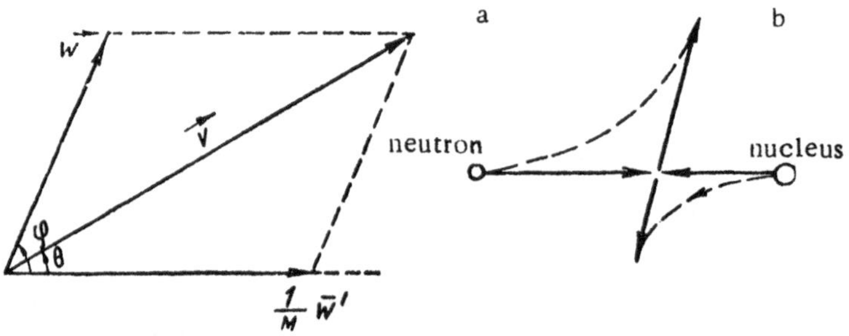

Fig. 1. Coordinate systems: a) laboratory coordinate system; b) center-of-mass coordinate system.

Let us consider the neutron-nucleus system. For convenience, we shall use two coordinate systems, the stationary one fixed on the target nucleus (called the laboratory coordinate system), and the moving center-of-mass system (Fig. 1).

Let $\vec{v}'$ be the neutron velocity before collision in the laboratory system. Then in this system the velocity of the center-of-mass is

$$\vec{v}_c' = \frac{1}{1+M}\vec{v}', \qquad (2.1)$$

where M is the mass of the nucleus (we take the neutron mass to be equal to unity).

Obviously, in the moving center-of-mass system the neutron velocity will be

$$\vec{w}' = \vec{v}' - \vec{v}_c' = \frac{M}{1+M}\vec{v}', \qquad (2.2)$$

while the velocity of the nucleus will be

*Such a classification of reactors is, of course, quite arbitrary.

$$\vec{w}_0' = \vec{v}_0' - \vec{v}_c' = -\vec{v}_c = -\frac{1}{1+M}\vec{v}'. \tag{2.3}$$

We have thus been able to express these three velocities in the two coordinate systems in terms of the neutron velocity before collision in the laboratory system.

We have now to find the corresponding velocities after elastic collision of the neutron with the nucleus. For convenience, we shall denote them by the same letters, but without primes.

To find the velocities of the neutron and target nucleus in the center-of-mass system after collision, we shall make use of momentum and energy conservation, namely

$$\left.\begin{aligned} \vec{w} + M\vec{w}_0 &= \vec{w}' + M\vec{w}_0', \\ \frac{w^2}{2} + \frac{Mw_0^2}{2} &= \frac{w'^2}{2} + \frac{Mw_0'^2}{2}. \end{aligned}\right\} \tag{2.4}$$

Using (2.2) and (2.3), we can bring (2.4) into the form

$$\left.\begin{aligned} \vec{w} + M\vec{w}_0 &= 0, \\ w^2 + Mw_0^2 &= \frac{M}{1+M}v'^2. \end{aligned}\right\} \tag{2.5}$$

The first of Equations (2.5) gives

$$\vec{w} = -M\vec{w}_0,$$

which means that

$$w = Mw_0. \tag{2.6}$$

We now insert the value of w given by (2.6) into the second of (2.5). We then arrive at

$$w_0 = \frac{1}{1+M}v' = w_0'. \tag{2.7}$$

Inserting the value of w_0 from (2.7) into (2.6), we have

$$w = \frac{M}{1+M}v' = w'. \tag{2.8}$$

According to (2.7) and (2.8) the magnitudes of the velocities of both the neutron and the target nucleus are not changed by the collision process in the center-of-mass system.

Now we have only to find the angle by which the neutron velocity is changed as a result of the collision. To do this, we write the relative neutron velocity after the collision in the form

$$\vec{w} = \vec{v} - \vec{v}_c = \vec{v} - \vec{v}_c',$$

whence we obtain

$$\vec{v} = \vec{w} + \vec{v}_c' = \vec{w} + \frac{1}{1+M}\vec{v}'. \tag{2.9}$$

Using (2.2), we can write (2.9) in the form

$$\vec{v} = \vec{w} + \frac{1}{M}\vec{w}'. \tag{2.10}$$

From (2.10) it follows that

$$v^2 = w^2 + \frac{1}{M^2} w'^2 + \frac{2}{M} ww' \cos \varphi, \tag{2.11}$$

where

$$\cos \varphi = \frac{(\vec{w} \cdot \vec{w}')}{w \cdot w'}.$$

Solving (2.11) for $\cos \varphi$ and using (2.8), we obtain

$$\cos \varphi = 1 - \frac{(1+M)^2}{2M}\left[1 - \left(\frac{v}{v'}\right)^2\right]. \tag{2.12}$$

From (2.10) we can obtain an expression for

$$\cos \theta = \frac{(\vec{v}, \vec{w}')}{v \cdot w'}.$$

To do this, we project the velocity parallelogram on the $\vec{w}'$ axis. We then have

$$v \cdot \cos \theta = w \cos \varphi + \frac{1}{M} w'. \tag{2.13}$$

Now using (2.8) and (2.12) we can write (2.13) in the form

$$\cos \theta = \frac{1+M}{2}\frac{v}{v'} - \frac{M-1}{2}\frac{v'}{v}. \tag{2.14}$$

Equation (2.14) gives us the important relation

$$1 \leqslant \frac{v'}{v} \leqslant \frac{M+1}{M-1}. \tag{2.15}$$

To go on it is convenient to transform to the new independent variable u, called the lethargy, defined by

$$u = \ln \frac{E_0}{E}, \tag{2.16}$$

where E is the kinetic energy of the neutron, and E_0 is some fixed energy which we shall henceforth take at 2 Mev. Bearing in mind that $E = v^2/2$, we have

$$u = 2 \ln \frac{v_0}{v}. \tag{2.17}$$

We now use (2.17) to rewrite (2.12), (2.14) and (2.15) in the forms

$$\cos \varphi = 1 - \frac{(1+M)^2}{2M}\left[1 - e^{-(u-u')}\right], \tag{2.18}$$

$$\cos \theta = \frac{1+M}{2} e^{-\frac{u-u'}{2}} - \frac{M-1}{2} e^{\frac{u-u'}{2}}, \tag{2.19}$$

$$u - r \leqslant u' \leqslant u, \tag{2.20}$$

where $r = \ln\left(\frac{M+1}{M-1}\right)^2$ and $u' = 2\ln\frac{v_0}{v'}$. Analysis of (2.18), (2.19) and (2.20) shows that the change in lethargy or the logarithmic energy decrement $u - u'$ has a maximum value r. In addition, the cosine of the collision angle θ (as well as that of φ) is uniquely determined by the logarithmic energy decrement. In what follows these facts will be used to derive the neutron transport equations.

3. Kinetic Slowing-Down Equation

The space-energy distribution of the neutrons slowing down as a result of elastic collisions with nuclei is described by Boltzmann's kinetic equation [118, 121], which we shall now derive.

Let $\vec{r}$ be the radius vector of the neutron at time t, and $\vec{v}$ be its vector velocity, which is equal to $v\vec{\Omega}$, where v is the magnitude and $\vec{\Omega}$ is a unit vector (Fig. 2).

Let us now consider the neutron balance in an element of volume $d\vec{r}$ about the point M, for neutrons moving in the direction $\vec{\Omega}$. To do this we describe a cylinder of volume $d\vec{r}$ at M, oriented parallel to $\vec{\Omega}$.

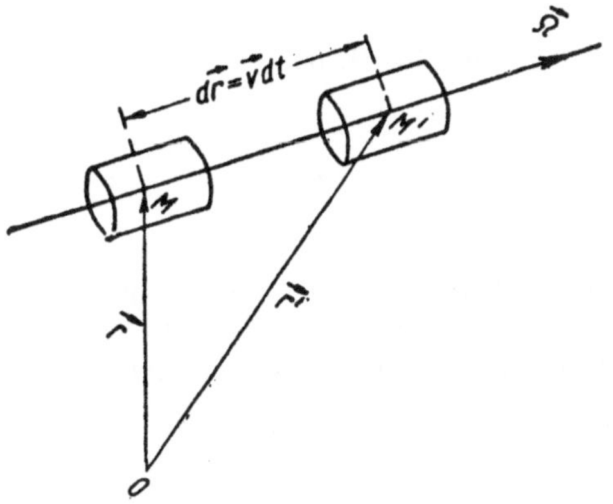

Fig. 2. Neutron path in the medium. $\vec{\Omega}$ is the unit vector along the neutron velocity; $\vec{r}$ is the neutron radius vector at time t; $\vec{r}'$ is the neutron radius vector at time $t + dt$.

After a time dt any neutron in the cylinder $d\vec{r}$ about M will move to a position close to M' if it has not been scattered by a moderator nucleus, and the distance MM' is given by:

$$ds = v\,dt.$$

Let us calculate the total change of the number of neutrons moving in the direction $\vec{\Omega}$ in the time dt.

Let $n(\vec{r}, \vec{v}, t)$ be the number of neutrons per unit volume of phase space at the point $(\vec{r}, \vec{v})$ at time t. Then the number of neutrons in the volume $d\vec{r}$ about M whose velocity lies in the interval $(\vec{v}, \vec{v} + d\vec{v})$ will be

$$n(\vec{r}, \vec{v}, t)\, d\vec{r}\, d\vec{v}.$$

At time t + dt the number of such neutrons will be

$$n(\vec{r} + \vec{v}dt, \vec{v}, t + dt)\, d\vec{r}\, d\vec{v}.$$

Thus the total change in the number of neutrons in the volume $d\vec{r}$ in a time dt is given by

$$dN = [n(\vec{r} + \vec{v}dt, \vec{v}, t + dt) - n(\vec{r}, \vec{v}, t)]\, d\vec{r}\, d\vec{v}. \tag{3.1}$$

We can expand the expression in square brackets in (3.1) in a Taylor's series, and keeping just the linear terms we have

$$dN = \left[\frac{\partial n}{\partial t}dt + \frac{\partial n}{\partial x}dx + \frac{\partial n}{\partial y}dy + \frac{\partial n}{\partial z}dz\right] d\vec{r}\, d\vec{v} \tag{3.2}$$

or

$$dN = \left(\frac{\partial n}{\partial t} + \vec{v}\,\nabla n\right) d\vec{r}\, d\vec{v}\, dt,$$

where

$$\vec{v}\,\nabla n = v_x\frac{\partial n}{\partial x} + v_y\frac{\partial n}{\partial y} + v_z\frac{\partial n}{\partial z} \tag{3.3}$$

and

$$v_x = \frac{dx}{dt}, \quad v_y = \frac{dy}{dt}, \quad v_z = \frac{dz}{dt}.$$

Let us now consider the processes which change the number of neutrons in the element of phase space ($d\vec{r}$, $d\vec{v}$, dt). First let us find the number which leave this element. It consists of the neutrons captured and scattered by the medium.*

Let Σ_c and Σ_s be the macroscopic capture and scattering cross sections, where $\Sigma_c = 1/\lambda_c$ and $\Sigma_s = 1/\lambda_s$, and let λ_c and λ_s be the corresponding mean free paths. Then

$$\frac{v\,dt}{\lambda_c} = \frac{ds}{\lambda_c} \quad \text{and} \quad \frac{v\,dt}{\lambda_s} = \frac{ds}{\lambda_s}$$

describe the number of neutron collisions per time dt leading to capture and scattering.

If the element of phase space contains $n\,d\vec{r}\,d\vec{v}$ neutrons, the total number of neutrons captured and scattered out of this element will be

$$\Sigma_c nv\,d\vec{r}\,d\vec{v}\,dt \quad \text{and} \quad \Sigma_s nv\,d\vec{r}\,d\vec{v}\,dt.$$

Thus, the total number of neutrons leaving the element ($d\vec{r}$, $d\vec{v}$) in the time dt will be

$$dP = \Sigma\,nv\,d\vec{r}\,d\vec{v}\,dt, \tag{3.4}$$

where $\Sigma = \Sigma_c + \Sigma_s$ is the total cross section.

Let us now calculate the number of neutrons dR which enter our element ($d\vec{r}$, $d\vec{v}$) of phase space in the time dt. This number is composed of the number of neutrons in the volume $d\vec{r}$ which, before scattering into the volume $d\vec{r}$, had velocity $\vec{v}'$ and for which the probability is $F(\vec{v}, \vec{v}')$ that after scattering they have velocity $\vec{v}$, plus the sources

$$S(\vec{r}, \vec{v}, t)\,d\vec{r}\,d\vec{v}\,dt.$$

In other words,

$$dR = \int d\vec{v}'\,\Sigma_s nv(\vec{r}, \vec{v}', t)\,F(\vec{v}, \vec{v}')\,d\vec{r}\,d\vec{v}\,dt + S(\vec{r}, \vec{v}, t)\,d\vec{r}\,d\vec{v}\,dt. \tag{3.5}$$

Here the integration is over all possible values of $\vec{v}'$.

We note that $F(\vec{v}, \vec{v}')$ must be normalized so that

$$\int d\vec{v}F(\vec{v}, \vec{v}') = 1. \tag{3.6}$$

Using (3.2), (3.4) and (3.5), we can write the neutron balance equation per element of phase space in the form

$$dN = -dP + dR. \tag{3.7}$$

Let us now insert the expressions for the differentials into (3.7) and divide by $d\vec{r}\,d\vec{v}\,dt$. We then arrive at Boltzmann's integro-differential equation

*For scattered neutrons, in general, one must account for changes both in the speed and in the direction $\vec{\Omega}$, so that a scattered neutron actually leaves the element ($d\vec{r}$, $d\vec{v}$) of phase space.

$$\frac{1}{v}\frac{d\varphi}{dt} = -\Sigma\varphi + \int d\vec{v}'\Sigma_s\varphi(\vec{r}, \vec{v}', t) F(\vec{v}', \vec{v}) + S(\vec{r}, \vec{v}, t). \tag{3.8}$$

Here we have introduced the notation

$$\varphi = nv.$$

In addition, in (3.8) we have used the relation

$$\frac{d\varphi}{dt} = \frac{\partial\varphi}{\partial t} + \vec{v}\nabla\varphi. \tag{3.9}$$

Let us now make use of the fact that

$$\vec{v} = \vec{\Omega}v, \quad \vec{\Omega} = \Omega_x\vec{i} + \Omega_y\vec{j} + \Omega_z\vec{k},$$

and go to a spherical coordinate system (Fig. 3). Then we have

$$\left.\begin{aligned} v_x &= \Omega_x v, & \Omega_x &= \sin\vartheta\cos\psi, \\ v_y &= \Omega_y v, & \Omega_y &= \sin\vartheta\sin\psi, \\ v_z &= \Omega_z v, & \Omega_z &= \cos\vartheta, \\ d\vec{v} &= v^2\sin\vartheta\, d\psi\, d\vartheta\, dv = v^2\, d\Omega\, dv \end{aligned}\right| \tag{3.10}$$

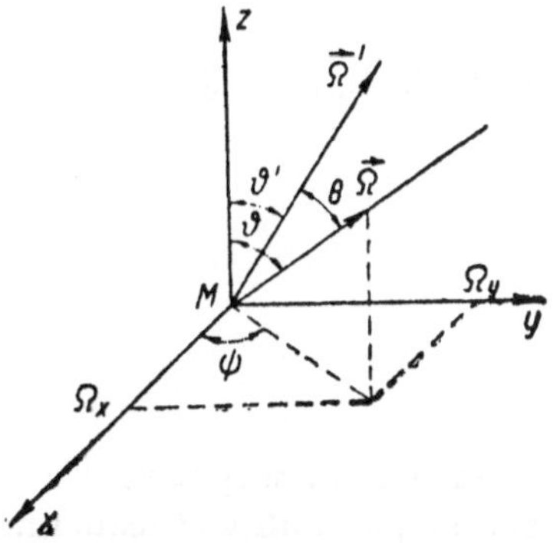

Fig. 3. Coordinate system fixed on a stationary nucleus. $\vec{\Omega}$ is the unit vector in the direction of the neutron velocity before the collision; $\vec{\Omega}'$ is the unit vector after collision; θ is the angle between $\vec{\Omega}$ and $\vec{\Omega}'$; ϑ is the polar angle, and ψ the azimuth angle.

and therefore (3.8) can be written

$$\frac{1}{v}\frac{d\varphi}{dt} = -\Sigma\varphi + \int dv' \int d\Omega'\Sigma_s\varphi(\vec{r}, \vec{\Omega}', v', t) F(\vec{\Omega}, \vec{\Omega}'; v, v') + S(\vec{r}, \vec{\Omega}, v, t), \tag{3.11}$$

where

$$\frac{1}{v}\frac{d\varphi}{dt} = \frac{1}{v}\frac{\partial\varphi}{\partial t} + \Omega\nabla\varphi. \tag{3.12}$$

The transition from the functions $\varphi(\vec{r}, \vec{v})$, $F(\vec{v}, \vec{v}')$ and $S(\vec{r}, \vec{v})$ defined in $(\vec{r}, \vec{v})$ phase space to the functions $\varphi(\vec{r}, \vec{\Omega}, v)$, $F(\vec{\Omega}, \vec{\Omega}', v, v')$ and $S(\vec{r}, \vec{\Omega}, v)$ defined in $(\vec{r}, \vec{\Omega}, v)$ space is made using the Jacobian of the transformation and the relations:

$$\left.\begin{aligned} |\vec{v}|^2\varphi(\vec{r}, \vec{v}) &= \varphi(\vec{r}, \vec{\Omega}, v), \\ |\vec{v}|^2 F(\vec{v}, \vec{v}') &= F(\vec{\Omega}, \vec{\Omega}'; v, v'), \\ |\vec{v}|^2 S(\vec{r}, \vec{v}) &= S(\vec{r}, \vec{\Omega}, v). \end{aligned}\right\} \tag{3.13}$$

Equation (3.6) can be used to show easily that the new F function is also normalized to unity; that is, that

$$\int dv \int d\Omega\, F(\vec{\Omega}, \vec{\Omega}'; v, v') = 1. \tag{3.14}$$

Finally, let us go to the new variable u in (3.11), using the expression

$$u = 2\ln\frac{v_0}{v}.$$

Then taking account of the Jacobian of the transformation, we arrive at the basic kinetic Boltzmann equation:

$$\frac{1}{v}\frac{\partial\varphi}{\partial t}+\vec{\Omega}\nabla\varphi=-\Sigma\varphi+\int\limits_{u-r}^{u}du'\int d\Omega'\,\Sigma_s\varphi(\vec{r},\vec{\Omega}',u',t)\,F(\vec{\Omega},\vec{\Omega}',u,u')+$$

$$+S(\vec{r},\vec{\Omega},u,t), \tag{3.15}$$

where the transition from $(\vec{r},\vec{\Omega},v)$ phase space to $(\vec{r},\vec{\Omega},u)$ space is made using the relations

$$\frac{v}{2}\varphi(\vec{r},\vec{\Omega},v)=\varphi(\vec{r},\vec{\Omega},u),$$

$$\frac{v}{2}F(\vec{\Omega},\vec{\Omega}';v,v')=F(\vec{\Omega},\vec{\Omega}';u,u'),$$

$$\frac{v}{2}S(\vec{r},\vec{\Omega},v)=S(\vec{r},\vec{\Omega},u),$$

and $F(\vec{\Omega},\vec{\Omega}';u,u')$ is a scattering function which is also normalized in the variables $\vec{\Omega},u$; that is,

$$\int\limits_{u'}^{u'+r}du\int d\Omega\,F(\vec{\Omega},\vec{\Omega}';u,u')=1. \tag{3.16}$$

In addition, (3.15) takes account of the fact that u' varies within the limits given by (2.20), namely:

$$u-r\leq u'\leq u.$$

Let us now obtain an explicit expression for the probability density

$$F(\vec{\Omega},\vec{\Omega}';u,u').$$

For this purpose, let us first consider the probability distribution law in the center-of-mass system. We shall assume that the neutrons are scattered isotropically in the system, that is, that the probability of scattering by a nucleus into a given angle does not depend on the direction, but only on the solid angle. Taking account of the normalization, we have

$$\frac{d\Omega}{4\pi}=\frac{1}{4\pi}\sin\varphi\,d\varphi\,d\beta, \tag{3.17}$$

where β is the azimuth. This means that the probability density is $1/4\pi$.

Noting that there is a functional relation between u and φ, we differentiate (2.18) to obtain

$$\sin\varphi\,d\varphi=\frac{(1+M)^2}{2M}e^{-(u-u')}du. \tag{3.18}$$

From the assumption, therefore, that the scattering is isotropic in the center-of-mass system, we obtain the neutron energy distribution. The probability density of this distribution is

$$\frac{(1+M)^2}{8\pi M}e^{-(u-u')}. \tag{3.19}$$

Thus, the lethargy distribution of the neutrons is given by (3.19).

Let us now fix u' and attempt to find the angular distribution of the neutrons. It is clear that the existence of the relation

$$\cos\theta=\frac{M+1}{2}e^{-\frac{u-u'}{2}}-\frac{M-1}{2}e^{\frac{u-u'}{2}}, \tag{3.20}$$

where

$$\cos\theta = (\vec{\Omega}, \vec{\Omega}') = \cos\vartheta\cos\vartheta' + \sin\vartheta\sin\vartheta'\cos(\psi - \psi'),$$

completely determines the angle θ. This means that there is no independent neutron angular distribution, and therefore we are dealing with a two-dimensional, rather than a three-dimensional, probability distribution law.

In particular, Equation (3.20) shows that a neutron can collide with a nucleus only if

$$\cos\theta - \left(\frac{M+1}{2} e^{-\frac{u-u'}{2}} - \frac{M-1}{2} e^{\frac{u-u'}{2}}\right) = 0. \tag{3.21}$$

As for cases for which

$$\cos\theta - \left(\frac{M+1}{2} e^{-\frac{u-u'}{2}} - \frac{M-1}{2} e^{\frac{u-u'}{2}}\right) \neq 0, \tag{3.22}$$

their probability distribution vanishes, since for these cases a neutron and nucleus do not collide. Bearing this in mind, we can formally write out a three-dimensional distribution law using a δ-function. According to (3.19) the probability distribution law is

$$F(\vec{\Omega}, \vec{\Omega}'; u, u') =$$

$$= \frac{(1+M)^2}{8\pi M} e^{-(u-u')} \delta\left[\mu_0 - \left(\frac{M+1}{2} e^{-\frac{u-u'}{2}} - \frac{M-1}{2} e^{\frac{u-u'}{2}}\right)\right], \tag{3.23}$$

where

$$\mu_0 = \cos\theta.$$

It is easily seen by performing the necessary integration that $F(\vec{\Omega}, \vec{\Omega}'; u, u')$ is normalized to unity. Finally, bearing in mind the structure of the function in (3.23), it is convenient to use the new notation

$$F(\vec{\Omega}, \vec{\Omega}'; u, u') = f(\mu_0 u - u'). \tag{3.24}$$

Let us now turn to the kinetic equation (3.15). For the stationary case it becomes, using (3.24):

$$\vec{\Omega}\nabla\varphi + \Sigma\varphi = \int_{u-r}^{u} du' \int d\Omega' \Sigma_s \varphi(\vec{r}, \vec{\Omega}', u') f(\mu_0, u - u') + S(\vec{r}, \vec{\Omega}, u). \tag{3.25}$$

Let us now introduce another very important characteristic, namely the differential neutron slowing-down density. This is defined as the number of neutrons which are slowed down through the lethargy u per unit time per unit solid angle about the direction $\vec{\Omega}$. We denote this by $q(\vec{r}, \vec{\Omega}, u)$.

We obtain an expression for q in the following way. The number of neutrons travelling in all possible $\vec{\Omega}'$ directions which are elastically scattered from the interval (u', u' + du') into the interval (u", u" + du") and into the $\vec{\Omega}$ direction (Fig. 4) is

$$du'' du' \int d\Omega' \Sigma_s \varphi(\vec{r}, \vec{\Omega}', u') f(\mu_0, u'' - u'). \tag{3.26}$$

Therefore the number of neutrons leaving the interval (u', u'+ du') and entering the interval (u, u' + r) is found simply by integration, i.e.:

$$du' \int_{u}^{u'+r} du'' \int d\Omega' \Sigma_s \varphi(\vec{r}, \vec{\Omega}', u') f(\mu_0, u'' - u'). \tag{3.27}$$

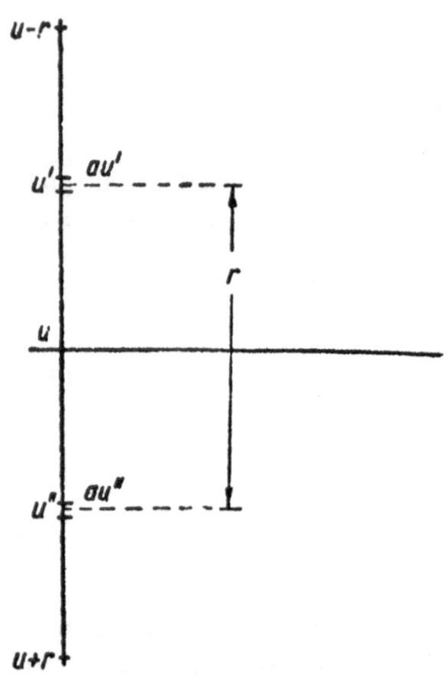

Fig. 4. Change of neutron lethargy in a collision with a nucleus. u' is the neutron lethargy before collision; u is the neutron lethargy after collision; r = max (u − u').

But the total number of neutrons going from the interval $u - r \le u' \le u$ to lethargies less than u will be

$$q(\vec{r}, \vec{\Omega}, u) = \int_{u-r}^{u} du' \int_{u}^{u'+r} du'' \int d\Omega' \times$$

$$\times \Sigma_s \varphi(\vec{r}, \vec{\Omega}', u') f(\mu_0, u'' - u'). \tag{3.28}$$

This expression can be written

$$q(\vec{r}, \vec{\Omega}, u) = \int_{u-r}^{u} du' \int d\Omega' \times$$

$$\times \Sigma_s \varphi(\vec{r}, \vec{\Omega}', u') F(\mu_0, u, u'), \tag{3.29}$$

where

$$F(\mu_0, u, u') = \int_{u}^{u'+r} f(\mu_0, u'' - u') du'' \tag{3.30}$$

is the probability that as a result of elastic collision a neutron will drop below u from a unit interval about the lethargy u'.

Let us now integrate $q(\vec{r}, \vec{\Omega}, u)$ over the unit sphere. We then obtain

$$q_0(\vec{r}, u) = \int_{u-r}^{u} du' \int d\Omega' \Sigma_s \varphi(\vec{r}, \vec{\Omega}', u') F_0(\vec{\Omega}', u, u'), \tag{3.31}$$

where

$$F_0(\vec{\Omega}', u, u') = \int d\Omega F(\mu_0, u, u'). \tag{3.32}$$

Obviously $q_0(\vec{r}, u)$, which is called the slowing-down density, is the total number of neutrons passing through lethargy u per unit time.

It is helpful to note than in Cartesian coordinates $\vec{\Omega}\nabla\varphi$ becomes

$$\vec{\Omega}\nabla\varphi = \sin\vartheta\cos\phi\frac{\partial\varphi}{\partial x} + \sin\vartheta\sin\phi\frac{\partial\varphi}{\partial y} + \cos\vartheta\frac{\partial\varphi}{\partial z}. \tag{3.33}$$

To obtain this expression we have made use of the fact that

$$\vec{\Omega} = \sin\vartheta\cdot\cos\phi\cdot\vec{i} + \sin\vartheta\cdot\sin\phi\cdot\vec{j} + \cos\vartheta\cdot\vec{k},$$
$$\nabla\varphi = \frac{\partial\varphi}{\partial x}\cdot\vec{i} + \frac{\partial\varphi}{\partial y}\vec{j} + \frac{\partial\varphi}{\partial z}\vec{k}. \tag{3.34}$$

In one-dimensional geometries the expression for $\vec{\Omega}\nabla\varphi$ simplifies somewhat and becomes:

$$\vec{\Omega}\nabla\varphi = \begin{cases} \mu\frac{\partial\varphi}{\partial z} & \text{for plane geometry} \\ \mu\frac{\partial\varphi}{\partial r} + \frac{1-\mu^2}{r}\frac{\partial\varphi}{\partial\mu} & \text{for spherical geometry} \\ \sin\phi\left(\mu\frac{\partial\varphi}{\partial\rho} + \frac{1-\mu^2}{\rho}\frac{\partial\varphi}{\partial\mu}\right) & \text{for cylindrical geometry} \end{cases} \tag{3.35}$$

where $\mu = \cos\vartheta$ is $(\vec{\Omega}, \vec{k})$ for the plane geometry, $(\vec{\Omega}, \vec{r}^0)$ for the spherical one, and $(\vec{\Omega}^0, \vec{\rho}^0)$ for the cylindrical one; we have written $\vec{\Omega}^0$ for the projection, in the cylindrical geometry, of $\vec{\Omega}$ onto the z = 0 plane,

$$\rho = \sqrt{x^2 + y^2}, \quad \text{and} \quad \cos\psi = (\vec{\Omega}, \vec{k}).$$

Here $\vec{k}, \vec{r}^0$, and $\vec{\rho}^0$ are the unit vectors in the $\vec{z}, \vec{r}$ and $\vec{\rho}$ directions. We now note that $\vec{\Omega}\nabla\varphi$ can be written in the Ω direction in the form

$$\vec{\Omega}\nabla\varphi = \frac{\partial\varphi}{\partial s},$$

where s is the distance along $\vec{\Omega}$ measured from an arbitrary point.

In conclusion we note that if the moderator consists of a mixture of nuclei with different atomic numbers, the kinetic slowing-down equation (3.25) becomes

$$\vec{\Omega}\nabla\varphi + \Sigma\varphi = \sum_k \int_{u-r_k}^{u} du' \int d\Omega' \, \Sigma_{sk}\varphi f_k(\mu_0, u-u') + S(\vec{r}, \vec{\Omega}, u), \tag{3.36}$$

where the index k is used to denote quantities associated with the element whose mass is M_k,

$$\Sigma_{sk} = \rho_k\sigma_{sk}, \quad \Sigma_{ck} = \rho_k\sigma_{ck}, \quad \Sigma = \sum_k (\rho_k\sigma_{sk} + \rho_k\sigma_{ck}),$$

ρ_k is the number of nuclei of the k-th kind per cm^3 of moderator, and σ_{sk} and σ_{ck} are the microscopic cross sections for scattering and capture by this nucleus.

4. Kinetic Equation of the Thermal Neutron Group

In the energy range in which thermal motion of the moderator nuclei, chemical bonds, and crystal effects are of importance to the neutron spectrum, the model of elastic scattering of neutrons by stationary nuclei requires further refinement. We shall assume that the neutron scattering law and therefore the scattering function are known. On this assumption the kinetic equation can be written very simply.

To be specific, let us consider neutron scattering by nuclei of a monatomic gas moderator, taking into account the thermal motion of the moderator nuclei (when the neutron spectrum is established under the influence of thermal motion of the nuclei, molecular bonds, and crystal effects, we shall call this process thermalization). In this case, as is well known, the moderator nuclei have a Maxwell distribution [23, 33, 80, 92]

$$M(v) = 4\pi \frac{v^2}{(2\pi kT)^{3/2}} \cdot e^{-\frac{M_j v^2}{2kT}}, \tag{4.1}$$

where T is the temperature of the medium, M_j is the mass of the nucleus whose atomic number is j, in units of the neutron mass, and k is Boltzmann's constant.

In the center-of-mass coordinate system of the neutron and nucleus, all the laws of an elastic collision are valid for energies below E_C. This means that in a single neutron-nucleus collision, the previously established scattering laws will be valid. Let us denote the corresponding scattering function by $p(\vec{v}', \vec{v}; \vec{v}_0)$, where $\vec{v}'$ is the neutron velocity before the collision, $\vec{v}$ is its velocity after collision, and $\vec{v}_0$ is the velocity of the nucleus. If the velocity of the nucleus is much less than that of the incident neutron, so that it may be considered stationary, the scattering function $p(\vec{v}', \vec{v}; \vec{v}_0)$ becomes the function $F(\vec{v}', \vec{v})$, which we considered in the previous section.

Let us now go on to constructing $g(\vec{\Omega}, \vec{\Omega}'; v, v')$. This function, also called the scattering function, is defined by [73]

$$g(\vec{\Omega}, \vec{\Omega}'; v, v') = \frac{1}{4\pi}\int_0^{2\pi}\int_{-1}^{1}\int_0^{\infty} v_R \Sigma_s(v_R) M(v_0)\, p(\vec{v}' \to \vec{v}; \vec{v}_0)\, dv_0\, d\mu'\, d\varphi', \tag{4.2}$$

where

$$p(\vec{v}' \to \vec{v}; \vec{v}_0) = \begin{cases} 0 & (v < v_m), \\ \frac{1}{2\pi}\frac{2v}{(v_M^2 - v_m^2)}\delta\left(\frac{A+1}{2}\cdot\frac{v}{v_M} - \frac{A-1}{2}\frac{v_M}{v} - \mu_0\right) & (v_m < v < v_M), \\ 0 & (v > v_M), \end{cases}$$

$$v_M = \frac{M_j}{M_j+1}v_R + v_c, \quad v_m = \left|\frac{M_j}{M_j+1}v_R - v_c\right|, \tag{4.3}$$

$$\vec{v}_c = \frac{\vec{v}' + M_j\vec{v}_0}{1+M_j}, \quad \vec{v}_R = \vec{v}' - \vec{v}_0,$$

$$A = \frac{M_j}{M_j+1}\cdot\frac{v_R}{v_c},$$

$$\mu' = \cos(\widehat{\vec{v}', \vec{v}_0}), \quad \mu_0 = \cos(\widehat{\vec{v}, \vec{v}_c}).$$

If the scattering cross section can be written as the sum of exponentials in the form

$$\Sigma_s(v_R) = \sum_i A_i e^{-\alpha_i v_R^2},$$

integration of (4.3) gives

$$g(\vec{\Omega}, \vec{\Omega}'; v, v') = \frac{(M_j+1)^2}{4M_j}\frac{\beta}{\sqrt{M_j\pi^3}}\frac{v^2}{|\vec{v} - \vec{v}'|}\times$$

$$\times\sum_i A_i\tau_i^2 e^{-\alpha_i\tau_i^2 v'^2 - \frac{(\alpha_i + M_j\beta^2)}{4}\left[\frac{|\vec{v}-\vec{v}'|}{M_j}\lambda_i - \tau_i^2\frac{v'^2 - v^2}{|\vec{v}-\vec{v}'|}\right]^2} \tag{4.3'}$$

where

$$\beta = \frac{1}{\sqrt{2\kappa T}}, \quad \tau_i^2 = \frac{M_j\beta^2}{M_j\beta^2 + \alpha_i}, \quad \lambda_i = 1 + M_j(1 - \tau_i^2).$$

Equation (4.3')was obtained by L.V. Maiorov and V.V. Smelov.

Having found the scattering function, it is not difficult to derive the kinetic equation. The resulting equation is

$$\frac{dn}{dt} = -[\gamma(\vec{r}, v, t) + w(\vec{r}, v, t)]\cdot n +$$
$$+\int dv' \int d\Omega' n(\vec{r}, \vec{\Omega}', v', t)\, g(\vec{\Omega}, \vec{\Omega}'; v, v') + S(\vec{r}, v, t), \tag{4.4}$$

where $n(\vec{r}, \vec{\Omega}, v, t)$ is the neutron density in $(\vec{r}, \vec{\Omega}, v, t)$ phase space, $\gamma(v)$ is the probability per unit time that the neutron with velocity $v = |\vec{v}|$ will be captured, and $w(v)$ is the probability that a neutron will be scattered by the moderator.* We shall further write this equation in the form

* It should be noted that $g(\vec{\Omega}, \vec{\Omega}'; v, v')$ depends parametrically on $\vec{r}$ and t.

$$\frac{1}{v}\frac{d\varphi}{dt} = -\frac{\gamma(\vec{r}, v, t) + w(\vec{r}, v, t)}{v}\varphi + \int_0^{v_C} dv' \int d\Omega' \varphi(\vec{r}, \vec{\Omega}', v', t) \frac{g(\vec{\Omega}, \vec{\Omega}'; v, v')}{v'} + S(\vec{r}, \vec{\Omega}, v, t), \quad (4.5)$$

where

$$\varphi = nv, \text{ and } v_C = v(E_C).$$

The transition from $g(\vec{v}, \vec{v}')$, the function given in $(\vec{r}, \vec{v}, t)$ phase space, to $g(v, v'; \vec{\Omega}, \vec{\Omega}')$ in $(\vec{r}, \vec{\Omega}, v, t)$ space is accomplished using the Jacobian of the transformation and the relation

$$|\vec{v}|^2 g(\vec{v}, \vec{v}') = g(\vec{\Omega}, \vec{\Omega}'; v, v').$$

Let us now integrate Equation (4.5) over all velocities $\underline{v}$ between the limits given by $0 \leq v \leq v_C$. We then obtain

$$\frac{\partial}{\partial t}\left(\frac{\Phi}{v_T}\right) + \vec{\Omega}\nabla\Phi = -\Sigma_T\Phi + \int_0^{v_C} dv' \int d\Omega' \varphi(\vec{r}, \Omega', v', t) \frac{G(\vec{\Omega}, \vec{\Omega}'; v')}{v'} + S(\vec{r}, \vec{\Omega}, t), \quad (4.6)$$

where

$$\Phi = \int_0^{v_C} \varphi\, dv, \quad v_T = \frac{\int_0^{v_C} \varphi\, dv}{\int_0^{v_C} \varphi \frac{dv}{v}},$$

$$\Sigma_T = \frac{\int_0^{v_C} [\gamma(\vec{r}, v, t) + w(\vec{r}, v, t)]\varphi \frac{dv}{v}}{\int_0^{v_C} \varphi\, dv},$$

$$G(\vec{\Omega}, \vec{\Omega}', v') = \int_0^{v_C} g(\vec{\Omega}, \vec{\Omega}'; v, v')\, dv,$$

$$S(\vec{r}, \vec{\Omega}, t) = \int_0^{v_C} S(\vec{r}, \vec{\Omega}, v, t)\, dv.$$

If we now assume the rigorous validity of

$$\varphi(\vec{r}, \vec{\Omega}', v', t) = \Phi(\vec{r}, \vec{\Omega}', t) M(v'), \quad (4.7)$$

which holds when there is no leakage or capture [20, 30, 80], Equation (4.6) can be written

$$\frac{1}{v_T}\frac{\partial\Phi}{\partial t} + \vec{\Omega}\nabla\Phi = -\Sigma_T\Phi + \int d\Omega' \Sigma_{sT}\Phi(\vec{r}, \vec{\Omega}', t) g(\vec{\Omega}, \vec{\Omega}') + S(\vec{r}, \vec{\Omega}, t), \quad (4.8)$$

where

$$\left.\begin{aligned} \Sigma_{sT} &= \int_0^{v_C} w(\vec{r}, v, t) M(v) \frac{dv}{v}, \\ g(\vec{\Omega}, \vec{\Omega}') &= \frac{1}{\Sigma_{sT}} \int_0^{v_C} dv' \frac{G(\vec{\Omega}, \vec{\Omega}', v')}{v'} M(v'). \end{aligned}\right\} \quad (4.9)$$

It should be noted further that in our case v_T and Σ_T are of the form

$$v_T = \frac{1}{\int\limits_0^{v_C} M(v)\,\frac{dv}{v}}\,,$$

$$\Sigma_T = \int\limits_0^{v_C} [\gamma(v) + w(v)]\, M(v)\,\frac{dv}{v}\,.$$

It is assumed here that the neutron spectrum is normalized according to

$$\int\limits_0^{v_C} M(v)\,dv = 1.$$

In addition, in (4.9) we have used the fact that $g(\vec{\Omega}, \vec{\Omega}')$ is normalized to unity according to

$$\int d\Omega\, g(\vec{\Omega}, \vec{\Omega}') = 1.$$

We have now to find the form of $S(\vec{r}, \vec{\Omega}, t)$. To do this we will assume that the external sources all arise as a result of neutrons slowed down from energies greater than E_C. It is easily seen that in this case we have

$$S(\vec{r}, \vec{\Omega}, t) = q_T(\vec{r}, \vec{\Omega}, t),$$

where q_T is the differential slowing-down density for $E = E_C$ (or $u = u_T$), and

$$q_T(\vec{r}, \vec{\Omega}, t) = \int\limits_{u_T - r}^{u_T} du' \int d\Omega' \Sigma_s \varphi(\vec{r}, \vec{\Omega}', u', t)\, F(\mu_0, u'), \tag{4.10}$$

$$F(\mu_0, u') = \int\limits_{u_T}^{u' + r} f(\mu_0, u - u')\, du.$$

We shall make one other very important assumption concerning the way g depends on $\vec{\Omega}$ and $\vec{\Omega}'$. This is that

$$g(\vec{\Omega}, \vec{\Omega}') = g(\mu_0),$$

where

$$\mu_0 = (\vec{\Omega}\vec{\Omega}') = \cos\theta.$$

Consider Equation (4.8). For stationary processes it can be rewritten

$$\vec{\Omega}\nabla\Phi + \Sigma_T\Phi = \int d\Omega' \Sigma_{sT}\Phi(\vec{r}, \vec{\Omega}')\, g(\mu_0) +$$

$$+ \int\limits_{u_T - r}^{u_T} du' \int d\Omega' \Sigma_s \varphi(\vec{r}, \vec{\Omega}', u')\, F(\mu_0, u'). \tag{4.11}$$

This will be the basic equation of neutron transport for the thermal group.

In conclusion it is useful to note that if $\underline{k}$ different types of nuclei take part in neutron scattering, the kinetic equation (4.11) is written

$$\vec{\Omega}\nabla\Phi + \Sigma_T\Phi = \sum_k \Big\{ \int d\Omega' \Sigma_{sk}\Phi g(\mu_0) + \int_{u_T - r}^{u_T} du' \int d\Omega' \Sigma_{sk}\varphi F(\mu_0, u') \Big\}, \tag{4.12}$$

where

$$\Sigma_T = \sum_k (\rho_k \sigma_{skT} + \rho_k \sigma_{ckT}).$$

The one-group model of thermal neutron scattering described in this section is satisfactory when the neutrons have an approximately Maxwell distribution. This is so, as was already mentioned, in most thermal reactors. If neutron capture in the reactor becomes more important, the neutron distribution deviates from the Maxwellian. In intermediate reactors, the neutron spectrum is not Maxwellian, and therefore one must analyze in more detail the process of neutron slowing-down in the thermal energy range. As a rule in such reactors, however, the thermal neutron yield in fission is so small that even a large error in the determination of the actual thermal neutron spectrum hardly matters when calculating the fundamental characteristics of the reactor.

It is particularly important to know the actual neutron spectrum in reactors in which neutrons are captured to a significant degree at energies above the thermal. This type of reactor is transitional between the thermal and intermediate categories. In calculating the spectrum of the neutron flux in such reactors one must use more accurate theoretical models. In the simplest case of a monatomic gas moderator, this problem has been treated in detail by several authors [33, 73, 109]. This situation is, however, usually far removed from reality, since it takes no account of chemical bonds and crystal effects which may be extremely important in these processes.

5. Basic Reactor Equations

In treating processes taking place in nuclear reactors, it should be borne in mind that the neutron source is the stationary self-maintaining fission chain reaction of the fuel nuclei.

If Σ_f is the macroscopic fission cross section at lethargy u, and Σ_{fT} is the fission cross section averaged over the thermal neutron spectrum, the number of neutrons produced per unit volume as a result of fission in the fuel is given by*

$$Q(\vec{r}) = \nu_f \int d\Omega \Big(\int_{-\infty}^{u_T} \Sigma_f \varphi \, du + \Sigma_{fT}\Phi \Big), \tag{5.1}$$

where ν_f is the effective neutron yield per fission.

In Equation (5.1) we assume, in addition, that the neutrons are produced isotropically in fission.

The full set of basic reactor equations with (5.1) is then written

$$\left.\begin{aligned} \vec{\Omega}\nabla\varphi + \Sigma\varphi &= \int_{u-r}^{u} du' \int d\Omega' \Sigma_s \varphi(\vec{r}, \vec{\Omega}', u') f(\mu_0, u-u') + \\ &+ \frac{\nu_f \chi(u)}{4\pi} \int d\Omega \Big(\int_{-\infty}^{u_T} \Sigma_f \varphi \, du + \Sigma_{fT}\Phi \Big), \\ \vec{\Omega}\nabla\Phi + \Sigma_T\Phi &= \int d\Omega' \Sigma_{sT}\Phi(\vec{r}, \vec{\Omega}') g(\mu_0) + \\ &+ \int_{u_T - r}^{u_T} du' \int d\Omega' \Sigma_s \varphi(\vec{r}, \vec{\Omega}', u') F(\mu_0, u'), \end{aligned}\right\} \tag{5.2}$$

*Here the integral is taken from the lethargy $u = -\infty$ corresponding to $E = \infty$, to u_T corresponding to E_C.

where $\chi(u)$ is the "spectrum" of neutrons produced by fission.* In terms of E, as is well known, χ can be written approximately in the form [23]

$$\chi(E) = 0{,}484 e^{-E} \operatorname{sh} \sqrt{2E}. \tag{5.3}$$

The necessary boundary conditions for Equations (5.2) are

$$\left.\begin{aligned} \varphi(\vec{r}, \vec{\Omega}, u) &= 0 \\ \Phi(\vec{r}, \vec{\Omega}) &= 0 \end{aligned}\right\} \text{ on } (S) \text{ for } (\vec{\Omega}\vec{n}^0) < 0, \tag{5.4}$$

where $\vec{n}^0$ is the outer normal to the concave surface S bounding the reactor volume G. Equation (5.4) states that there is no neutron flux through S directed from the outside into G.

Finally, we shall assume that the solution of Equations (5.2) can be found in the class of functions continuous in

$$G \times \Omega \times U,$$

where **U** is the lethargy interval $-\infty \le u \le u_T$, and Ω is the interval of all possible angles ϑ and ψ.

For notational convenience, we shall write the solution of Equations (5.2) in the form of a vector function f

$$f = \begin{vmatrix} \varphi \\ \Phi \end{vmatrix}, \tag{5.5}$$

whose components are $\varphi(\vec{r}, \vec{\Omega}, u)$ and $\Phi(\vec{r}, \vec{\Omega})$.

Further, let **L** be the operator which transforms f into:

$$\mathbf{L}f = \begin{vmatrix} \vec{\Omega}\nabla\varphi + \Sigma\varphi - \int\limits_{u-r}^{u} du' \int d\Omega' \Sigma_s \varphi(\vec{r}, \vec{\Omega}', u') f(\mu_0, u-u') - \\ \qquad - \frac{\nu_f \chi(u)}{4\pi} \int d\Omega \left(\int\limits_{-\infty}^{u_T} \Sigma_f \varphi\, du + \Sigma_{fT}\Phi \right) \\ \vec{\Omega}\nabla\Phi + \Sigma_T\Phi - \int d\Omega' \Sigma_{sT}\Phi(\vec{r}, \vec{\Omega}') g(\mu_0) - \\ \qquad - \int\limits_{u_T - r}^{u_T} du' \int d\Omega' \Sigma_s \varphi(\vec{r}, \vec{\Omega}', u') F(\mu_0, u') \end{vmatrix} \tag{5.6}$$

Then Equation (5.2) can be written

$$\mathbf{L}f = 0. \tag{5.7}$$

We shall make use of this notation in what follows.

6. Adjoint Reactor Equations

It is known that to every linear operator **L** defined for a domain **G** bounded by a surface S, there corresponds an adjoint operator, in the sense of Lagrange, which is defined by the functional equation [37, 42, 76]

---

*In general χ depends on $\vec{\Omega}, \vec{\Omega}'$; u, u'. For simplicity we shall assume that the fission neutrons are isotropically distributed and depend only on u.

$$(f^*, \mathbf{L}f) = (f, \mathbf{L}^* f^*). \tag{6.1}$$

where f and f^* are functions belonging to sets $\{f\}$ and $\{f^*\}$ having certain definite properties, and

$$(\varphi, \varphi^*) = \int_G \varphi\varphi^* \, d\vec{r} \tag{6.2}$$

is the scalar product of two functions φ and φ^*.

As a simple example of this property, let us consider the Sturm-Liouville problem [74, 76, 78]. Consider the second order differential equation

$$\frac{d}{dx} D \frac{d\varphi}{dx} + (\lambda - \Sigma)\varphi = 0, \tag{6.3}$$

where λ is a parameter, and D and Σ are piecewise smooth functions ($\Sigma \geq 0$). We supplement Equation (6.3) with boundary conditions, which we shall take for simplicity to be

$$\varphi(0) = \varphi(1) = 0. \tag{6.4}$$

It follows from this that the region **G** consists of the line segments $0 \leq x \leq 1$, and the boundary S consists of the two points $x = 0$ and $x = 1$.

The problem is to find those values of λ for which there exist nonvanishing solutions of (6.3) which are continuous throughout **G** up to the boundary S together with the current $D\frac{d\varphi}{dx}$, for which $\frac{d}{dx}D\frac{d\varphi}{dx}$ is piecewise continuous, and which satisfy the boundary conditions (6.4).

Consider the operator

$$\mathbf{L} = \frac{d}{dx} D \frac{d}{dx} + (\lambda - \Sigma). \tag{6.5}$$

Let us consider the set of functions $\{f\}$ acted upon by **L**. This set consists of continuous functions whose current $D\frac{df}{dx}$ is continuous, for which the derivative of the current $\frac{d}{dx}D\frac{df}{dx}$ is piecewise continuous, and which satisfy the conditions

$$f(0) = f(1) = 0. \tag{6.6}$$

We note that for any general element of the set $\{f\}$ we may have

$$\mathbf{L}f \neq 0,$$

while this vanishes only if f is a solution of (6.3).

We now introduce another set of functions $\{f^*\}$ defined in **G** up to the surface S, whose properties we shall establish below.

Now let us use (6.1) to find the adjoint operator $\mathbf{L}^*$, at the same time establishing the properties of $\{f^*\}$.

Consider the functional

$$(f^*, \mathbf{L}f) = \int_0^1 f^* \left[\frac{d}{dx} D \frac{df}{dx} + (\lambda - \Sigma) f \right] dx. \tag{6.7}$$

Integrating the first term in the integrand by parts, we obtain

$$(f^*, Lf)=\left(Df^*\frac{df}{dx}-Df\frac{df^*}{dx}\right)\Bigg|_0^1+\int_0^1 f\left[\frac{d}{dx}D\frac{df^*}{dx}+(\lambda-\Sigma)f^*\right]dx. \tag{6.8}$$

Note that we have already used the continuity of f^* and $D\frac{df^*}{dx}$ on the segment $0 \leq x \leq 1$. If we now assume that

$$f^*(0)=f^*(1), \tag{6.9}$$

we can make use of (6.6) to rewrite (6.8) in the form

$$(f^*, Lf)=\int_0^1 f\left[\frac{d}{dx}D\frac{df^*}{dx}+(\lambda-\Sigma)f^*\right]dx=(f, L^*f^*). \tag{6.10}$$

We have thus obtained the adjoint operator

$$L^*=\frac{d}{dx}D\frac{d}{dx}+(\lambda-\Sigma). \tag{6.11}$$

It is obvious that $L^* = L$ and that $\{f^*\}$ is the same as $\{f\}$. As is known, such operators are called self-adjoint in the sense of Lagrange.

If we now choose those functions f of the set $\{f\}$ which satisfy solutions (6.3), writing $f = \varphi$, we obtain

$$L\varphi=\frac{d}{dx}D\frac{d\varphi}{dx}+(\lambda-\Sigma)\varphi=0. \tag{6.12}$$

If we now assume that

$$L^*\varphi^*=\frac{d}{dx}D\frac{d\varphi^*}{dx}+(\lambda-\Sigma)\varphi^*=0, \tag{6.13}$$

we find that (6.10) is satisfied. Thus, up to an arbitrary factor, the solutions of (6.12) and (6.13) are the same.

Let us now consider the case of an operator which is not self-adjoint. We wish to solve the problem

$$\left.\begin{aligned}\frac{\partial}{\partial x}D\frac{\partial\varphi}{\partial x}-\Sigma\varphi=\frac{\partial\varphi}{\partial u}-\lambda S(u)\int_{-\infty}^{\infty}\sigma\varphi\,du,\\ \varphi(0, u)=\varphi(1, u)=0,\end{aligned}\right\} \tag{6.14}$$

where D, Σ and σ are given functions of $\underline{x}$ and $\underline{u}$, and S(u) is a continuous function of $\underline{u}$ normalized so that

$$\int_{-\infty}^{\infty} S(u)\,du=1.$$

Equation (6.14) is parabolic. Let us define the set $\{f\}$ in which we wish to find a solution of (6.14) in the following way. We shall assume that f and the current $D\frac{\partial f}{\partial x}$ are continuous over the region **G** (that is, $0 \leq x \leq$

≤ 1), that $\frac{\partial}{\partial x} D \frac{\partial f}{\partial x}$ is piecewise continuous, that it is continuously differentiable in u, and that it satisfies the boundary conditions $f(0, u) = f(1, u) = 0$ and initial conditions $f(-\infty, x) = 0$. In addition, we must assume that

$$\int_{-\infty}^{\infty} \sigma |f| \, du < \infty.$$

Then the operator **L** on the set $\{f\}$ is

$$\mathbf{L} = \frac{\partial}{\partial x} D \frac{\partial}{\partial x} - \Sigma - \frac{\partial}{\partial u} + \lambda S(u) \int_{-\infty}^{\infty} \sigma du.$$

We now consider the adjoint function φ^* belonging to the set $\{f^*\}$, whose properties will be established below, satisfying the equation

$$\mathbf{L}^* \varphi^* = 0.$$

where $\mathbf{L}^*$ is the adjoint operator. We shall find this operator using the functional equation

$$(f^*, \mathbf{L} f) = (f, \mathbf{L}^* f^*), \tag{6.15}$$

where the scalar product must be defined by

$$(\varphi, \varphi^*) = \int_{-\infty}^{\infty} du \int_0^1 dx \, \varphi \cdot \varphi^*.$$

Then

$$(f^*, \mathbf{L} f) = \int_{-\infty}^{\infty} du \int_0^1 dx f^* \left[\frac{\partial}{\partial x} D \frac{\partial f}{\partial x} - \Sigma f - \frac{\partial f}{\partial u} + \lambda S(u) \int_{-\infty}^{\infty} \sigma f \, du \right]. \tag{6.16}$$

If the functions of $\{f^*\}$ are continuous in $\mathbf{G} \times \mathbf{U}$ together with the current $D \frac{\partial f^*}{\partial x}$, and if the derivative of the current $\frac{\partial}{\partial x} D \frac{\partial f^*}{\partial x}$ is piecewise continuous, if they are continuously differentiable in u, and if they satisfy the conditions $f^*(0) = f^*(1) = 0$, $f^*(\infty, x) = 0$ and $\int_{-\infty}^{\infty} S(u) |f^*(u)| \, du < \infty$, then integrating by parts over x and u in (6.16) we arrive at

$$(f^*, \mathbf{L} f) = \int_{-\infty}^{\infty} du \int_0^1 dx f \left[\frac{\partial}{\partial x} D \frac{\partial f^*}{\partial x} - \Sigma f^* + \frac{\partial f^*}{\partial u} + \lambda \sigma \int_{-\infty}^{\infty} S(u) f^* du \right] = (f, \mathbf{L}^* f^*), \tag{6.17}$$

where

$$\mathbf{L}^* = \frac{\partial}{\partial x} D \frac{\partial}{\partial x} - \Sigma + \frac{\partial}{\partial u} + \lambda \sigma \int_{-\infty}^{\infty} S(u) \, du.$$

If now in (6.17) we choose f to be the solution of (6.14), writing $f = \varphi$, we arrive at the adjoint equation

$$\left.\begin{array}{l}\frac{\partial}{\partial x}D\frac{\partial\varphi^*}{\partial x}-\Sigma\varphi^*+\frac{\partial\varphi^*}{\partial u}+\lambda_J\int\limits_{-\infty}^{\infty}S(u)\,\varphi^*\,du=0,\\ \varphi^*(0,u)=\varphi^*(1,u)=0.\qquad \varphi^*(\infty,x)=0.\end{array}\right\}\tag{6.18}$$

In this problem the operator $\mathbf{L}^*$ is not the same as $\mathbf{L}$, so that it is no longer self-adjoint.*

In many cases we must deal not with single equations, but rather with sets of differential equations. It is found that in this case it is also possible to obtain the adjoint equations by using the concept of vector functions. We will also illustrate this method by simple example.

First consider the problem given by (6.3) and (6.4), where we reduce (6.3) to a set of second-order equations:

$$\left.\begin{array}{l}\frac{d\varphi_2}{dx}+(\lambda-\Sigma)\,\varphi_1=0,\\ D\frac{d\varphi_1}{dx}-\varphi_2=0,\\ \varphi_1(0)=\varphi_1(1)=0.\end{array}\right\}\tag{6.19}$$

Let us consider the vector function

$$f=\begin{vmatrix}f_1\\ f_2\end{vmatrix}\tag{6.20}$$

with components f_1 and f_2, this vector function belonging to the set of continuous vector functions $\{f\}$, differentiable and satisfying the conditions

$$f_1(0)=f_1(1)=0.\tag{6.21}$$

Further, let $\mathbf{L}$ be the operator which transforms a function f of the set $\{f\}$ into:

$$\mathbf{L}f=\begin{vmatrix}\frac{df_2}{dx}+(\lambda-\Sigma)\,f_1\\ D\frac{df_1}{dx}-f_2\end{vmatrix}\tag{6.22}$$

We will find $\mathbf{L}^*$ using the functional relation (6.1). To do this we first define the scalar product (f, f^*) by:

$$(f,f^*)=\int\limits_G(f_1f_1^*+f_2f_2^*)\,d\vec{r},\tag{6.23}$$

where f_1^* and f_2^* are the components of the vector function f^*, defined as

$$f^*=\begin{vmatrix}f_1^*\\ f_2^*\end{vmatrix}.\tag{6.24}$$

*It is interesting to note that if D and Σ are independent of $\underline{u}$, while $\sigma = 1$, integration of (6.18) and (6.25) over $\underline{u}$ gives

$$\frac{d}{dx}D\frac{d\varphi}{dx}+(\lambda-\Sigma)\,\varphi=0\ \text{and}\ \frac{d}{dx}D\frac{d\varphi^*}{dx}+(\lambda-\Sigma)\,\varphi^*=0,$$

where

$$\varphi(0)=\varphi(1)=0,\qquad \varphi^*(0)=\varphi^*(1)=0$$

We then have

$$(f^*,Lf)=\int_0^1 \left[f_1^* \left(\frac{df_2}{dx}+(\lambda-\Sigma)f_1\right)+f_2^*\left(D\frac{df_1}{dx}-f_2\right)\right] dx. \tag{6.25}$$

Integrating (6.25) by parts, we arrive at

$$(f^*,Lf)=(f_1^*f_2+Df_1f_2^*)\Big|_0^1+\int_0^1\left[f_1\left(-\frac{dDf_2^*}{dx}+(\lambda-\Sigma)f_1^*\right)+f_2\left(-\frac{df_1^*}{dx}-f_2^*\right)\right]dx. \tag{6.26}$$

We have here used in addition the continuity of f^*. If, further, we assume that

$$f_1^*(0)=f_1^*(1)=0, \tag{6.27}$$

Equation (6.26) can be written

$$(f^*,\ Lf)=(f,\ L^*f^*), \tag{6.28}$$

where

$$L^*f^*=\left|\begin{matrix} -\dfrac{dDf_2^*}{dx}+(\lambda-\Sigma)f_1^* \\ -\dfrac{df_1^*}{dx}-f_2^* \end{matrix}\right| . \tag{6.29}$$

Assuming, finally, f to be a solution of (6.19), we have

$$Lf=0,$$

so that from (6.28) we have

$$L^*f^*=0.$$

We thus arrive at the adjoint set of equations

$$\left.\begin{matrix} -\dfrac{dD\varphi_2^*}{dx}+(\lambda-\Sigma)\varphi_1^*=0, \\ -\dfrac{d\varphi_1^*}{dx}-\varphi_2^*=0,\ \varphi_1^*(0)=\varphi_1^*(1)=0. \end{matrix}\right\} \tag{6.30}$$

This set is equivalent to the second-order equation

$$\left.\begin{matrix} \dfrac{d}{dx}D\dfrac{d\varphi^*}{dx}+(\lambda-\Sigma)\varphi^*=0, \\ \varphi^*(0)=\varphi^*(1)=0 \end{matrix}\right\} \tag{6.31}$$

We have in this way arrived at (6.13).

Thus, when one is dealing with a set of partial differential equations, the mathematical tools of vector function theory can be used to obtain the corresponding set of adjoint equations.

In conclusion it should be noted that if f and f^* are nonzero solutions of a set of basic and adjoint equations, they have the same eigenvalue spectrum of λ. That these eigenvalue spectra are the same is extremely important and will be used in what follows in order to control the accuracy of the calculations. Further, the adjoint equations can be used to develop an effective perturbation theory. Finally, the solution of the adjoint equation can be used in general to analyze the kinetics of a nuclear reactor [82].

Having completed this introduction, let us go on to a derivation of the set of adjoint reactor equations. We shall do this using vector function methods.

Let $\boldsymbol{f}$ belong to $\{\boldsymbol{f}\}$, the set of vector functions defined in $\mathbf{G} \times \boldsymbol{\Omega} \times \mathbf{U}$ for which the integro-differential operators of (5.2) have meaning. Let us further assume that the functions $\boldsymbol{f}$ satisfy the boundary conditions of (5.4).*

Consider the adjoint vector $\boldsymbol{f}^*$ belonging to $\{f^*\}$, a set of vector functions defined in $\mathbf{G} \times \boldsymbol{\Omega} \times \mathbf{U}$, which we shall further define later. Let us write $\boldsymbol{f}^*$ in the form

$$f^* = \begin{vmatrix} \varphi^* \\ \Phi^* \end{vmatrix}, \tag{6.32}$$

where Φ^* is a function only of $\vec{r}$ and $\vec{\Omega}$, and φ is a function of $\vec{r}$, $\vec{\Omega}$ and $\underline{u}$. Then the adjoint operator is defined by the functional equation

$$(f^*, \mathbf{L}f) = (f, \mathbf{L}^*f^*), \tag{6.33}$$

where $(\boldsymbol{f}, \boldsymbol{f}^*)$ is the scalar product in $\mathbf{G} \times \boldsymbol{\Omega} \times \mathbf{U}$, defined in the following way.

We write

$$f = \begin{vmatrix} f_1 \\ f_2 \end{vmatrix}, \quad f^* = \begin{vmatrix} f_1^* \\ f_2^* \end{vmatrix},$$

and if

$$f_1 = f_1(\vec{r}, \vec{\Omega}, u), \quad f_2 = f_2(\vec{r}, \vec{\Omega}),$$

and

$$f_1^* = f_1^*(\vec{r}, \vec{\Omega}, u), \quad f_2^* = f_2^*(\vec{r}, \vec{\Omega}),$$

then

$$(f, f^*) = \int_G d\vec{r} \int d\Omega \left(\int_{-\infty}^{u_T} f_1 f_1^* \, du + f_2^* f_2 \right). \tag{6.34}$$

Using the definition (6.34) of the scalar product and Equation (5.6), we write

$$\begin{aligned}(f^*, \mathbf{L}f) = \int_G d\vec{r} \int d\Omega \Bigg\{ \int_{-\infty}^{u_T} du\varphi^* \Big[\vec{\Omega}\nabla\varphi + \Sigma\varphi - \\ - \int_{-\infty}^{u_T} du' \int d\Omega' \Sigma_s \varphi f(\mu_0, u - u') - \nu_f \frac{\chi(u)}{4\pi} \int d\Omega \Big(\int_{-\infty}^{u_T} \Sigma_f \varphi \, du + \Sigma_{f_T}\Phi \Big) \Big] + \\ + \Phi^* \Big[\vec{\Omega}\nabla\Phi + \Sigma_T\Phi - \int d\Omega' \Sigma_{sT} \Phi g(\mu_0) - \\ - \int_{-\infty}^{u_T} du' \int d\Omega' \Sigma_s \varphi F(\mu_0, u') \Big] \Bigg\},\end{aligned} \tag{6.35}$$

*For the one-velocity kinetic equations, these conditions have been formulated by V.S. Vladimirov [13].

where for convenience we have taken the region of integration over all of U, and

$$f(\mu_0, u-u')=0, \quad \text{outside of} \quad u-r\leqslant u'\leqslant u$$
$$F(\mu_0, u')=0, \quad \text{»} \quad \text{»} \quad u_T-r\leqslant u'\leqslant u_T. \tag{6.36}$$

Let us now transform the right side of (6.35). Consider the integral

$$\int d\vec{r}\varphi^*\vec{\Omega}\nabla\varphi. \tag{6.37}$$

Recalling that according to Gauss' theorem

$$\int_G d\vec{r}(\vec{\Omega}\nabla)\varphi\cdot\varphi^* = \int_S dS\,\varphi\cdot\varphi^*(\vec{\Omega}, \vec{n}^0), \tag{6.38}$$

where $\vec{n}^0$ is the unit vector normal to the surface S, and $\vec{\Omega}$ is independent of the variables of integration, we arrive immediately at the relation

$$\int_G d\vec{r}\varphi^*\vec{\Omega}\nabla\varphi = -\int_G d\vec{r}\varphi\vec{\Omega}\nabla\varphi^* + \int_S dS\varphi\cdot\varphi^*(\vec{\Omega}, \vec{n}^0). \tag{6.39}$$

If we now require that

$$\varphi^*(\vec{r}, \vec{\Omega}, u)=0 \text{ on } S \text{ for } (\vec{\Omega}, \vec{n}^0)>0, \tag{6.40}$$

and recall (5.4), we have

$$\int_S dS\varphi\cdot\varphi^*(\vec{\Omega}, \vec{n}^0)=0. \tag{6.41}$$

It should be noted that (6.41) remains valid for all possible directions $\vec{\Omega}$.

Thus,(6.39) becomes

$$\int_G d\vec{r}\varphi^*\vec{\Omega}\nabla\varphi = -\int_G d\vec{r}\varphi\vec{\Omega}\nabla\varphi^*. \tag{6.42}$$

Similarly, we can obtain

$$\int_G d\vec{r}\Phi^*\vec{\Omega}\nabla\Phi = -\int_G d\vec{r}\Phi\vec{\Omega}\nabla\Phi^* \tag{6.43}$$

on the condition that

$$\Phi^*(\vec{r}, \vec{\Omega})=0 \text{ on } S \text{ for } (\vec{\Omega}, \vec{n}^0)>0. \tag{6.44}$$

As a result, we have

$$\int_G d\vec{r}\int d\Omega\left[\int_{-\infty}^{u_T} du\varphi^*(\vec{\Omega}\nabla\varphi+\Sigma\varphi)+\Phi^*(\vec{\Omega}\nabla\Phi+\Sigma_T\Phi)\right]=$$

$$=\int_G d\vec{r}\int d\Omega\left[\int_{-\infty}^{u_T} du\varphi(-\vec{\Omega}\nabla\varphi^*+\Sigma\varphi^*)+\Phi(-\vec{\Omega}\nabla\Phi^*+\Sigma_T\Phi^*)\right]. \tag{6.45}$$

Let us further consider the expression

$$\int_{-\infty}^{u_T} du\varphi^*\int_{-\infty}^{u_T} du'\Sigma_s\varphi(\vec{r},\vec{\Omega}',u')f(\mu_0,u-u'). \tag{6.46}$$

Changing the order of integration in (6.46), we obtain

$$\int_{-\infty}^{u_T} du'\Sigma_s\varphi(\vec{r},\vec{\Omega}',u')\int_{-\infty}^{u_T} du\varphi^*(\vec{r},\vec{\Omega},u)f(\mu_0,u-u') \tag{6.47}$$

and therefore

$$\int_G d\vec{r}\int_{-\infty}^{u_T} du\int d\Omega\varphi^*(\vec{r},\vec{\Omega},u)\int_{-\infty}^{u_T} du'\int d\Omega'\Sigma_s\varphi(\vec{r},\vec{\Omega}',u')f(\mu_0,u-u')=$$

$$=\int_G d\vec{r}\int_{-\infty}^{u_T} du'\int d\Omega'\Sigma_s\varphi(\vec{r},\vec{\Omega}',u')\int_{-\infty}^{u_T} du\int d\Omega\varphi^*(\vec{r},\vec{\Omega},u)f(\mu_0,u-u'). \tag{6.48}$$

Similarly, we obtain

$$\int_G d\vec{r}\int d\Omega\Phi^*(\vec{r},\vec{\Omega})\int d\Omega'\Sigma_{sT}\Phi(\vec{r},\vec{\Omega}')g(\mu_0)=$$

$$=\int_G d\vec{r}\int d\Omega'\Phi(\vec{r},\vec{\Omega}')\int d\Omega\Phi^*(\vec{r},\vec{\Omega})g(\mu_0). \tag{6.49}$$

We now perform the transformation

$$\nu_f\int_G d\vec{r}\int_{-\infty}^{u_T} du\int d\Omega\varphi^*(\vec{r},\vec{\Omega},u)\chi(u)\int d\Omega'\left(\int_{-\infty}^{u_T}\Sigma_f\varphi du'+\Sigma_{fT}\Phi\right)=$$

$$=\nu_f\int_G d\vec{r}\int_{-\infty}^{u_T} du'\int d\Omega'\Sigma_f\varphi(\vec{r},\vec{\Omega}',u')\int d\Omega\int_{-\infty}^{u_T}\chi(u)\varphi^*(\vec{r},\vec{\Omega},u)du+$$

$$+\nu_f\int_G d\vec{r}\int d\Omega'\Sigma_{fT}\Phi(\vec{r},\vec{\Omega}')\int d\Omega\int_{-\infty}^{u_T}\chi(u)\varphi^*(\vec{r},\vec{\Omega},u)du. \tag{6.50}$$

Finally, let us consider the last expression in (6.35), changing the order of integration so that we may write it

$$\int_G d\vec{r}\int d\Omega\Phi^*(\vec{r},\vec{\Omega})\int_{-\infty}^{u_T} du'\int d\Omega'\Sigma_s\varphi(\vec{r},\vec{\Omega}',u')F(\mu_0,u')=$$

$$=\int_G d\vec{r}\int_{-\infty}^{u_T} du'\int d\Omega'\Sigma_s\varphi(\vec{r},\vec{\Omega}',u')\int d\Omega\Phi^*(\vec{r},\vec{\Omega})F(\mu_0,u'). \tag{6.51}$$

We can now use (6.45) and (6.48-6.51) to rewrite (6.35) in the form

$$(f^*, \mathbf{L}f) = \int\limits_G d\vec{r} \int d\Omega' \Big\{ \int\limits_{-\infty}^{u_T} du' \varphi \Big[-\vec{\Omega}\nabla\varphi^* + \Sigma\varphi^* - \\ - \Sigma_s \int\limits_{-\infty}^{u_T} du \int d\Omega \varphi^* f(\mu_0, u-u') - \frac{\nu_f \Sigma_f}{4\pi} \int\limits_{-\infty}^{u_T} du \int d\Omega \chi(u) \varphi^* - \\ - \Sigma_s \int d\Omega \Phi^* (\vec{r}, \vec{\Omega}) F(\mu_0, u') \Big] + \Phi \Big[-\vec{\Omega}\nabla\Phi^* + \Sigma_T \Phi^* - \\ - \Sigma_{sT} \int d\Omega \Phi^* g(\mu_0) - \frac{\nu_f \Sigma_{fT}}{4\pi} \int\limits_{-\infty}^{u_T} du \int d\Omega \chi(u) \varphi^* \Big] \Big\}. \tag{6.52}$$

It follows from (6.52) that

$$(f^*, \mathbf{L}f) = (f, \mathbf{L}^* f^*), \tag{6.53}$$

where $\mathbf{L}^* f^*$ is

$$\mathbf{L}^* f^* = \left| \begin{matrix} -\vec{\Omega}\nabla\varphi^* + \Sigma\varphi^* - \Sigma_s \int\limits_{-\infty}^{u_T} du \int d\Omega' \varphi^* f(\mu_0, u-u') - \\ - \frac{\nu_f \Sigma_f}{4\pi} \int\limits_{-\infty}^{u_T} du \int d\Omega \chi(u) \varphi^* - \Sigma_s \int d\Omega \Phi^* F(\mu_0, u) \\ -\vec{\Omega}\vec{\nabla}\Phi^* + \Sigma_T \Phi^* - \Sigma_{sT} \int d\Omega \Phi^* g(\mu_0) - \\ - \frac{\nu_f \Sigma_{fT}}{4\pi} \int\limits_{-\infty}^{u_T} du \int d\Omega \chi(u) \varphi^* \end{matrix} \right| \tag{6.54}$$

Let us now assume that f is a solution of (5.7). It is then sufficient to assume that f^* satisfies

$$\mathbf{L}^* f^* = 0. \tag{6.55}$$

We then arrive at the set of adjoint equations*

$$\left. \begin{aligned} -\vec{\Omega}\nabla\varphi^* + \Sigma\varphi^* &= \Sigma_s \int\limits_u^{u+r} du' \int d\Omega' \varphi^* (\vec{r}, \vec{\Omega}', u') f(\mu_0, u-u') + \\ &+ \frac{\nu_f \Sigma_f}{4\pi} \int\limits_{-\infty}^{u_T} du \int d\Omega \chi(u) \varphi^* + \Sigma_s \int d\Omega \Phi^* F(\mu_0, u), \\ -\vec{\Omega}\nabla\Phi^* + \Sigma_T \Phi^* &= \Sigma_{sT} \int d\Omega' \Phi^* (\vec{r}, \vec{\Omega}') g(\mu_0) + \\ &+ \frac{\nu_f \Sigma_{fT}}{4\pi} \int\limits_{-\infty}^{u_T} du \int d\Omega \chi(u) \varphi^*. \end{aligned} \right\} \tag{6.56}$$

The derivation of (6.56) also implies certain properties for the functions whose solutions we wish to find.

*This set of adjoint reactor equations was first obtained in the Soviet Union by L.N. Usachev [82]. He also analyzed the physical meaning of each of the terms involved.

Thus the function f^* must be such that the integro-differential operators of (6.56) have meaning. In addition, they must satisfy the condition

$$\left.\begin{aligned} &f^* = 0 \text{ on } S \text{ for } (\vec{\Omega}, \vec{n}^0) > 0. \\ &\varphi^* (\vec{r}, \vec{\Omega}, u_T) = \Phi^* (\vec{r}, \vec{\Omega}). \end{aligned}\right\} \tag{6.57}$$

If nuclei of more than one kind take part in the slowing down and diffusion, the set of adjoint equations becomes

$$\left.\begin{aligned} -\vec{\Omega}\nabla\varphi^* + \Sigma\varphi^* &= \sum_k \Sigma_{sk} \left[\int_u^{u+r} du' \int d\Omega' \varphi^* f_k (\mu_0, u - u') + \right. \\ &\left. + \int d\Omega' \Phi^* (\vec{r}, \vec{\Omega}') F (\mu_0, u) \right] + \frac{\nu_f \Sigma_f}{4\pi} \int_{-\infty}^{u_T} du \int d\Omega \chi (u) \varphi^*, \\ -\vec{\Omega}\nabla\Phi^* + \Sigma_T\Phi^* &= \sum_k \Sigma_{sk} \int d\Omega' \Phi^* g (\mu_0) + \\ &+ \frac{\nu_f \Sigma_{fT}}{4\pi} \int_{-\infty}^{u_T} du \int d\Omega \chi (u) \varphi^*. \end{aligned}\right\} \tag{6.58}$$

7. Definition of the Method of Successive Approximations

The set of basic reactor equations (5.2) and the adjoint set (6.56) are homogeneous. This means that we are dealing with an eigenvalue problem. It is known that nonvanishing solutions of a set of integro-differential equations cannot exist for all values of the parameters, but only for certain well-defined values. We assume, for instance, that all the parameters in Equations (5.2) and (6.26) are fixed, except for ν_f which, for convenience, we shall denote by λ_1.

We wish to find those values of λ for which there exist nonzero solutions of the equations. Let these values be $\lambda_1, \lambda_2, \ldots$*

Let us further assume that the eigenfunctions of the basic and adjoint equations which belong to the first eigenvalue $\lambda_1 = \lambda$ are nonnegative. Then these functions will describe the neutron flux and importance** in a reactor which is critical when

$$\lambda_1 = \lambda. \tag{7.1}$$

Consider Relation (7.1). The left side of this equation contains a quantity which is defined only by the properties of the fissile substance (for U^{235}, for instance, $\nu_f \cong 2.46$). The parameter λ on the right side of (7.1), however, is defined by the reactor geometry, as well as by the scattering properties and absorbing properties of the material filling the reactor volume G.

Of course, for any chosen reactor parameters and dimensions, λ is not necessarily equal to ν_f. Therefore, into the computational problem must enter the choice of just such combinations of physical parameters or reactor dimensions which will give $\lambda = \nu_f$. Only in this case will the reactor be critical.

We shall later find it convenient to introduce a new reactor characteristic K_{eff}, defined by

$$K_{eff} = \frac{\nu_f}{\lambda}. \tag{7.2}$$

We shall call K_{eff} the effective neutron multiplication factor of the reactor.

---

*The basic and adjoint equations have the same eigenvalue spectrum.

**The term "importance" has been discussed by L.N. Usachev [82].

Using (7.2), the condition of (7.1) for a critical reactor becomes

$$K_{eff} = 1. \tag{7.3}$$

It has been shown by Usachev [82] that the effective multiplication factor of a system indicates how many secondary neutrons are produced in the reactor (by each primary neutron) after a complete cycle of transformations which starts with the neutron slowing down and culminates in the production of the secondary neutrons. It follows from this that if $K_{eff} < 1$, the chain reaction decays and the reactor is thus subcritical, whereas if $K_{eff} > 1$, the chain reaction progresses and the reactor is supercritical.

Thus, in treating the basic and adjoint reactor equations, the physical quantity ν_f should be replaced by

$$\lambda = \nu_f / K_{eff}. \tag{7.4}$$

In this way the eigenvalue problem for K_{eff} becomes a complete problem.

The direct solution for the first eigenvalue of the set of basic and adjoint equations is a very difficult problem. This problem is usually effectively solved by using the method of successive approximations, which transforms the eigenvalue problem into a succession of computationally simpler Cauchy problems.*

The method of successive approximations or, as it is sometimes called, the "method of source iteration," for a set of basic reactor equations is formulated in the following way [101, 138].

We are given a certain source distribution

$$Q(\vec{r}) = \nu_f \int d\Omega \left(\int_{-\infty}^{u_T} \Sigma_f \varphi \, du + \Sigma_{f_T} \Phi \right). \tag{7.5}$$

and by successive solutions of Equations (5.2) we find new approximate values. This process should be continued until the ratio of two successive values of $Q^{(l)}(\vec{r})$ is equal to the same constant, to the necessary accuracy, for all points in the core. This ratio is the desired eigenvalue of the problem, that is,

$$K_{eff} = \lim_{l \to \infty} \frac{Q^{(l)}(\vec{r})}{Q^{(l-1)}(\vec{r})}. \tag{7.6}$$

Thus, the method of successive approximations gives the first eigenvalue of the problem. It should be noted at the same time that the corresponding eigenvalue problem can be formulated in a somewhat different way.

Indeed, consider Equations (5.2), into which we introduce K_{eff} explicitly, writing

$$\left.\begin{aligned}
\vec{\Omega}\nabla\varphi + \Sigma\varphi &= \int_{u-r}^{u} du' \int d\Omega' \, \Sigma_s \varphi(\vec{r}, \vec{\Omega}', u') f(\mu_0, u-u') + \\
&+ \frac{\nu_f}{4\pi K_{eff}} \chi(u) \int d\Omega \left(\int_{-\infty}^{u_T} \Sigma_f \varphi \, du + \Sigma_{f_T} \Phi \right), \\
\vec{\Omega}\nabla\Phi + \Sigma_T \Phi &= \int d\Omega' \, \Sigma_{sT} \Phi(\vec{r}, \vec{\Omega}') g(\mu_0) + \\
&+ \int_{u_T - r}^{u_T} du' \int d\Omega' \, \Sigma_s \varphi(\vec{r}, \vec{\Omega}', u') F(\mu_0, u').
\end{aligned}\right\} \tag{7.7}$$

*For a definition of the method of successive approximations and its convergence, see L.N. Usachev [82].

This set of equations determines the first eigenvalue $\lambda = \nu_f/K_{eff}$ which will give an everywhere nonnegative solution of the problem. This formulation of the problem, with the explicit dependence of the solution on the desired parameter K_{eff}, will be used later.

It should be noted that the two above formulations of the eigenvalue problem are equivalent.

The method of successive approximations for the solution of the set of adjoint equations (6.56) can be formulated in the following way.

We are given a distribution function

$$Q^*(\vec{r}) = \nu_f \int d\Omega \int_{\infty}^{u_T} du\, \chi(u)\, \varphi^*(\vec{r}, \vec{\Omega}, u). \tag{7.8}$$

We then solve Equations (6.56) to obtain a new approximation, etc. Then the eigenvalue is obtained from the ratio of two successive values $Q^{*(l)}(\vec{r})$, that is,

$$K_{eff} = \lim_{l\to\infty} \frac{Q^{*(l)}(\vec{r})}{Q^{*(l-1)}(\vec{r})}. \tag{7.9}$$

Just as we did for the basic reactor equations, we can formulate the corresponding eigenvalue problem for the adjoint equations, in which the dependence of the solution on the characteristic parameter $\lambda = 1/K_{eff}$ is exhibited explicitly in the integro-differential equations:

$$\left.\begin{aligned}
&-\vec{\Omega}\nabla\varphi^* + \Sigma\varphi^* = \Sigma_s \int_u^{u+r} du' \int d\Omega' \varphi^*(\vec{r}, \vec{\Omega}', u') f(\mu_0, u - u') + \\
&+ \frac{\nu_f}{4\pi K_{eff}} \Sigma_f \int_{-\infty}^{u_T} du \int d\Omega \chi(u) \varphi^* + \Sigma_s \int d\Omega \Phi^* F(\mu_0, u), \\
&-\vec{\Omega}\nabla\Phi^* + \Sigma_T \Phi^* = \Sigma_{sT} \int d\Omega' \Phi^*(\vec{r}, \vec{\Omega}') g(\mu_0) + \\
&+ \frac{\nu_f}{4\pi K_{eff}} \Sigma_{fT} \int_{-\infty}^{u_T} du \int d\Omega \chi(u) \varphi^*.
\end{aligned}\right\} \tag{7.10}$$

It is quite clear that the K_{eff} values obtained from the basic and adjoint equations will be the same.

It should be recalled that in defining the method of successive approximations, we must assume the existence of a first eigenvalue K_{eff} and the convergence of the successive approximation method. These two propositions have not yet been proved in general. V.S. Vladimirov [11, 12, 15] has proven the validity of the successive approximation method in general for one-velocity problems. In Chapter V we shall prove the convergence of the successive approximation method for a space-energy problem in a reactor of any shape without a reflector. The proof will be within the framework of age theory.

Chapter II

DIFFUSION APPROXIMATION

It is extremely difficult to obtain exact solutions of Equations (5.2) and (6.56) in order to calculate the critical mass and space-energy distribution of the neutron flux and importance. In most cases, therefore, one attempts to find only approximate solutions. Unique among the methods for attaining these is the so-called diffusion approximation. Essentially, this method consists of attempting to obtain a solution of the kinetic equations in the form of a series of spherical functions, using only the first two terms of the expansion.

The diffusion approximation can be used for a more or less satisfactory description of processes taking place in reactors whose dimensions are much greater than the scattering mean free path of the neutrons. This is not a very restrictive requirement, and as a rule it is fulfilled for a large class of problems of practical interest.

The present chapter will give a derivation of the diffusion equations using the kinetic reactor equations.[*]

8. The Diffusion Approximation. Set of Basic Reactor Equations

Let us consider the neutron slowing-down equation. As has been established in Chapter I, this equation is of the form

$$\vec{\Omega}\nabla\varphi + \Sigma\varphi = \int_{u-r}^{u} du' \int d\Omega' \Sigma_s \varphi f(\mu_0, u-u') + S(\vec{r}, \vec{\Omega}, u). \tag{8.1}$$

Let us rewrite (8.1) in the form

$$\frac{d\varphi}{ds} + \Sigma\varphi = I(\vec{r}, \vec{\Omega}, u),$$

where

$$I(\vec{r}, \vec{\Omega}, u) = \int_{u-r}^{u} du' \int d\Omega' \Sigma_s \varphi f(\mu_0, u-u') + S(\vec{r}, \vec{\Omega}, u).$$

Consider a point M in an infinite homogeneous medium, and let us choose some distribution $I(\vec{r}, \vec{\Omega}, u)$ of slowing-down sources. It is easily seen that the number of neutrons of lethargy $\underline{u}$ scattered into the element of solid angle $d\Omega$ about the direction $\vec{\Omega}$ at the point M will be

$$\varphi d\Omega = \int_{0}^{\infty} I e^{-\Sigma l} dl \, d\Omega.$$

For convenience, we have taken M as the origin of coordinates. Let us assume for simplicity that $I(\vec{r}, \vec{\Omega}, u) = I(u)$. Then

$$\varphi d\Omega = I(u) \int_{0}^{\infty} e^{-\Sigma l} dl \, d\Omega = \frac{I(u)}{\Sigma} d\Omega.$$

---

[*] In deriving this equation we shall follow Marchak, Brooks and Hurwitz [118].

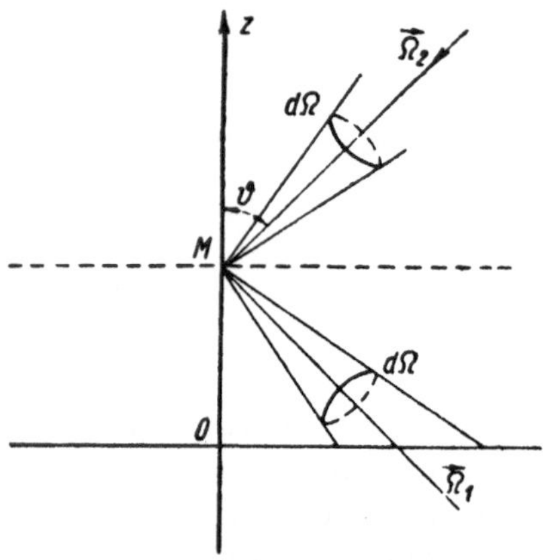

Fig. 5. Neutron paths in the vicinity of a point M in the semi-infinite volume $z \geq 0$. $\vec{\Omega}_1$ and $\vec{\Omega}_2$ are unit vectors along the neutron velocity; $d\Omega$ is the solid angle.

Thus, in this case the neutron flux φ at M does not depend on $\vec{\Omega}$, but only on $d\Omega$, so that it is isotropic.

If I is a function of $\vec{r}$ and of the point $\vec{\Omega}$, the flux φ is no longer isotropic.

This is most easily seen from the example of the semi-infinite space $z \geq 0$ bordering on the vacuum. Let us now assume that $I = I(u)$ for $z \geq 0$, and that $I = 0$ for $z < 0$. Let us calculate the number of neutrons scattered into an element of solid angle in different directions $\vec{\Omega}$ (Fig. 5).

It is easily seen that

$$\varphi d\Omega = \begin{cases} \frac{I}{\Sigma} d\Omega & \text{for} \quad 0 \leqslant \vartheta \leqslant \frac{\pi}{2}, \\ \frac{I}{\Sigma}\left(1 - e^{-\frac{\Sigma z}{\cos\vartheta}}\right) d\Omega & \text{for} \quad \frac{\pi}{2} \leqslant \vartheta \leqslant \pi. \end{cases}$$

Thus, the neutron flux is found to be anisotropic in this case. The anisotropy is, of course, small for $\Sigma_z \gg 1$, and it vanishes completely as $z \to \infty$. Thus, far from the interface the flux is almost isotropic. As $z \to 0$, on the other hand, the anisotropy approaches a maximum, since

$$\varphi d\Omega = \begin{cases} \frac{I}{\Sigma} d\Omega & \text{for} \quad 0 \leqslant \vartheta \leqslant \frac{\pi}{2}, \\ 0 & \text{for} \quad \frac{\pi}{2} \leqslant \vartheta \leqslant \pi. \end{cases}$$

We have thus analyzed qualitatively the limiting case, in which $I(\vec{r}, \vec{\Omega}, u)$ is a piecewise constant function of $\vec{r}$. The situation is substantially the same when $I(\vec{r}, \vec{\Omega}, u)$ is a continuous but locally rapidly varying function of $\vec{r}$ and $\vec{\Omega}$. Then, in the neighborhood of a strong local inhomogeneity, the anisotropy of φ also increases.

Thus, on the basis of the above remarks we may assert that the anisotropy of the neutron flux will be small if

$$\frac{d\varphi}{ds} \ll \Sigma\varphi \tag{8.2}$$

for any directions $\vec{\Omega}$ about M.

We may attempt to find a solution of (8.1) in the form of a series of spherical functions [74, 78]:

$$\varphi(\vec{r}, \vec{\Omega}, u) = \frac{1}{4\pi} \sum_{l=0}^{\infty} \left[\varphi_l P_l(\mu) + \sum_{m=1}^{l} (\varphi_l^m \sin m\psi + \varphi_l^{-m} \cos m\psi) P_l^m(\mu) \right], \tag{8.3}$$

where $\varphi_l(\vec{r}, u)$, $\varphi_l^m(\vec{r}, u)$ and $\varphi_l^{-m}(\vec{r}, u)$ are the desired functions, and $P_l(\mu)$ and $P_l^m(\mu)$ are Legendre polynomials and associated Legendre polynomials in which $\mu = \cos\vartheta$.

Obviously, if (8.2) is fulfilled, we need keep only the terms with $l = 0$ and $l = 1$ in (8.3).

Then taking account of the form of $P_l^m(\mu)$, the solution can be written approximately in the form

$$\varphi(\vec{r}, \vec{\Omega}, u) = \frac{1}{4\pi}[\varphi_0(\vec{r}, u) + 3\vec{\Omega}\vec{\varphi}_1(\vec{r}, u)] \tag{8.4}$$

where

$$\vec{\varphi}_1 = \varphi_1^{(1)}\vec{i} + \varphi_1^{(-1)}\vec{j} + \varphi_1\vec{k},$$

and

$$\vec{\Omega} = \sin\vartheta \cdot \cos\psi \cdot \vec{i} + \sin\vartheta \cdot \sin\psi \vec{j} + \cos\vartheta \cdot \vec{k}.$$

It is easily shown that

$$\varphi_0 (\vec{r}, u) = \int d\Omega \varphi (\vec{r}, \vec{\Omega}, u), \quad \vec{\varphi}_1 (\vec{r}, u) = \int d\Omega \cdot \vec{\Omega} \varphi (\vec{r}, \vec{\Omega}, u).$$

In what follows we shall make use of vector notation for writing (8.4). Let the collision function $f(\mu_0, u)$ be written approximately in the form

$$f(\mu_0, u) = \frac{1}{4\pi} [f_0(u) + 3\mu_0 f_1(u)], \tag{8.5}$$

where

$$f_0(u) = \int d\Omega f(\mu_0, u), \quad f_1(u) = \int d\Omega \mu_0 f(\mu_0, u).$$

Bearing in mind that $f(\mu_0, u)$ is of the form

$$f(\mu_0, u) = \frac{(M+1)^2}{8\pi M} e^{-u} \delta \left[\mu_0 - \left(\frac{M+1}{2} e^{-\frac{u}{2}} - \frac{M-1}{2} e^{\frac{u}{2}} \right) \right], \tag{8.6}$$

we see that according to (8.5)

$$f_0(u) = \frac{(M+1)^2}{4M} e^{-u}, \quad f_1(u) = f_0(u) \mu_0(u), \tag{8.7}$$

where

$$\mu_0(u) = \frac{M+1}{2} e^{-\frac{u}{2}} - \frac{M-1}{2} e^{\frac{u}{2}}.$$

Inserting (8.4) and (8.5) into (8.1) and integrating over $d\Omega$, we obtain an equation relating the new unknown functions φ_0 and $\vec{\varphi}_1$. In order to obtain the necessary second equation, we again insert (8.4) and (8.5) into (8.1), this time multiplying the result by $\vec{\Omega}$ and integrating over $\vec{\Omega}$.

As a result we arrive at the integro-differential slowing-down equations

$$\begin{aligned} \nabla\vec{\varphi}_1 + \Sigma\varphi_0 &= \int_{u-r}^{u} du' \Sigma_s \varphi_0 f_0 (u-u') + S(\vec{r}, u), \\ \frac{1}{3} \nabla\varphi_0 + \Sigma\vec{\varphi}_1 &= \int_{u-r}^{u} du' \Sigma_s \vec{\varphi}_1 f_1 (u-u') \\ S(\vec{r}, u) &= \int d\Omega S(\vec{r}, \vec{\Omega}, u). \end{aligned} \tag{8.8}$$

Here we have used the notation $\nabla\vec{A} = \operatorname{div}\vec{A}$, and $\nabla A = \operatorname{grad} A$. It must be noted that in deriving (8.8) we used the easily verified relations

$$\begin{aligned} &\int d\Omega = \int_0^{2\pi} d\varphi \int_0^{\pi} \sin\vartheta \, d\vartheta = 4\pi, \quad \int d\Omega \, \vec{\Omega} = 0, \\ &\int \vec{\Omega} \nabla (\vec{\Omega} \vec{\varphi}_1) \, d\Omega = \frac{4\pi}{3} \nabla\vec{\varphi}_1, \quad \int d\Omega \, \vec{\Omega} (\vec{\Omega}, \vec{\Omega}') = \frac{4\pi}{3} \vec{\Omega}', \end{aligned} \tag{8.9}$$

where, as before,

$$\vec{\Omega} = \sin\vartheta\cdot\cos\psi\vec{i} + \sin\vartheta\cdot\sin\psi\cdot\vec{j} + \cos\vartheta\cdot\vec{k}.$$

Let us now consider the kinetic equation for the thermal neutron group, namely

$$\vec{\Omega}\nabla\Phi + \Sigma_T\Phi = \int d\Omega'\Sigma_{sT}\Phi g(\mu_0) + \int_{u_T-r}^{u_T} du' \int d\Omega'\Sigma_s\varphi F(\mu_0, u'). \tag{8.10}$$

Let us write $g(\mu_0)$ in the form

$$g(\mu_0) = \frac{1}{4\pi}[1 + 3\mu_0 g_1], \tag{8.11}$$

where

$$g_1 = \bar{\mu}_{0T} = \text{const}. \tag{8.12}$$

We shall attempt to find a solution to (8.10) in the form

$$\Phi(\vec{r}, \vec{\Omega}) = \frac{1}{4\pi}[\Phi_0(\vec{r}) + 3\vec{\Omega}\vec{\Phi}_1(\vec{r})], \tag{8.13}$$

where

$$\Phi_0(\vec{r}) = \int d\Omega\Phi(\vec{r}, \vec{\Omega}), \quad \vec{\Phi}_1(\vec{r}) = \int d\Omega\vec{\Omega}\Phi(\vec{r}, \vec{\Omega}.) \tag{8.14}$$

We now insert (8.5), (8.4), (8.11) and (8.13) into (8.10) and perform roughly the same mathematical operations as we did in deriving Equations (8.8). The resulting equations for Φ_0 and $\vec{\Phi}_1$ are

$$\left.\begin{aligned} \nabla\vec{\Phi}_1 + \Sigma_{cT}\Phi_0 &= \int_{u_T-r}^{u_T} du'\Sigma_s\varphi_0 F_0(u'), \\ \frac{1}{3}\nabla\Phi_0 + \Sigma_{trT}\vec{\Phi}_1 &= \int_{u_T-r}^{u_T} du'\Sigma_s\vec{\varphi}_1 F_1(u'), \end{aligned}\right\} \tag{8.15}$$

where

$$\left.\begin{aligned} F_0(u') = \int_{u_T}^{u'+r} f_0(u-u')\,du, \quad F_1(u') = \int_{u_T}^{u'+r} f_1(u-u')\,du, \\ \Sigma_{trT} = \Sigma_{sT}(1 - \bar{\mu}_{0T}) + \Sigma_{cT}. \end{aligned}\right\} \tag{8.16}$$

Finally, we must consider the expression for $S(\vec{r}, u)$, namely

$$S(\vec{r}, u) = \frac{\nu_f}{K_{eff}}\chi(u)\int d\Omega\left[\int_{-\infty}^{u_T}\Sigma_f\varphi\,du + \Sigma_{fT}\Phi\right]. \tag{8.17}$$

According to (8.4) and (8.13), $S(\vec{r}, u)$ can be written

$$S(\vec{r}, u) = \frac{\nu_f}{K_{eff}}\chi(u)\left[\int_{-\infty}^{u_T}\Sigma_f\varphi_0\,du + \Sigma_{fT}\Phi_0\right]. \tag{8.18}$$

Then the set of reactor equations becomes

$$\left.\begin{aligned}
&\nabla\vec{\varphi}_1 + \Sigma\varphi_0 = \int\limits_{u-r}^{u} du'\Sigma_s\varphi_0 f_0(u-u') + \\
&+ \frac{\nu_f}{K_{\text{eff}}}\chi(u)\left[\int\limits_{-\infty}^{u_T}\Sigma_f\varphi_0 du + \Sigma_{fT}\Phi_0\right], \\
&\frac{1}{3}\nabla\varphi_0 + \Sigma\vec{\varphi}_1 = \int\limits_{u-r}^{u} du'\Sigma_s\vec{\varphi}_1 f_1(u-u'), \\
&\nabla\vec{\Phi}_1 + \Sigma_{cT}\Phi_0 = \int\limits_{u_T-r}^{u_T} du'\Sigma_s\varphi_0 F_0(u'), \\
&\frac{1}{3}\nabla\Phi_0 + \Sigma_{trT}\vec{\Phi}_1 = \int\limits_{u_T-r}^{u_T} du'\Sigma_s\vec{\varphi}_1 F_1(u').
\end{aligned}\right\} \quad (8.19)$$

If the moderator consists of several different types of nuclei, the set of reactor equations is

$$\left.\begin{aligned}
&\nabla\vec{\varphi}_1 + \Sigma\varphi_0 = \sum_k \int\limits_{u-r_k}^{u} du'\Sigma_{sk}\varphi_0 f_{0k}(u-u') + \\
&+ \frac{\nu_f}{K_{\text{eff}}}\chi(u)\left[\int\limits_{-\infty}^{u_T}\Sigma_f\varphi_0 du + \Sigma_{fT}\Phi_0\right], \\
&\frac{1}{3}\nabla\varphi_0 + \Sigma\vec{\varphi}_1 = \sum_k \int\limits_{u-r_k}^{u} du'\Sigma_{sk}\vec{\varphi}_1 f_{1k}(u-u'), \\
&\nabla\vec{\Phi}_1 + \Sigma_{cT}\Phi_0 = \sum_k \int\limits_{u_T-r_k}^{u_T} du'\Sigma_{sk}\varphi_0 F_0(u'), \\
&\frac{1}{3}\nabla\Phi_0 + \Sigma_{trT}\vec{\Phi}_1 = \sum_k \int\limits_{u_T-r_k}^{u_T} du'\Sigma_{sk}\vec{\varphi}_1 F_1(u').
\end{aligned}\right\}. \quad (8.20)$$

9. The Diffusion Approximation. Set of Adjoint Reactor Equations

In order to find the set of adjoint reactor equations in the diffusion approximation, let us consider Equations (7.10), namely:

$$\left.\begin{aligned}
&-\vec{\Omega}\nabla\varphi^* + \Sigma\varphi^* = \Sigma_s\int\limits_{u}^{u+r} du'\int d\Omega'\varphi^* f(\mu_0, u-u') + \\
&+ \Sigma_s\int d\Omega'\Phi^* F(\mu_0, u) + \frac{\nu_f}{4\pi K_{\text{eff}}}\Sigma_f\int\limits_{-\infty}^{u}\chi(u)\,du\int d\Omega\varphi^*, \\
&-\vec{\Omega}\nabla\Phi^* + \Sigma_T\Phi^* = \Sigma_{sT}\int d\Omega'\Phi^* g(\mu_0) + \frac{\nu_f}{4\pi K_{\text{eff}}}\Sigma_{fT}\int\limits_{-\infty}^{u_T}\chi(u)\,du\int d\Omega\varphi^*.
\end{aligned}\right\} \quad (9.1)$$

We shall attempt to obtain a solution of (9.1), taking account of (8.5), in the form

$$\varphi^*(\vec{r}, \vec{\Omega}, u) = \frac{1}{4\pi}[\varphi_0^*(\vec{r}, u) + 3\vec{\Omega}\vec{\varphi}_1^*(\vec{r}, u)],$$
$$\Phi^*(\vec{r}, \vec{\Omega}) = \frac{1}{4\pi}[\Phi_0^*(\vec{r}) + 3\vec{\Omega}\vec{\Phi}_1^*(\vec{r})]. \tag{9.2}$$

We now insert (8.5) and (9.2) into (9.1), and integrate over $\vec{\Omega}$. We then again insert (8.5) and (9.2) into (9.1), but this time take the scalar product with $\vec{\Omega}$ and integrate over $\vec{\Omega}$. After some simple mathematical operations similar to those of Section 8, we arrive at the diffusion equations

$$\left.\begin{aligned}
&-\nabla\vec{\varphi}_1^* + \Sigma\varphi_0^* = \Sigma_s \int_u^{u+r} du'\varphi_0^* f_0(u'-u) + \Sigma_s\Phi_0^* F_0(u) + \\
&\quad + \frac{\nu_f}{K_{\text{eff}}}\Sigma_f \int_{-\infty}^{u_T} \chi(u)\varphi_0^* du, \\
&-\frac{1}{3}\nabla\varphi_0^* + \Sigma\vec{\varphi}_1^* = \Sigma_s \int_u^{u+r} du'\vec{\varphi}_1^* f_1(u'-u) + \Sigma_s\vec{\Phi}_1^* F_1(u) \\
&-\nabla\vec{\Phi}_1^* + \Sigma_{cT}\Phi_0^* = \frac{\nu_f}{K_{\text{eff}}}\Sigma_{fT}\int_{-\infty}^{u_T}\chi(u)\varphi_0^* du, \\
&-\frac{1}{3}\nabla\Phi_0^* + \Sigma_{trT}\vec{\Phi}_1^* = 0.
\end{aligned}\right\} \tag{9.3}$$

If the moderator consists of several different types of nuclei, the set of adjoint reactor equations becomes

$$\left.\begin{aligned}
&-\nabla\vec{\varphi}_1^* + \Sigma\varphi_0^* = \sum_h \left[\Sigma_{sh}\int_u^{u+r_h} du'\varphi_0^* f_{0h}(u'-u) + \Sigma_{sh}\Phi_0^* F_0(u)\right] + \\
&\quad + \frac{\nu_f}{K_{\text{eff}}}\Sigma_f \int_{-\infty}^{u_T}\chi(u)\varphi_0^* du, \\
&-\frac{1}{3}\nabla\varphi_0^* + \Sigma\vec{\varphi}_1^* = \sum_h\left[\Sigma_{sh}\int_u^{u+r_h} du'\vec{\varphi}_1^* f_{1h}(u'-u) + \Sigma_{sh}\vec{\Phi}_1^* F_1(u)\right], \\
&-\nabla\vec{\Phi}_1^* + \Sigma_T\Phi_0^* = \frac{\nu_f}{K_{\text{eff}}}\Sigma_{fT}\int_{-\infty}^{u_T}\chi(u)\varphi_0^* du, \\
&-\frac{1}{3}\nabla\Phi_0^* + \Sigma_{trT}\vec{\Phi}_1^* = 0.
\end{aligned}\right\} \tag{9.4}$$

10. Boundary Conditions

For a complete definition of the eigenvalue problem for Equations (8.19), we must include the boundary conditions on the surface (S).

Consider, for instance, the boundary condition for $\varphi(\vec{r}, \vec{\Omega}, u)$, namely:

$$\varphi(\vec{r}, \vec{\Omega}, u) = 0 \quad \text{on } S \quad \text{for} \quad (\vec{\Omega}, \vec{n}^0) < 0, \tag{10.1}$$

which means that the neutrons leaving the reactor region **G** cannot be reflected by the vacuum and return to the reactor.

To obtain the approximate boundary conditions corresponding to the diffusion approximation, we replace (10.1) by the integral condition

$$\int_{\Omega} d\Omega\,|\Omega_n|\,\varphi(\vec{r}, \vec{\Omega}, u) = 0 \text{ on } S, \tag{10.2}$$

where

$$\Omega_n = -\cos\vartheta,$$

and the integration is taken over the unit hemisphere swept out by the ends of the $\vec{\Omega}$ vectors directed into **G**, and therefore bounded by the tangent plane to the convex surface S. Equation (10.2) states that the integrated neutron flux through S, from the vacuum into **G**, vanishes.

Let us direct the z axis along the external normal to S at the point whose radius vector is $\vec{r}$. Then (10.2) can be written

$$\int_0^{2\pi} d\psi \int_{\frac{\pi}{2}}^{\pi} \varphi(\vec{r}, \vec{\Omega}, u)\cos\vartheta\sin\vartheta\, d\vartheta = 0 \text{ on } S. \tag{10.3}$$

Let us now insert (8.4) into (10.3). Then integration gives

$$2(\vec{\varphi}_1)_{\vec{n}^0} - \varphi_0 = 0, \tag{10.4}$$

where $(\vec{\varphi}_1)_{\vec{n}^0}$ is the projection of $\vec{\varphi}_1$ onto the external normal $\vec{n}^0$.

Similarly, we find the boundary conditions in the diffusion approximation for Φ to be

$$2(\vec{\Phi}_1)_{\vec{n}^0} - \Phi_0 = 0 \text{ on } S. \tag{10.5}$$

Equations (10.4) and (10.5) are our necessary boundary conditions for (8.19).

Let us now go on to establishing the boundary conditions for the set of adjoint equations (9.3). First consider the boundary conditions for $\varphi^*(\vec{r}, \vec{\Omega}, u)$. The exact boundary condition is

$$\varphi^*(\vec{r}, \vec{\Omega}, u) = 0 \text{ on } S \text{ for } (\vec{\Omega}, \vec{n}^0) > 0, \tag{10.6}$$

but we shall replace it by the approximation

$$\int_{\Omega} d\vec{\Omega}\,|\Omega_n|\,\varphi^*(\vec{r}, \vec{\Omega}, u) = 0 \text{ on } S, \tag{10.7}$$

where the integration is taken over the hemisphere directed outwardly from the reactor. Let us write (10.7) in the form

$$\int_0^{2\pi} d\psi \int_{\frac{\pi}{2}}^{\pi} \varphi^*(\vec{r}, \vec{\Omega}, u)\cos\vartheta\cdot\sin\vartheta\cdot d\vartheta = 0 \text{ on } S. \tag{10.8}$$

We now insert (9.2) into (10.8) and perform the necessary integration. The resulting boundary condition we obtain for φ^* is

$$2(\vec{\varphi}_1^*)_{\vec{n}^0} + \varphi_0^* = 0 \text{ on } S. \tag{10.9}$$

The boundary condition we obtain similarly for Φ^* is

$$2(\vec{\Phi}_1^*)_{\vec{n}_0} + \Phi_0^* = 0 \quad \text{on} \quad S. \tag{10.10}$$

11. Region of Existence of a Solution

Let us consider the vector function

$$f = \begin{Vmatrix} \varphi_0 \\ \vec{\varphi}_1 \\ \Phi_0 \\ \vec{\Phi}_1 \end{Vmatrix}, \tag{11.1}$$

whose components are the scalar functions $\varphi_0(\vec{r}, u)$ and $\Phi_0(\vec{r})$ and the vector functions $\vec{\varphi}_1(\vec{r}, u)$ and $\vec{\Phi}_1(\vec{r})$.

Then in terms of f the set of basic reactor equations is

$$\mathbf{L}f = \begin{Vmatrix} \nabla\vec{\varphi}_1 + \Sigma\varphi_0 - \int\limits_{u-r}^{u} du'\Sigma_s\varphi_0 f_0(u-u') - \frac{1}{K_{\text{eff}}}\chi(u)Q \\ \nabla\varphi_0 + 3\Sigma\vec{\varphi}_1 - 3\int\limits_{u-r}^{u} du'\Sigma_s\vec{\varphi}_1 f_1(u-u') \\ \nabla\vec{\Phi}_1 + \Sigma_{cT}\Phi_0 - \int\limits_{u_T-r}^{u_T} du'\Sigma_s\varphi_0 F_0(u') \\ \nabla\Phi_0 + 3\Sigma_{trT}\vec{\Phi}_1 - 3\int\limits_{u_T-r}^{u_T} du'\Sigma_s\vec{\varphi}_1 F_1(u') \end{Vmatrix} = 0, \tag{11.2}$$

where

$$Q = \nu_f\left(\int\limits_{-\infty}^{u_T} \Sigma_f\varphi_0\, du + \Sigma_{fT}\Phi_0\right). \tag{11.3}$$

We assume further that the components of f satisfy conditions (10.4) and (10.5).

Let us further say that the solution of (11.2) belongs to $\{f\}$, the class of functions for which the operator $\mathbf{L}$ has meaning and whose components vanish at the external extrapolated reactor surface.

Let us further consider the adjoint vector function f^* given by

$$f^* = \begin{Vmatrix} \varphi_0^* \\ \vec{\varphi}_1^* \\ \Phi_0^* \\ \vec{\Phi}_1^* \end{Vmatrix}, \tag{11.4}$$

which is a solution of

$$
\mathbf{L}^* \boldsymbol{f}^* = \left| \begin{array}{l}
-\nabla\vec{\varphi}_1^* + \Sigma\varphi_0^* - \Sigma_s \int\limits_u^{u+r} \varphi_0^* f_0 (u' - u)\, du' - \\
\qquad - \Sigma_s \Phi_0^* F_0 (u) - \dfrac{1}{K_{\text{eff}}} \Sigma_f Q^* \\
-\nabla\varphi_0^* + 3\Sigma\vec{\varphi}_1^* - 3\Sigma_s \int\limits_u^{u+r} \vec{\varphi}_1^* f_1 (u' - u)\, du' - \\
\qquad - 3\Sigma_s \vec{\Phi}_1^* F_1 (u) \\
-\nabla\vec{\Phi}_1^* + \Sigma_{c\mathrm{T}} \Phi_0^* - \dfrac{1}{K_{\text{eff}}} \Sigma_{f\mathrm{T}} Q^* \\
-\nabla\Phi_0^* + 3\Sigma_{tr\mathrm{T}} \vec{\Phi}_1^*
\end{array} \right| = 0, \tag{11.5}
$$

where

$$
Q^* = \nu_f \int\limits_{-\infty}^{u_{\mathrm{T}}} \varphi_0^* \chi (u)\, du. \tag{11.6}
$$

It is easily shown that Equations (11.5), which were derived in Section 9, are indeed adjoint to the basic equations of (11.2) if the vector function $\boldsymbol{f}^*$ belongs to the set $\{\boldsymbol{f}^*\}$ whose elements satisfy boundary conditions (10.9) and (10.10) and for which the operator $\mathbf{L}^*$ has meaning.

Chapter III

DIFFUSION-AGE APPROXIMATION

The preceding chapter gave the transition to the diffusion approximation, which simplifies the mathematical statement of the problem. The solution of the basic and adjoint reactor equations in the diffusion approximation is also, however, very difficult. The computational difficulties are particularly great when the moderator consists of a mixture of elements. These difficulties are due primarily to the complicated structure of the energy operators involved in the solution. Hence, in order to develop an effective method for reactor calculations, we must further simplify the mathematical statement of the problem. This time the simplification must be based on assumptions as to the type of energy dependence of the solution. The final result of these assumptions is that the exact energy operators of the problem are replaced by approximate ones. The present chapter shall deal with this subject.

12. The Diffusion-Age Approximation for the Basic Set of Reactor Equations

Consider the set of neutron slowing-down equations in the diffusion approximation, that is,

$$\left.\begin{aligned} \nabla \vec{\varphi}_1 + \Sigma\varphi_0 &= \int_{u-r}^{u} du'\Sigma_s\varphi_0 f_0(u-u') + S, \\ \frac{1}{3}\nabla\varphi_0 + \Sigma\vec{\varphi}_1 &= \int_{u-r}^{u} du'\Sigma_s\vec{\varphi}_1 f_1(u-u') \end{aligned}\right\} \tag{12.1}$$

Let us assume that over the range of variation $\underline{r}$ of the lethargy, the functions $\Sigma_s\varphi_0$ and $\Sigma_s\vec{\varphi}_1$ vary only weakly* so that they can be represented with sufficient accuracy by the first few terms of a Taylor's series:

$$\left.\begin{aligned} \Sigma_s\varphi_0(\vec{r}, u') &= \Sigma_s\varphi_0(\vec{r}, u) + \frac{\partial\Sigma_s\varphi_0}{\partial u}(u'-u) + \frac{1}{2}\frac{\partial^2\Sigma_s\varphi_0}{\partial u^2}(u'-u)^2, \\ \Sigma_s\vec{\varphi}_1(\vec{r}, u') &= \Sigma_s\vec{\varphi}_1(\vec{r}, u) + \frac{\partial\Sigma_s\vec{\varphi}_1}{\partial u}(u'-u). \end{aligned}\right\} \tag{12.2}$$

Let us now insert (12.2) into the right side of (12.1) and integrate. We then obtain

$$\left.\begin{aligned} \nabla\vec{\varphi}_1 + \Sigma_c\varphi_0 &= -\xi\frac{\partial\Sigma_s\varphi_0}{\partial u} + \frac{\overline{\xi^2}}{2}\frac{\partial^2\Sigma_s\varphi_0}{\partial u^2} + S, \\ \frac{1}{3}\nabla\varphi_0 + \Sigma_{tr}\vec{\varphi}_1 &= -\overline{\Delta u\mu_0}\frac{\partial\Sigma_s\vec{\varphi}_1}{\partial u}, \end{aligned}\right\} \tag{12.3}$$

where

$$\left.\begin{aligned} \Sigma_{tr} &= \Sigma_s(1-\bar{\mu}_0) + \Sigma_c, & \xi &= \int_0^r t f_0(t)\,dt, \\ \overline{\xi^2} &= \int_0^r t^2 f_0(t)\,dt, & \bar{\mu}_0 &= \int_0^r \mu_0(t) f_0(t)\,dt, \end{aligned}\right\} \tag{12.4}$$

---

*Such an assumption is usually valid enough when the atomic number M of the moderator is much greater than 1.

$$\overline{\Delta u\mu_0} = \int_0^r t\mu_0(t) f_0(t)\,dt, \qquad \mu_0(t) = \frac{M+1}{2} e^{-\frac{t}{2}} - \frac{M-1}{2} e^{\frac{t}{2}}, \tag{12.5}$$

and

$$f_0(t) = \frac{(M+1)^2}{4M} e^{-t} \tag{12.6}$$

is the neutron lethargy distribution function normalized to unity. As is implied by the structure of (12.4), we shall call ξ the mean lethargy loss, $\overline{\xi^2}$ the mean square lethargy loss, $\bar{\mu}_0$ the mean cosine of the scattering angle, and $\overline{\Delta u \mu_0}$ the mean product of the lethargy loss and the cosine of the scattering angle.

Let us estimate the order of magnitude of the quantities appearing in (12.3). To do this, we go from the independent variables $\vec{r}$ and u to new variables $\vec{r}'$ and u' according to

$$\vec{r} = L\vec{r}', \qquad u = Uu', \tag{12.7}$$

where L and U are quantities characterizing the scale of the processes, and $\vec{r}'$ and u' are new dimensionless variables. Further, let L be chosen so that

$$\Sigma_s = \frac{1}{L}\Sigma_s', \qquad \Sigma_c = \frac{1}{L}\Sigma_c', \qquad S = \frac{1}{L}S', \tag{12.8}$$

where $1/L$ is the maximum of $\Sigma_s(\vec{r}, u)$ and $\Sigma_c(\vec{r}, u)$ in the interval $(u - r, u)$. We note finally that φ_0 and $|\vec{\varphi}_1|$ are of the same order of magnitude.

Let us now insert (12.7) and (12.8) into (12.3), multiplying term by term by L. We then obtain

$$\left.\begin{aligned} \nabla'\vec{\varphi}_1 + \Sigma_c'\varphi_0 &= -\frac{\xi}{U}\frac{\partial\Sigma_s'\varphi_0}{\partial u'} + \frac{\overline{\xi^2}}{2U^2}\frac{\partial^2\Sigma_s'\varphi_0}{\partial u'^2} + S' \\ \frac{1}{3}\nabla'\varphi_0 + \Sigma_{tr}'\vec{\varphi}_1 &= -\frac{\overline{\Delta u\mu_0}}{U}\frac{\partial\Sigma_s'\vec{\varphi}_1}{\partial u'}, \end{aligned}\right\} \tag{12.9}$$

where we have used the relation

$$\nabla = \frac{1}{L}\nabla'.$$

Let us now consider the first of (12.9). Obviously, the terms on the left side of this equation characterize the decrease of the number of neutrons per unit volume of $(\vec{r}', \vec{u}')$ phase space due to leakage and capture, respectively. Therefore, they can be compensated only by a corresponding influx of neutrons due to slowing-down and external sources, which is described by the terms on the right. This means that taken as a whole, the order of magnitude of the terms on the left and the right of the first of Equations (12.9) must be the same.

In view of the fact that neutron slowing down is described primarily by the first term on the right of (12.9), we must require that

$$\xi/U = 1, \tag{12.10}$$

from which it follows that $U = \xi$. With this fact in mind, Equations (12.9) can be written

$$\left.\begin{aligned} \nabla'\vec{\varphi}_1 + \Sigma_c'\varphi_0 &= -\frac{\partial\Sigma_s'\varphi_0}{\partial u'} + \varepsilon\frac{\partial^2\Sigma_s'\varphi_0}{\partial u'^2} + S', \\ \frac{1}{3}\nabla'\varphi_0 + \Sigma_{tr}'\vec{\varphi}_1 &= -\delta^2\frac{\partial\Sigma_s'\vec{\varphi}_1}{\partial u'}, \end{aligned}\right\} \tag{12.11}$$

where

$$\varepsilon = \frac{\overline{\xi^2}}{2\xi^2}, \quad \delta^2 = \frac{\Delta u \overline{\mu_0}}{\xi}. \tag{12.12}$$

It is easily shown that ϵ is of the first order of smallness, while δ^2 is of the second. This means that in the first approximation we may neglect terms of order ϵ and δ^2 in (12.11). With this in mind we return from the dimensionless variables to the original ones, arriving at the following set of equations for neutron slowing down in the reactor:

$$\left.\begin{aligned} \nabla\vec{\varphi}_1 + \Sigma_c\varphi_0 &= -\frac{\partial\xi\Sigma_s\varphi_0}{\partial u} + S, \\ \frac{1}{3}\nabla\varphi_0 + \Sigma_{tr}\vec{\varphi}_1 &= 0. \end{aligned}\right\} \tag{12.13}$$

Let us now write (12.13) as the single second-order equation

$$\nabla D\nabla\varphi - \Sigma_c\varphi = \frac{\partial\xi\Sigma_s\varphi}{\partial u} - S, \tag{12.14}$$

where $D = 1/3\Sigma_{tr}$, and we have dropped the index 0 on the unknown function. Equation (12.14) will now be considered the basic slowing-down equation in the diffusion-age approximation.*

Let us return to the set of equations for the thermal group, namely

$$\left.\begin{aligned} \nabla\vec{\Phi}_1 + \Sigma_{cT}\Phi_0 &= \int_{u_T - r}^{u_T} du'\Sigma_s\varphi_0 F_0(u'), \\ \frac{1}{3}\nabla\Phi_0 + \Sigma_{trT}\vec{\Phi}_1 &= \int_{u_T - r}^{u_T} du'\Sigma_s\vec{\varphi}_1 F_1(u'). \end{aligned}\right\} \tag{12.15}$$

Let us write out the integrals on the right side of (12.15) using the expansion (12.2). We then obtain

$$\left.\begin{aligned} \nabla\vec{\Phi}_1 + \Sigma_{cT}\Phi_0 &= \xi\Sigma_s\varphi_0 - \frac{\overline{\xi^2}}{2}\frac{\partial\Sigma_s\varphi_0}{\partial u}, \\ \frac{1}{3}\nabla\Phi_0 + \Sigma_{trT}\vec{\Phi}_1 &= \delta_1^2\xi\Sigma_s\vec{\varphi}_1, \end{aligned}\right\} \tag{12.16}$$

where

$$\delta_1^2 = \frac{1}{\xi}\int_{u_T - r}^{u_T} (u_T - u')F_1(u')\,du'.$$

Equations (12.16) can be written in terms of the dimensionless variables. We then obtain

$$\left.\begin{aligned} \nabla'\vec{\Phi}_1 + \Sigma'_{cT}\Phi_0 &= \xi\Sigma'_s\varphi_0 - \varepsilon\frac{\partial\xi\Sigma'_s\varphi_0}{\partial u'}, \\ \frac{1}{3}\nabla'\Phi_0 + \Sigma'_{trT}\vec{\Phi}_1 &= \delta_1^2\xi\Sigma'_s\vec{\varphi}_1. \end{aligned}\right\} \tag{12.17}$$

A study of the order of magnitude of the quantities in (12.17) shows that the term containing the parameter ϵ is of the first order of smallness, while δ_1^2 is of the second. Therefore, keeping only the principal terms in the

*For a discussion of the limits of applicability of diffusion-age theory, see Galanin [20].

equations and going back to the original variables, we obtain the following set of equations for the thermal group:

$$\left.\begin{aligned} &\nabla\vec{\Phi}_1 + \Sigma_{cT}\Phi_0 = \xi\Sigma_s\varphi_0, \\ &\frac{1}{3}\nabla\Phi_0 + \Sigma_{trT}\vec{\Phi}_1 = 0. \end{aligned}\right\} \tag{12.18}$$

This set of equations can be written in the form of a single diffusion equation

$$\nabla D_T \nabla\Phi - \Sigma_{cT}\Phi = -\xi\Sigma_s\varphi(\vec{r}, u_T), \tag{12.19}$$

where

$$D_T = \frac{1}{3\Sigma_{trT}}, \quad \Phi = \Phi_0, \quad \varphi = \varphi_0.$$

Thus, in the diffusion-age approximation the set of basic reactor equations becomes

$$\left.\begin{aligned} &\nabla D\nabla\varphi - \Sigma_c\varphi = \frac{\partial\xi\Sigma_s\varphi}{\partial u} - \frac{\nu_f}{K_{\text{eff}}}\chi(u)\left(\int_{-\infty}^{u_T} du\Sigma_f\varphi + \Sigma_{fT}\Phi\right), \\ &\nabla D_T\nabla\Phi - \Sigma_{cT}\Phi = -\xi\Sigma_s\varphi(\vec{r}, u_T). \end{aligned}\right\} \tag{12.20}$$

One should now note that if the moderator consists of several different kinds of nuclei, we must attempt to find a solution of the set of basic reactor equations in the form

$$\left.\begin{aligned} &\Sigma_{sk}\varphi_0(\vec{r}, u') = \Sigma_{sk}\varphi_0(\vec{r}, u) + \frac{\partial\Sigma_{sk}\varphi_0}{\partial u}(u' - u) + \dots, \\ &\Sigma_{sk}\vec{\varphi}_1(\vec{r}, u') = \Sigma_{sk}\vec{\varphi}_1(\vec{r}, u) + \dots. \end{aligned}\right\} \tag{12.21}$$

Operations similar to those performed above will again lead to Equations (12.20) with the substitutions

$$\left.\begin{aligned} &\Sigma_s = \sum_k \rho_k\sigma_{sk}, \quad \Sigma_c = \sum_k \rho_k\sigma_{ck}, \quad \Sigma_{tr} = \Sigma_s(1 - \bar{\mu}_0) + \Sigma_c, \\ &\Sigma_f = \sum_k \rho_k\sigma_{fk}, \quad \bar{\mu}_0 = \frac{1}{\Sigma_s}\sum_k \bar{\mu}_{0k}\rho_k\sigma_{sk}, \quad \xi = \frac{1}{\Sigma_s}\sum_k \xi_k\rho_k\sigma_{sk}, \end{aligned}\right\} \tag{12.22}$$

where σ_{sk}, σ_{ck} and σ_{fk} are the microscopic scattering, capture, and fission cross sections for nuclei of mass M_k, and ρ_k is the number of nuclei of this kind per unit volume of moderator.

13. The Diffusion-Age Approximation for the Adjoint Set of Equations

Let us now go on to obtaining the adjoint set of reactor equations. We start with

$$\left.\begin{aligned} &-\nabla\vec{\varphi}_1^* + \Sigma\varphi_0^* = \Sigma_s\int_u^{u+r} du'\varphi_0^* f_0(u' - u) + \Sigma_s\Phi_0^* F_0(u) + \\ &\qquad + \frac{1}{K_{\text{eff}}}\Sigma_f Q^*(\vec{r}), \\ &-\frac{1}{3}\nabla\varphi_0^* + \Sigma\vec{\varphi}_1^* = \Sigma_s\int_u^{u+r} du'\vec{\varphi}_1^* f_1(u' - u) + \Sigma_s\vec{\Phi}_1^* F_1(u), \end{aligned}\right\} \tag{13.1}$$

where

$$Q^*(\vec{r}) = \nu_f \int\limits_{-\infty}^{u_T} \chi(u)\,\varphi_0^*\,du. \tag{13.2}$$

We shall attempt to find a solution of (13.1) in the form

$$\left.\begin{aligned} \varphi_0^*(\vec{r}, u') &= \varphi_0^*(\vec{r}, u) + \frac{\partial \varphi_0^*}{\partial u}(u' - u) + \frac{1}{2}\frac{\partial^2 \varphi_0^*}{\partial u^2}(u' - u)^2, \\ \vec{\varphi}_1^*(\vec{r}, u') &= \vec{\varphi}_1^*(\vec{r}, u) + \frac{\partial \vec{\varphi}_1^*}{\partial u}(u' - u). \end{aligned}\right\} \tag{13.3}$$

Let us now insert (13.3) into (13.1). We then obtain

$$\left.\begin{aligned} -\nabla\vec{\varphi}_1^* + \Sigma_c\varphi_0^* &= \xi\Sigma_s\frac{\partial \varphi_0^*}{\partial u} + \frac{\overline{\xi^2}}{2}\Sigma_s\frac{\partial^2\varphi_0^*}{\partial u^2} + \Sigma_s\Phi_0^*F_0(u) + \\ &\quad + \frac{1}{K_{\text{eff}}}\Sigma_f Q^*(\vec{r}), \\ -\frac{1}{3}\nabla\varphi_0^* + \Sigma_{tr}\vec{\varphi}_1^* &= \overline{\Delta u\mu_0}\,\Sigma_s\frac{\partial\vec{\varphi}_1^*}{\partial u} + \Sigma_s\vec{\Phi}_1^*F_1(u). \end{aligned}\right\} \tag{13.4}$$

Let us now evaluate the orders of magnitude of the quantities in (13.4). To do this let us again go over to the dimensionless variables $\vec{r}'$ and u' using (12.7). Then,recalling (12.8), we obtain

$$\left.\begin{aligned} -\nabla'\vec{\varphi}_1^* + \Sigma_c'\varphi_0^* &= \Sigma_s'\frac{\partial\varphi_0^*}{\partial u'} + \varepsilon\Sigma_s'\frac{\partial^2\varphi_0^*}{\partial u'^2} + \Sigma_s'\Gamma\Phi_0^*F_0(u) + \\ &\quad + \frac{1}{K_{\text{eff}}}\Sigma_f' Q^*, \\ -\frac{1}{3}\nabla'\varphi_0^* + \Sigma_{tr}'\vec{\varphi}_1^* &= \delta^2\Sigma_s'\frac{\partial\vec{\varphi}_1^*}{\partial u'} + \Sigma_s'\vec{\Phi}_1^*F_1(u), \end{aligned}\right\} \tag{13.5}$$

where

$$\Sigma_f' = L\Sigma_f.$$

Let us now eliminate the terms of first and second order of smallness from (13.5), and again go to the original variables. We then have

$$\left.\begin{aligned} -\nabla\vec{\varphi}_1^* + \Sigma_c\varphi_0^* &= \xi\Sigma_s\frac{\partial\varphi_0^*}{\partial u} + \Sigma_s\Phi_0^*F_0(u) + \frac{1}{K_{\text{eff}}}\Sigma_f Q^*, \\ -\frac{1}{3}\nabla\varphi_0^* + \Sigma_{tr}\vec{\varphi}_1^* &= \Sigma_s\vec{\Phi}_1^*F_1(u). \end{aligned}\right\} \tag{13.6}$$

Let us solve the second of (13.6) for $\vec{\varphi}_1^*$, and insert this into the first. This gives

$$\begin{aligned} \nabla D\nabla\varphi_0^* - \Sigma_c\varphi_0^* = &- \xi\Sigma_s\frac{\partial\varphi_0^*}{\partial u} - \Sigma_s\Phi_0^*F_0(u) - m\nabla\vec{\Phi}_1^*F_1(u) - \\ &- \frac{1}{K_{\text{eff}}}\Sigma_f Q^*, \end{aligned} \tag{13.7}$$

where

$$m = \frac{\Sigma_s}{\Sigma_{tr}} \approx 1.$$

We note that $F_0(u)$ and $F_1(u)$ are nonzero only in the interval $u_T - r \le u \le u_T$.

For M ≫ 1, Equation (13.7) can be simplified to some extent. Indeed, if this is the case we can write

$$F_0(u) = \xi\mathcal{E}(u - u_T), \qquad F_1(u) = \bar{\mu}_0\delta(u - u_T).$$

We have here used the relations

$$\int_{u_T - r}^{u_T} F_0(u)\,du = \xi, \qquad \int_{u_T - r}^{u_T} F_1(u)\,du = \bar{\mu}_0.$$

Then (13.7) becomes

$$\nabla D \nabla \varphi^* - \Sigma_c \varphi^* = -\xi\Sigma_s \frac{\partial \varphi^*}{\partial u} - (\xi\Sigma_s \Phi_0^* + \bar{\mu}_0 m \nabla \vec{\Phi}_1^*)\,\delta(u - u_T) -$$
$$- \frac{1}{K_{\text{eff}}}\Sigma_f Q^*, \tag{13.8}$$

where

$$\varphi^* = \varphi_0^*.$$

We now integrate (13.8) over the interval $u_T - \epsilon \leq u \leq u_T + \epsilon$. Then setting $\varphi^*(\vec{r}, u_T + \epsilon) = 0$, we obtain

$$\int_{u_T - \varepsilon}^{u_T + \varepsilon} \left[(\nabla D \nabla \varphi^* - \Sigma_c \varphi^*) + \frac{1}{K_{\text{eff}}}\Sigma_f Q^* \right] du = \xi\Sigma_{sT}\varphi^*(\vec{r}, u_T - \varepsilon) -$$
$$- (\xi\Sigma_{sT}\Phi_0^* - \bar{\mu}_0 m \nabla \vec{\Phi}_1^*), \tag{13.9}$$

where

$$\xi\Sigma_{sT} = \xi\Sigma_s \Big|_{u = u_T}.$$

We now let ϵ approach zero and assume that the integrand in (13.9) remains bounded. We then arrive at

$$\varphi^*(\vec{r}, u_T) = \Phi_0^*(\vec{r}) - \frac{\bar{\mu}_0}{\xi\Sigma_{sT}} \nabla \vec{\Phi}_1^*.$$

The second term on the right side of this equation is usually small, so that we may write

$$\varphi^*(\vec{r}, u_T) = \Phi_0^*(\vec{r}). \tag{13.10}$$

We shall henceforth consider (13.10) an "initial" condition. Using it, we can rewrite (13.8) in the form

$$\nabla D \nabla \varphi^* - \Sigma_c \varphi^* = -\xi\Sigma_s \frac{\partial \varphi^*}{\partial u} - \frac{1}{K_{\text{eff}}}\Sigma_f Q^*. \tag{13.11}$$

Let us now go on to a consideration of the adjoint equations for the thermal group. To do this, consider the last two equations of (9.3), namely:

$$\left.\begin{aligned} -\nabla\vec{\Phi}_1^* + \Sigma_{cT}\Phi_0^* &= \frac{1}{K_{\text{eff}}}\Sigma_{fT} Q^*(\vec{r}), \\ -\frac{1}{3}\nabla\Phi_0^* + \Sigma_{tr\,T}\vec{\Phi}_1^* &= 0, \end{aligned}\right\} \tag{13.12}$$

where

$$Q^*(\vec{r}) = \nu_f \int_{-\infty}^{u_T} \chi(u)\,\varphi^*\,du. \tag{13.13}$$

Equations (13.12) can be reduced to the single diffusion equation

$$\nabla D_T \nabla \Phi^* - \Sigma_{cT}\Phi^* = -\frac{1}{K_{\text{eff}}}\,\Sigma_{fT} Q^*, \tag{13.14}$$

where

$$\Phi^* = \Phi_0^*.$$

Thus, in the diffusion-age approximation the set of adjoint reactor equations becomes

$$\left.\begin{aligned} \nabla D \nabla \varphi^* - \Sigma_c \varphi^* &= -\xi\Sigma_s \frac{\partial \varphi^*}{\partial u} - \frac{\nu_f}{K_{\text{eff}}}\,\Sigma_f \int_{-\infty}^{u_T} \chi(u)\,\varphi^*\,du, \\ \nabla D_T \nabla \Phi^* - \Sigma_{cT}\Phi^* &= -\frac{\nu_f}{K_{\text{eff}}}\,\Sigma_{fT} \int_{-\infty}^{u_T} \chi(u)\,\varphi^*\,du, \\ \varphi^*(\vec{r}, u_T) &= \Phi^*(\vec{r}). \end{aligned}\right\} \tag{13.15}$$

When the moderator consists of a mixture of several different kinds of nuclei, we can again obtain (13.15) as indicated above, except that the quantities Σ_s, Σ_c, $\bar{\mu}_0$ and ξ must be replaced by their effective values as given by (12.22).

Finally, one must mention that when the moderator consists of a mixture of nuclei, the effective constants for the adjoint equations will have the same values as those for the basic reactor equations, namely the values given by (12.22).

14. Extrapolated Reactor Boundary

Let us now consider the boundary conditions in the diffusion-age approximation. We start with the condition given by (10.4), namely:

$$2(\vec{\varphi}_1)_{\vec{n}0} - \varphi_0 = 0 \quad \text{on } S. \tag{14.1}$$

Bearing in mind that in the diffusion-age approximation (12.13) is satisfied

$$\vec{\varphi}_1 = -D\nabla\varphi_0, \tag{14.2}$$

we can rewrite (14.1) in the form:

$$\varphi_0 + d\,\frac{\partial \varphi_0}{\partial s} = 0 \text{ for } s = 0, \tag{14.3}$$

where

$$d = 2D = \frac{2}{3}\lambda_{tr},$$

in which λ_{tr} is the transport mean free path, and $\underline{s}$ is the distance from the boundary along the normal to the surface. We rewrite (14.3) in the form

$$-\frac{\varphi_0}{\frac{\partial \varphi_0}{\partial s}} = d. \tag{14.4}$$

We shall call d the extrapolation distance. Its meaning becomes clear from Equation (14.4) if we assume that φ_0 is a linear function of s, writing:

$$\varphi_0 = c_1 + c_2 s. \tag{14.5}$$

Inserting (14.5) into (14.4), we then obtain

$$c_1 = - c_2 d, \tag{14.6}$$

which means that close to S the function φ_0 is of the form

$$\varphi_0 = c(s - d). \tag{14.7}$$

This last equation allows us to introduce the so-called extrapolated boundary S_e of the reactor, which lies at a distance d from S.

It follows from (14.7) that the boundary condition on S_e can be written

$$\varphi_0 (\vec{r}, u) = 0. \tag{14.8}$$

Similarly, we could have introduced the extrapolated boundary of the reactor by considerations involving Φ_0. It should be borne in mind, however, that the extrapolated reactor boundary is related to the mean free path λ_{tr}, and is therefore a function of the energy.

For reactors whose dimensions are much greater than the mean free path we can assume a certain extrapolated boundary independent of the energy.

When the reactor dimensions are not large, the energy dependence of d can no longer be neglected, and we must make direct use of the boundary conditions in the form of (14.2).

Let us now go on to a consideration of the boundary condition for the adjoint equations. To do this we use the diffusion boundary conditions of (10.9), namely

$$2 (\vec{\varphi}_1^*)_{\vec{n}0} + \varphi_0^* = 0, \tag{14.9}$$

for which the relation between $\vec{\varphi}_1^*$ and φ_0^* is given by

$$\vec{\varphi}_1^* = D\nabla\varphi_0^*. \tag{14.10}$$

Using (14.10), Equation (14.9) becomes

$$\varphi_0^* + d\frac{\partial \varphi_0^*}{\partial s} = 0 \quad \text{for} \quad s = 0. \tag{14.11}$$

In view of the fact that (14.11) is identical with (14.3), the extrapolated boundaries S_e will be the same for φ_0 and φ_0^*. The same is true of Φ^*. A more accurate expression for the extrapolation distance can be obtained by using the asymptotic solutions of the kinetic equations [96, 130, 144].

15. The Region of Definition of the Solution in the Diffusion-Age Approximation

Let us consider the vector function f whose components are given by the column

$$f = \begin{vmatrix} \varphi_0 \\ \vec{\varphi}_1 \\ \Phi_0 \\ \vec{\Phi}_1 \end{vmatrix}$$

Then Equations (12.20) can be written

$$\mathbf{L}f = \begin{vmatrix} \nabla\vec{\varphi}_1 + \Sigma_c\varphi_0 + \dfrac{\partial\xi\Sigma\varphi_0}{\partial u} - \dfrac{\chi(u)}{K_{eff}}Q \\ \nabla\varphi_0 + 3\Sigma_{tr}\vec{\varphi}_1 \\ \nabla\vec{\Phi}_1 + \Sigma_{cT}\Phi_0 - \xi\Sigma_s\varphi_{0T} \\ \nabla\Phi_0 + 3\Sigma_{trT}\vec{\Phi}_1 \end{vmatrix} = 0. \tag{15.1}$$

The elements of the set $\{f\}$ are continuous, differentiable in $\mathbf{G} \times \mathbf{U}$, and satisfy the boundary conditions

$$\begin{matrix} \varphi_0 = 0 \\ \Phi_0 = 0 \end{matrix} \quad \text{on } S_e$$

Let us further consider the vector function f^* as defined by

$$f^* = \begin{vmatrix} \varphi_0^* \\ \vec{\varphi}_1^* \\ \Phi_0^* \\ \vec{\Phi}_1^* \end{vmatrix} \tag{15.2}$$

Then the set of adjoint equations can be written in the vector form

$$\mathbf{L}^*f^* = \begin{vmatrix} -\nabla\vec{\varphi}_1^* + \Sigma_c\varphi_0^* - \xi\Sigma_s\dfrac{\partial\varphi_0^*}{\partial u} - \dfrac{\Sigma_f}{K_{eff}}Q^* \\ -\nabla\varphi_0^* + 3\Sigma_{tr}\vec{\varphi}_1^* \\ -\nabla\vec{\Phi}_1^* + \Sigma_{cT}\Phi_0^* - \dfrac{\Sigma_{fT}}{K_{eff}}Q^* \\ -\nabla\Phi_0^* + 3\Sigma_{trT}\vec{\Phi}_1^* \end{vmatrix} = 0. \tag{15.3}$$

Here it is assumed that the elements of the set $\{f^*\}$ are continuous and differentiable functions in $\mathbf{G} \times \mathbf{U}$, and that their components satisfy the subsidiary requirements

$$\begin{matrix} \varphi_0^* = 0, \quad \Phi_0^* = 0 \quad \text{on } S_e \\ \varphi_T^* = \Phi^*(\vec{r}). \end{matrix} \tag{15.4}$$

It is easily shown that Equations (15.1) and (15.3) defined on the sets $\{f\}$ and $\{f^*\}$ are indeed mutually adjoint.

16. The Method of Successive Approximations

In Chapter I we defined the method of successive approximations for finding the eigenvalue K_{eff} of the problem both for the basic and adjoint reactor equations. It is convenient, however, to introduce a new definition of the method of successive approximations, namely:

$$K_{eff}^{(p,n)} = \frac{\int_G d\vec{r}\, Q^{*(p)}(\vec{r})\, Q^{(n)}(\vec{r})}{\int_G d\vec{r}\, Q^{*(p)}(\vec{r})\, Q^{(n-1)}(\vec{r})}, \tag{16.1}$$

where

$$\left.\begin{aligned} Q(\vec{r}) &= \nu_f\left(\int_{-\infty}^{u_T} \Sigma_f\, \varphi du + \Sigma_{fT}\Phi\right), \\ Q^*(\vec{r}) &= \nu_f \int_{-\infty}^{u_T} \chi(u)\, \varphi^*(\vec{r}, u)\, du, \end{aligned}\right\} \tag{16.2}$$

and the indices n and p denote the iteration number of the set of basic and adjoint equations. Generally speaking, it can be shown for a nonreflected reactor that the method of successive approximations defined in this way depends only on the total number of iterations of the basic and adjoint equations (see Chapter V).

In practice, the method defined by (16.1) has many advantages over that described in Chapter I, particularly when $Q(\vec{r})$ has a sudden rise at the boundary between the core and the reflector. In many systems the function $Q^*(\vec{r})$, on the other hand, decreases monotonically as one moves from the center of the reactor to its boundary. Therefore, if Equations (16.1) are used to calculate K_{eff}, the region near the boundary with the sudden rise has much less influence on the results. Because this region is responsible for very large errors due to the rapid space and energy fluctuations of the functions, this method of calculating the eigenvalue is considerably more accurate.

Finally, analysis shows that in calculating the eigenvalue equations (16.1) one need not obtain a solution of the adjoint equations, it being sufficient to have a solution of just the basic reactor equations. Recalling that in the central region of the reactor the solutions of both the basic and adjoint equations have roughly the same space dependence, corresponding to that of an eigenfunction of an equivalent nonreflected reactor, it is convenient to take the adjoint function $Q^*(\vec{r})$ as just this eigenfunction. In practice this procedure reduces to cutting off the sudden rise in $Q(\vec{r})$ at a level corresponding to the eigenfunction of the equivalent nonreflected reactor.

In calculating the critical mass of a reactor, it is more convenient to use the adjoint, rather than the basic equations, for their solution is simpler to approximate. This is because these solutions are monotonic functions of the coordinates and vary slowly with energy. For this reason it is possible, even with the most simple-minded finite-difference methods of computation, to obtain highly accurate solutions.

Chapter IV

REFINEMENT OF THE DIFFUSION-AGE THEORY

The diffusion-age theory usually can be used for a successful description of neutron slowing down by nuclei whose mass is much larger than unity. Even then, however, it is often insufficient. This happens, for instance, when dealing with strong absorption of slowing-down neutrons. It then becomes necessary to mold the diffusion-age theory to the specific peculiarities of the problem.

In this chapter we shall consider methods for refining the age equation both for problems involving space and energy, and for those involving energy alone.

17. Slowing Down of Neutrons in Infinite Homogeneous Media

Consider the slowing down of neutrons in a homogeneous infinite medium. It is clear that in this case the neutron flux φ is independent of $\vec{r}$, and of $\vec{\Omega}$. Therefore, φ will be isotropic and will depend only on u. Bearing this in mind and integrating the kinetic slowing-down equation over space and solid angle, we arrive at the integral equation

$$\Sigma\varphi = \int_{u-r}^{u} du' \Sigma_s \varphi(u') f_0(u-u') + S. \tag{17.1}$$

It is useful to note that this equation can be written as the set of integro-differential equations

$$\begin{gathered} \frac{dq}{du} + \Sigma_c \varphi = S, \\ q = \int_{u-r}^{u} du' \Sigma_s \varphi F_0(u, u'), \end{gathered} \tag{17.2}$$

where

$$F_0'(u, u') = \int_{u}^{u'+r} f_0(u-u'')\, du''. \tag{17.3}$$

In deriving (17.2), use is made of the identity

$$\frac{dq}{du} = \Sigma\varphi - \int_{u-r}^{u} du' \Sigma_s \varphi f_0(u-u'). \tag{17.4}$$

The function $q(u)$, called the slowing-down density, describes the total number of neutrons passing through lethargy u per unit time.

Let us now go on to a calculation of q. We shall do this by attempting to find a solution of (17.1) in the form

$$\Sigma_s\varphi(u') = \Sigma_s\varphi(u) + \frac{d\Sigma_s\varphi}{du}(u'-u) + \frac{1}{2}\frac{d^2\Sigma_s\varphi}{du^2}(u'-u)^2 + \\ + \frac{1}{6}\frac{d^3\Sigma_s\varphi}{du^3}(u'-u)^3 + \ldots \tag{17.5}$$

We now insert (17.5) into (17.1) and integrate. We then obtain

$$\Sigma_c\varphi = -\xi\frac{d\Sigma_s\varphi}{du} + \frac{\overline{\xi^2}}{2}\frac{d^2\Sigma_s\varphi}{du^2} - \frac{\overline{\xi^3}}{6}\frac{d^3\Sigma_s\varphi}{du^3} + \ldots + S, \tag{17.6}$$

where the

$$\overline{\xi^l} = \int\limits_0^r t^l f_0(t)\,dt, \quad \overline{\xi^1} = \xi \tag{17.7}$$

are the "moments" of the distribution function.

Let us write (17.6) in the form

$$\frac{d\psi}{du'} + \alpha\psi - S = \varepsilon\frac{d^2\psi}{du'^2} - \mu\frac{d^3\psi}{du'^3} + \ldots, \tag{17.8}$$

where

$$\psi = \Sigma_s\varphi, \quad \alpha = \frac{\Sigma_c}{\Sigma_s}, \quad u' = \frac{u}{\xi}, \\ \varepsilon = \frac{\overline{\xi^2}}{2\xi^2}, \quad \mu = \frac{\overline{\xi^3}}{6\xi^3}. \tag{17.9}$$

It can be shown that ϵ is of the first order of smallness, while μ is of the second.

Taking only the principal terms in (17.8), we arrive at

$$\frac{d\psi}{du'} + \alpha\psi - S = 0, \tag{17.10}$$

which can be written in the form

$$\frac{dq}{du} + \Sigma_c\varphi = S, \\ q = \xi\Sigma_s\varphi. \tag{17.11}$$

We have here made use of (17.9).

In this way we have arrived at the well-known age approximation.

Let us now keep the term in (17.8) which contains ϵ. We then arrive at

$$\frac{d\psi}{du'} + \alpha\psi - S = \varepsilon\frac{d^2\psi}{du'^2}. \tag{17.12}$$

It should be noted that in this equation the highest derivative is multiplied by a small parameter.

Let us now consider (17.12). First, we solve for the first derivative, obtaining

$$\frac{d\psi}{du'} = -\alpha\psi + S + \varepsilon\frac{d^2\psi}{du'^2}. \tag{17.13}$$

Let us now differentiate with respect to u'. This gives

$$\frac{d^2\psi}{du'^2} = -\frac{d\alpha\psi}{du'} + \frac{dS}{du'} + \varepsilon\frac{d^3\psi}{du'^3}. \tag{17.14}$$

We now insert this expression for the second derivative into the right side of (17.12). Then up to terms of order ϵ^2 we have

$$\frac{d\psi}{du'} + \alpha\psi - S = -\varepsilon\frac{d}{du'}(\alpha\psi - S). \tag{17.15}$$

This equation can be written in the form

$$\frac{dq}{du} + \Sigma_c\varphi = S, \tag{17.16}$$

where

$$q = \xi\Sigma_s(1 + \varepsilon\alpha)\varphi - \gamma S, \quad \gamma = \varepsilon\xi.$$

Here we have again used (17.9).

Relation (17.16) between q and φ was first obtained by Greuling and Goertzel [54].

Equations (17.16) thus improve age theory by taking into account the mean square logarithmic energy decrement.

It is of some interest to write out the equation for q itself. According to (17.16), as is easily shown, this is

$$\frac{dq}{du} + \frac{\Sigma_c}{\xi\Sigma_s + \gamma\Sigma_c} q = \frac{\xi\Sigma_s}{\xi\Sigma_s + \gamma\Sigma_c} S. \tag{17.17}$$

the solution of this equation is

$$q = \int_{-\infty}^{u} S(u') \frac{\xi\Sigma_s}{\xi\Sigma_s + \gamma\Sigma_c} e^{-\int_{u'}^{u} \frac{\Sigma_c du''}{\xi\Sigma_s + \gamma\Sigma_c}} du'. \tag{17.18}$$

If the moderator is made up of a mixture of different elements, γ is given by

$$\gamma = \frac{1}{\xi\Sigma_s}\sum_h \gamma_h\xi_h\Sigma_{sh}. \tag{17.19}$$

The other effective quantities in (17.18) are given by the usual formulas of age theory.

18. Age Theory With Resonance Absorption. Effective Resonance Integral

In the preceding section we refined age theory to include the second moment of energy. The equations thus obtained are essentially valid only when Σ_c is either small or when it hardly changes over the logarithmic energy decrement ξ.

In many important problems, on the other hand, one must deal with resonance absorption in energy intervals much smaller than that corresponding to the mean logarithmic decrement. In such cases inclusion of the second moment of energy adds nothing significant to slowing-down theory.

It was Wigner [54] who suggested a method for improving age theory for such problems. His procedure is the following.

Consider the basic slowing-down equation

$$\Sigma \varphi = \int_{u-r}^{u} \Sigma_s \varphi f_0 (u - u') \, du', \qquad (18.1)$$

where $\Sigma = \Sigma_c + \Sigma_s$ is the total macroscopic cross section, Σ_c is a macroscopic cross section for resonance capture, and $\Sigma_s = \Sigma_{sr} + \Sigma_{sn}$ is the macroscopic scattering cross section, composed of the resonance scattering term Σ_{sr} and the potential scattering term Σ_{sn}.

If we assume that the resonance width (Γ_j) is much lower than the maximum logarithmic energy decrement $\underline{r}$ and that $\Sigma_{sn} r \gg \Sigma_{sr} \Gamma_j$, we can replace Σ_s and the integrand of (18.1) by Σ_{sn}. This gives

$$\Sigma \varphi = \int_{u-r}^{u} \Sigma_{sn} \varphi f_0 (u - u') \, du' \qquad (18.1')$$

which also leads to the following expression for $\underline{q}$:

$$q = \int_{u-r}^{u} \Sigma_{sn} \varphi F_0 (u, u') \, du'. \qquad (18.2)$$

Now let us expand $\Sigma_{sn} \varphi$ in a Taylor's series. Writing

$$\Sigma_{sn} \varphi (u') = \Sigma_{sn} \varphi (u) + \dots, \qquad (18.3)$$

and using (18.2), we obtain

$$q = \xi \Sigma_{sn} \varphi. \qquad (18.4)$$

Solving (18.4) for $\Sigma_{sn} \varphi$, we have

$$\Sigma_{sn} \varphi = \frac{q}{\xi}. \qquad (18.5)$$

Finally, inserting (18.5) into the integrand of (18.1') we arrive at

$$(\Sigma_c + \Sigma_s) \varphi = \frac{q}{\xi},$$

and, therefore,

$$\varphi = \frac{q}{\xi (\Sigma_c + \Sigma_s)}. \qquad (18.6)$$

If we now use the first of Equations (17.2), which in this case is

$$- \Sigma_c \varphi = \frac{dq}{du} \qquad (18.7)$$

and use (18.6) to eliminate φ from this equation, we arrive at Wigner's equation

$$\frac{dq}{du} + \frac{\Sigma_c}{\xi (\Sigma_c + \Sigma_s)} \cdot q = 0, \qquad (18.8)$$

whose solution, for $q(0) = 1$, is

$$q = e^{-\int_0^u \frac{\Sigma_c du}{\xi(\Sigma_c + \Sigma_s)}} . \quad (18.9)$$

In terms of E this solution becomes

$$q(E) = e^{-\int_E^{E_0} \frac{\Sigma_c}{\xi(\Sigma_c+\Sigma_s)} \frac{dE}{E}} , \quad (18.10)$$

where E_0 is the energy corresponding to lethargy $u = 0$.

The functions q(u) and q(E) defined by (18.9) and (18.10) are the probabilities of avoiding resonance capture. Henceforth we shall denote them by $\langle\varphi\rangle$.

For a single resonance level, (18.10) can be written*

$$\langle\varphi\rangle_j = e^{-\frac{\rho}{\xi\Sigma_{sn}} J^j_{\text{eff}}} , \quad (18.11)$$

where

$$J^j_{\text{eff}} = \int_{E_j-\alpha}^{E_j+\alpha} \Sigma_c \frac{\Sigma_{sn}}{\Sigma} \frac{dE}{E} \quad (18.12)$$

is the effective resonance integral for the j-th resonance, ρ is the number of absorbing nuclei per cubic centimeter, and α is a number which is greater than the resonance width (in energy units).

Let us assume that close to resonance $\sigma_c(E)$ and $\sigma_{sr}(E)$ are given by the Breit-Wigner formula [3, 26]

$$\sigma_c(E) = \frac{\sigma^j_c}{1+\left(\frac{E-E_j}{\Gamma_j/2}\right)^2}, \quad \sigma_{sr}(E) = \frac{\sigma^j_{sr}}{1+\left(\frac{E-E_j}{\Gamma_j/2}\right)^2} , \quad (18.13)$$

where σ^j_c and σ^j_{sr} are the cross sections for absorption and resonance scattering at the maximum of the resonance.

Transforming to the new variable x defined by

$$x = \frac{E-E_j}{\Gamma_j/2} , \quad (18.14)$$

in (18.13), we obtain

$$\sigma_c = \frac{\sigma^j_c}{1+x^2} , \quad \sigma_{sr} = \frac{\sigma^j_{sr}}{1+x^2} . \quad (18.15)$$

Let us now insert (18.15) into (18.12). After some simple operations we arrive at

$$J^j_{\text{eff}} = \frac{\Gamma_j}{2E_j} \frac{\sigma^j_c}{\sigma^j_r} \int_{-\infty}^{\infty} \frac{\psi(x)\,dx}{\psi(x) + \frac{1}{h_j}} , \quad (18.16)$$

*If the absorbing substance is in a homogeneous mixture with the moderator, $\xi\Sigma_{sn}$ in (18.11) should be replaced by the appropriate quantity for the homogeneous mixture.

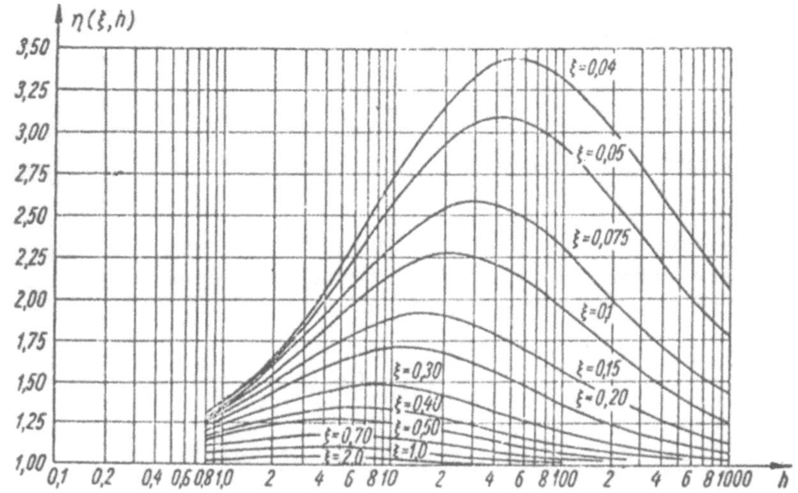

Fig. 6. The effect of Doppler broadening on resonance capture of neutrons in a homogeneous medium.

where

$$\psi(x)=\frac{1}{1+x^2},\quad h_j=\frac{\Sigma_r^j}{\Sigma_{sn}},\quad \sigma_r^j=\sigma_c^j+\sigma_{sr}^j. \tag{18.17}$$

We note further that the integration in (18.16) has been extended to infinity, which introduces no significant errors. Inserting (18.17) into (18.16), we obtain

$$J_{\text{eff}}^j=\frac{\pi\Gamma_j}{2E_j}\frac{\sigma_c^j}{\sqrt{1+h_j}}. \tag{18.16'}$$

We cannot yet use Expression (18.17) for $\psi(x)$ to calculate the probability of avoiding resonance capture, since this expression does not take account of the thermal motion of the moderator nuclei which, in the final analysis, causes Doppler line broadening [26]. To obtain the energy dependence of the observed resonance cross section, one must average it over all possible velocities of the neutron relative to the absorbing nuclei. One then obtains [26].

$$\sigma_c(x,\xi)=\sigma_c^j\psi(x,\xi),\quad \sigma_{sr}(x,\xi)=\sigma_{sr}^j\psi(x,\xi), \tag{18.18}$$

where

$$\left.\begin{aligned}\psi(x,\xi)&=\frac{\xi}{2\sqrt{\pi}}\int_{-\infty}^{\infty}\frac{e^{-\frac{\xi^2}{4}(x-y)^2}}{1+y^2}dy,\\ \xi&=\frac{\Gamma_j}{\Delta_j},\quad \Delta_j=2\sqrt{\frac{kT}{M}E_j},\end{aligned}\right\} \tag{18.19}$$

in which k is Boltzmann's constant, T is the absolute temperature of the medium, and M is the mass of the nucleus. Inserting (18.18) and (18.19) into (18.16), we obtain

$$J_{\text{eff}}^j(T)=J_{\text{eff}}^j\,\eta(\xi,h), \tag{18.20}$$

where

$$\eta(\xi,h)=\frac{2\sqrt{1+h}}{\pi h}\int_0^{\infty}\frac{\psi(x,\xi)}{\psi(x,\xi)+\frac{1}{h}}dx. \tag{18.21}$$

Figure 6 is a graph of $\eta(\xi, h)$ (see also Table I of Appendix F).

Analysis shows that $\eta(\xi, h)$ varies within the limits

$$1 \leqslant \eta < \sqrt{1+h}\,.$$

This function has been calculated for wide intervals of variation of ξ and h by Gordeev, Orlov and Sedel'nikov [24].

Thus, Equation (18.20) is used to compute the effective resonance integral, which, in turn, is then used to calculate the probability that a neutron will avoid capture by the given resonance level. This probability is given by

$$\langle\varphi\rangle_j = e^{-\frac{\rho}{\xi\Sigma_{sn}} J^j_{\text{eff}}(T)}. \tag{18.22}$$

19. Slowing Down of Neutrons in Inhomogeneous Media*

When the slowing down takes place in inhomogeneous or finite media, not only neutron absorption takes place, but also leakage. Starting with the diffusion approximation, diffusion-age theory for this case is based on the equations

$$\left.\begin{aligned} \nabla\vec{\varphi}_1 + \Sigma\varphi_0 &= \int\limits_{u-r}^{u} du'\, \Sigma_s\varphi_0 f_0(u-u') + S, \\ \frac{1}{3}\nabla\varphi_0 + \Sigma\vec{\varphi}_1 &= \int\limits_{u-r}^{u} du'\, \Sigma_s\vec{\varphi}_1 f_1(u-u'). \end{aligned}\right\} \tag{19.1}$$

We shall attempt to find a solution of (19.1) in the form

$$\left.\begin{aligned} \Sigma_s\varphi_0(\vec{r}, u') &= \Sigma_s\varphi_0(\vec{r}, u) + \frac{\partial\Sigma_s\varphi_0}{\partial u}(u'-u) + \frac{1}{2}\cdot\frac{\partial^2\Sigma_s\varphi_0}{\partial u^2}(u'-u)^2, \\ \Sigma_s\vec{\varphi}_1(\vec{r}, u') &= \Sigma_s\vec{\varphi}_1(\vec{r}, u) + \frac{\partial\Sigma_s\vec{\varphi}_1}{\partial u}(u'-u). \end{aligned}\right\} \tag{19.2}$$

To do this, we insert (19.2) into (19.1) and integrate. In this way one obtains Equations (12.3). Further, going over to the dimensionless variables $\vec{r}'$ and u' using (12.7), we obtain Equations (12.11) which, as we have seen, are

$$\left.\begin{aligned} \nabla'\vec{\varphi}_1 + \Sigma_c'\varphi_0 &= -\frac{\partial\Sigma_s'\varphi_0}{\partial u'} + \varepsilon\frac{\partial^2\Sigma_s'\varphi_0}{\partial u'^2} + S', \\ \frac{1}{3}\nabla'\varphi_0 + \Sigma_{tr}'\vec{\varphi}_1 &= -\delta^2\frac{\partial\Sigma_s'\vec{\varphi}_1}{\partial u'}, \end{aligned}\right\} \tag{19.3}$$

where by definition ϵ is a quantity of the first order of smallness, while δ^2 is of the second.

We shall improve diffusion-age theory by including terms of order ϵ. In so doing we shall neglect terms of order δ^2.

Then, using (19.3), we easily obtain the second-order equation

*If the reactor calculations are based on a theory which takes into account the second moment of energy, one can correct the neutron flux in the neighborhood of the boundary between the core and reflector. The flux in this region as given by age theory is usually too low. Such correction of the theory is particularly important for intermediate neutron reactors. The correction to the critical mass, on the other hand, is usually quite small.

$$\frac{\partial \Sigma_s \varphi}{\partial u} - \nabla D \nabla \varphi + \Sigma_c \varphi - S = \varepsilon \frac{\partial^2 \Sigma_s \varphi}{\partial u^2}, \tag{19.4}$$

where

$$\varphi = \varphi_0(\vec{r}, u).$$

Here we have omitted the primes and will continue to do so.

We can simplify Equation (19.4) by making use of the fact that ϵ multiplies the higher derivative. We first solve (19.4) for the first derivative, obtaining

$$\frac{\partial \Sigma_s \varphi}{\partial u} = \nabla D \nabla \varphi - \Sigma_c \varphi + S + \varepsilon \frac{\partial^2 \Sigma_s \varphi}{\partial u^2}. \tag{19.5}$$

We now differentiate (19.5) with respect to $\underline{u}$. This gives

$$\frac{\partial^2 \Sigma_s \varphi}{\partial u^2} = \frac{\partial}{\partial u}\left(\nabla D \nabla \varphi - \Sigma_c \varphi + S + \varepsilon \frac{\partial^2 \Sigma_s \varphi}{\partial u^2}\right). \tag{19.6}$$

We now insert (19.6) into the right side of (19.4), and maintain only first order terms. The result is

$$\frac{\partial}{\partial u}\left[(\Sigma_s + \varepsilon \Sigma_c)\varphi - \varepsilon \nabla D \nabla \varphi - \varepsilon S\right] - \nabla D \nabla \varphi + \Sigma_c \varphi = S. \tag{19.7}$$

Let us now return to the original variables. We then obtain*

$$\left.\begin{aligned} &\frac{\partial q}{\partial u} - \nabla D \nabla \varphi + \Sigma_c \varphi = S, \\ &q = (\xi \Sigma_s + \gamma \Sigma_c)\varphi - \gamma \nabla D \nabla \varphi - \gamma S. \end{aligned}\right\} \tag{19.8}$$

Let us now consider the equation for the thermal group. It is clear that in maintaining terms of order ϵ the diffusion equation becomes

$$\nabla D_T \nabla \Phi - \Sigma_{cT} \Phi = -q(\vec{r}, u_T), \tag{19.9}$$

where

$$q(\vec{r}, u_T) = (\xi \Sigma_s + \gamma \Sigma_c)\varphi - \gamma \nabla D \nabla \varphi \quad \text{for} \quad u = u_T.$$

Thus, in this approximation the basic equations will be

$$\begin{aligned} &\frac{\partial q}{\partial u} - \nabla D \nabla \varphi + \Sigma_c \varphi = S, \\ &\nabla D_T \nabla \Phi - \Sigma_{cT} \Phi = -q(\vec{r}, u_T), \\ &q = (\xi \Sigma_s + \gamma \Sigma_c)\varphi - \gamma \nabla D \nabla \varphi - \gamma S, \\ &S = \frac{\chi(u)}{K_{eff}} Q(\vec{r}), \quad Q(\vec{r}) = \nu_f\left(\int_{-\infty}^{u_T} \Sigma_f \varphi \, du + \Sigma_{fT} \Phi\right). \end{aligned} \tag{19.10}$$

Going through the same procedure for the adjoint equations, we arrive at

*Equation (19.8) was first obtained by Greuling and Goertzel [54].

$$
\begin{gathered}
\xi\Sigma_s \frac{\partial}{\partial u} \frac{q^*}{\xi\Sigma_s} + \nabla D \nabla \varphi^* - \Sigma_c \varphi^* = -S^*, \\
\nabla D_T \nabla \Phi^* - \Sigma_{cT} \Phi^* = -S_T^*, \\
q^* = (\xi\Sigma_s + \gamma\Sigma_c)\,\varphi^* - \gamma \nabla D \nabla \varphi^* - \gamma S^*, \\
S^* = \frac{\Sigma_f}{K_{eff}} Q^*(\vec{r}), \quad S_T^* = \frac{\Sigma_{fT}}{K_{eff}} \cdot Q^*(\vec{r}), \\
Q^* = \nu_f \int_{-\infty}^{u_T} \varphi^* \chi \, du.
\end{gathered}
\tag{19.11}
$$

If the moderator is a mixture of nuclei, γ in Equations (19.10) and (19.11) is given by (17.19), while the other quantities are given by the equations of age theory.

Chapter V

HOMOGENEOUS NONREFLECTED REACTOR

It is possible to obtain exact solutions of the basic and adjoint reactor equations of the diffusion-age approximation only in very rare cases. One such case is a homogeneous nonreflected reactor whose extrapolated boundary is independent of u.*

It is found that in this case the basic and adjoint equations can be solved in closed analytic form. The solution is unique for its simplicity and clarity, and at the same time it indicates the basic properties of the solution for a homogeneous reactor even with a reflector.

Indeed, far from the reflector in the core of a reactor with a reflector, the neutron flux and importance are established similarly as in a nonreflected reactor of some equivalent size. Therefore, to practically every reactor with a reflector there corresponds a nonreflected reactor with the same value of K_{eff} and the same neutron flux and importance distributions far from the boundary between the core and the reflector.

Finally, for nonreflected reactors one can establish the convergence of the method of successive approximations for finding K_{eff}, and one can estimate the rate of this convergence. This is particularly important because the method of successive approximations is just about the only effective computational scheme for reactors with a reflector.

For these reasons we shall now describe computational methods for this most simple case, the nonreflected reactor.

20. Neutron Density Spectrum in a Nonreflected Reactor

Consider a homogeneous nonreflected reactor of volume G bounded by a surface S_e. Then the set of basic reactor equations in the diffusion-age approximation becomes

$$\left.\begin{aligned} &D\nabla^2\varphi - \Sigma_c\varphi = \frac{\partial\xi\Sigma s\varphi}{\partial u} - \frac{\chi(u)}{K_{\text{eff}}}Q(\vec{r}),\\ &D_T\nabla^2\Phi - \Sigma_{cT}\Phi = -\xi\Sigma_s\varphi(\vec{r}, u_T),\\ &Q(\vec{r}) = \nu_f\left(\int\limits_{-\infty}^{u_T}\Sigma_f\varphi\,du + \Sigma_{fT}\Phi\right),\end{aligned}\right\} \tag{20.1}$$

where ∇^2 is the Laplacian.

The boundary conditions which must be added to (20.1) are, on the assumption that the extrapolated boundary is independent,

$$\left.\begin{aligned} \varphi(\vec{r}, u) &= 0,\\ \Phi(\vec{r}) &= 0\end{aligned}\right\} \quad \text{on} \quad S_e \tag{20.2}$$

Let us try to solve this equation by separating the variables. We thus write

---

*Romanovich and Pupko [67] have studied the theory of nonreflected reactors in great detail.

$$\varphi(\vec{r}, u) = \varphi_1(u) Y(\vec{r}),$$
$$\Phi(\vec{r}) = \Phi_1 Y(\vec{r}), \tag{20.3}$$

where $Y(r)$ satisfies

$$\nabla^2 Y + \varkappa^2 Y = 0 \tag{20.4}$$

with the condition

$$Y(\vec{r}) = 0 \quad \text{on} \quad S_e \tag{20.5}$$

The parameter κ^2 in (20.4) is found from the requirement that the problem have a nontrivial solution. With very general assumptions concerning the nature of S_e, the problem stated in (20.4) and (20.5) is known to give an infinite sequence of eigenvalues $\kappa_1^2, \kappa_2^2, \ldots, \kappa_n^2 \ldots$, with eigenfunctions $Y_1(\vec{r}), Y_2(\vec{r}), \ldots, Y_n(\vec{r}) \ldots$

Of all the $Y_n(\vec{r})$ functions, however, only $Y = Y_1$, that belonging to the maximum eigenvalue $\vec{\kappa}^2 = \kappa_1^2$, is non-negative, so that this is the function which describes the neutron flux density in the reactor.

Bearing this in mind, let us insert (20.3) into (20.1) and make use of (20.4). We then obtain

$$\left.\begin{aligned} (\varkappa^2 D + \Sigma_c)\varphi_1 &= -\frac{\partial \xi \Sigma_s \varphi_1}{\partial u} + \frac{1}{K_{\text{eff}}} \chi(u) Q_1, \\ (\varkappa^2 D_T + \Sigma_{cT})\Phi_1 &= \xi \Sigma_s \varphi_1(u_T), \\ Q_1 &= \nu_f \int_{-\infty}^{u_T} \Sigma_f \varphi_1 \, du + \Sigma_{fT} \Phi_1. \end{aligned}\right\} \tag{20.6}$$

Recalling that there are no sources with $u = -\infty$, which means that $\xi \Sigma_s \varphi_1(-\infty) = 0$, we can integrate the first of (20.6), obtaining

$$\xi \Sigma_s \varphi_1 = \frac{1}{K_{\text{eff}}} Q_1 \int_{-\infty}^{u} \chi(u') e^{-\int_{u'}^{u} \frac{\varkappa^2 D + \Sigma_c}{\xi \Sigma_s} du''} du'. \tag{20.7}$$

Setting $u = u_T$, we have

$$\xi \Sigma_s \varphi_1(u_T) = \frac{1}{K_{\text{eff}}} Q_1 \int_{-\infty}^{u_T} \chi(u') e^{-\int_{u'}^{u_T} \frac{\varkappa^2 D + \Sigma_c}{\xi \Sigma_s} du''} du',$$

so that

$$\Phi_1 = \frac{1}{K_{\text{eff}}} \frac{Q_1}{\varkappa^2 D_T + \Sigma_{cT}} \int_{-\infty}^{u_T} \chi(u') e^{-\int_{u'}^{u_T} \frac{\varkappa^2 D + \Sigma_c}{\xi \Sigma_s} du''} du'. \tag{20.8}$$

We now insert (20.7) and (20.8) into the last of (20.6). Then K_{eff} is given by*

$$K_{\text{eff}} = \nu_f \int_{-\infty}^{u_T} du \left[\frac{\Sigma_f}{\xi \Sigma_s} \int_{-\infty}^{u} \chi(u') e^{-\int_{u'}^{u} \frac{\varkappa^2 D + \Sigma_c}{\xi \Sigma_s} du''} du' + \frac{\Sigma_{fT} \chi(u)}{\varkappa^2 D_T + \Sigma_{cT}} e^{-\int_{u}^{u_T} \frac{\varkappa^2 D + \Sigma_c}{\xi \Sigma_s} du'} \right]. \tag{20.9}$$

---

* Equation (20.9) was first obtained by A.S. Romanovich [67].

Thus the eigenvalue K_{eff} for a nonreflected reactor is given by (20.9).

If, further, the arbitrary constant Q_1/K_{eff} is chosen equal to unity in (20.7) and (20.8), the neutron energy spectrum is finally given by

$$\left.\begin{aligned} \varphi_1 &= \frac{1}{\xi\Sigma_s} \int_{-\infty}^{u} \chi(u')\, e^{-\int_{u'}^{u} \frac{\varkappa^2 D+\Sigma_c}{\xi\Sigma_s} du''} du', \\ \Phi_1 &= \frac{1}{\varkappa^2 D_T + \Sigma_{cT}} \int_{-\infty}^{u_T} \chi(u')\, e^{-\int_{u'}^{u_T} \frac{\varkappa^2 D+\Sigma_c}{\xi\Sigma_s} du''} du'. \end{aligned}\right\} \tag{20.10}$$

These equations describe the neutron energy distribution from a unit source (that is, one for which $Q_1 = 1$) whose spectrum is $\chi(u)$.

Finally, a remark about the important case in which the reactor dimensions are allowed to increase to infinity. In this case, clearly, $\kappa_1^2 \to 0$, and we obtain the formula for K_{eff} and the neutron spectrum in an infinite homogeneous medium without neutron leakage.

21. Neutron Importance Spectrum in a Nonreflected Reactor

Let us now consider the adjoint equations (13.15), which for a homogeneous nonreflected reactor become

$$\left.\begin{aligned} D\nabla^2\varphi^* - \Sigma_c\varphi^* &= -\xi\Sigma_s \frac{\partial\varphi^*}{\partial u} - \frac{1}{K_{eff}} \Sigma_f Q^*, \\ D_T\nabla^2\Phi^* - \Sigma_{cT}\Phi^* &= -\frac{1}{K_{eff}} \Sigma_{fT} Q^*, \\ \varphi^*(\vec{r}, u_T) = \Phi^*(\vec{r}),\ Q^* &= \nu_f \int_{\infty-}^{u_T} \chi(u)\, \varphi^*\, du, \end{aligned}\right\} \tag{21.1}$$

with the condition

$$\left.\begin{aligned} \varphi^*(\vec{r}, u) &= 0, \\ \Phi^*(\vec{r}) &= 0 \end{aligned}\right\} \quad \text{on } S_e \tag{21.2}$$

We shall attempt to obtain a solution of (21.1) and (21.2) in the form

$$\left.\begin{aligned} \varphi^*(\vec{r}, u) &= \varphi_1^*(u)\, Y(\vec{r}), \\ \Phi^*(\vec{r}) &= \Phi_1^*\, Y(\vec{r}), \end{aligned}\right\} \tag{21.3}$$

where $Y(\vec{r})$ is again a solution of (20.4) with (20.5). Let us therefore insert (21.3) into (21.1). The equations thus obtained for φ_1^* and Φ_1^* are

$$\left.\begin{aligned} (\varkappa^2 D + \Sigma_c)\,\varphi_1^* &= \xi\Sigma_s \frac{d\varphi_1^*}{du} + \frac{1}{K_{eff}} \Sigma_f Q_1^*, \\ (\varkappa^2 D_T + \Sigma_{cT})\,\Phi_1^* &= \frac{1}{K_{eff}} \Sigma_{fT} Q_1^*, \\ \varphi_1^*(u_T) = \Phi_1^*,\ Q_1^* &= \nu_f \int_{-\infty}^{u_T} \chi(u)\, \varphi_1^*\ u. \end{aligned}\right\} \tag{21.4}$$

We shall find a solution of (21.4) in the form

$$\left.\begin{aligned} \Phi_1^* &= \frac{1}{K_{\text{eff}}}\frac{\Sigma_{fT}}{\varkappa^2 D_T+\Sigma_{cT}}Q_1^*, \\ \varphi_1^* &= \frac{1}{K_{\text{eff}}}Q_1^*\left[\frac{\Sigma_{fT}}{\varkappa^2 D_T+\Sigma_{cT}}e^{-\int_u^{u_T}\frac{\varkappa^2 D+\Sigma_c}{\xi\Sigma_s}du'}+\right. \\ &\left.+\int_u^{u_T}\frac{\Sigma_f}{\xi\Sigma_s}e^{-\int_u^{u'}\frac{\varkappa^2 D+\Sigma_c}{\xi\Sigma_s}du''}du'\right]. \end{aligned}\right\} \quad (21.5)$$

Let us insert the second of (21.5) into the last of (21.4). This gives

$$K_{\text{eff}} = \nu_f\int_{-\infty}^{u_T}du\left[\frac{\Sigma_{fT}\chi(u)}{\varkappa^2 D_T+\Sigma_{cT}}e^{-\int_u^{u_T}\frac{\varkappa^2 D+\Sigma_c}{\xi\Sigma_s}du'}+\right.$$

$$\left.+\chi(u)\int_u^{u_T}\frac{\Sigma_f}{\xi\Sigma_s}e^{-\int_u^{u'}\frac{\varkappa^2 D+\Sigma_c}{\xi\Sigma_s}du''}du'\right]. \quad (21.6)$$

Recalling the identity

$$\int_{-\infty}^{u_T}\chi(u)du\int_u^{u_T}\frac{\Sigma_f}{\xi\Sigma_s}e^{-\int_u^{u'}\frac{\varkappa^2 D+\Sigma_c}{\xi\Sigma_s}du''}du' = \int_{-\infty}^{u_T}\frac{\Sigma_f}{\xi\Sigma_s}du\int_{-\infty}^{u}\chi(u')e^{-\int_{u'}^{u}\frac{\varkappa^2 D+\Sigma_c}{\xi\Sigma_s}du''}du',$$

we can transform (21.6) to the form of (20.9).

We thus arrive at the obvious fact that the eigenvalues of the basic and adjoint equations are the same.

Consider the neutron importance spectrum. If we set $Q_1^*/K_{\text{eff}} = 1$, we arrive at

$$\left.\begin{aligned} \Phi_1^* &= \frac{\Sigma_{fT}}{\varkappa^2 D_T+\Sigma_{cT}}, \\ \varphi_1^* &= \frac{\Sigma_{fT}}{\varkappa^2 D_T+\Sigma_{cT}}e^{-\int_u^{u_T}\frac{\varkappa^2 D+\Sigma_c}{\xi\Sigma_s}du'}+\int_u^{u_T}\frac{\Sigma_f}{\xi\Sigma_s}e^{-\int_u^{u'}\frac{\varkappa^2 D+\Sigma_c}{\xi\Sigma_s}du''}du'. \end{aligned}\right\} \quad (21.7)$$

22. Refinement of the Computations for a Homogeneous Nonreflected Reactor

We have used the diffusion-age approximation to calculate the critical mass of nonreflected reactors. As was mentioned earlier, this approximation is sometimes inadequate and must be refined. In this section we will use diffusion-age approximation with the second moment of energy taken into account to obtain formulas for K_{eff}, the neutron spectrum, and the neutron importance in a nonreflected reactor. As before, we will assume that the reactor is homogeneous, so that the basic equations are

$$\left.\begin{aligned} D\nabla^2\varphi - \Sigma_c\varphi &= \frac{\partial q}{\partial u} - \frac{\chi(u)}{K_{\text{eff}}}Q(\vec{r}), \\ D_T\nabla^2\Phi - \Sigma_{cT}\Phi &= -q(\vec{r}, u_T), \\ q &= (\xi\Sigma_s+\gamma\Sigma_c)\varphi - \gamma D\nabla^2\varphi - \gamma S, \end{aligned}\right\} \quad (22.1)$$

where

$$Q = \nu_f\left(\int_{-\infty}^{u_T} \Sigma_f \varphi du + \Sigma_{fT}\Phi\right).$$

We shall attempt to find a solution of (22.1) with the boundary conditions

$$\left.\begin{array}{l} \varphi(\vec{r}, u) = 0 \\ \Phi(\vec{r}) = 0 \end{array} \text{ on } S_e \right\} \tag{22.2}$$

in the form

$$\left.\begin{array}{l} \varphi(\vec{r}, u) = \varphi_1(u) Y(\vec{r}), \\ \Phi(\vec{r}) = \Phi_1 \cdot Y(\vec{r}), \end{array}\right\} \tag{22.3}$$

where $Y(\vec{r})$ is the first eigenfunction of the problem given by (20.4) and (20.5).

Let us insert (22.3) into (22.1). We then arrive at

$$\left.\begin{array}{l} (\varkappa^2 D + \Sigma_c)\varphi_1 = -\frac{d}{du}(\xi\Sigma_s + \gamma\Sigma_c + \gamma\varkappa^2 D)\varphi_1 + \\ \quad + \frac{1}{K_{\text{eff}}}\left(\chi(u) + \gamma\frac{d\chi}{du}\right)Q_1, \\ (\varkappa^2 D_T + \Sigma_{cT})\Phi_1 = (\xi\Sigma_s + \gamma\Sigma_c + \gamma\varkappa^2 D)\varphi_1, \; u = u_T, \\ Q_1 = \nu_f\left(\int_{-\infty}^{u_T} \Sigma_f\varphi_1 du + \Sigma_{fT}\Phi_1\right). \end{array}\right\} \tag{22.4}$$

Solving (22.4), we obtain

$$\begin{aligned} K_{\text{eff}} &= \nu_f \int_{-\infty}^{u_T} du \left\{ \frac{\Sigma_f}{\xi\Sigma_s + \gamma\Sigma_c + \gamma\varkappa^2 D} \times \right. \\ &\times \int_{-\infty}^{u}\left[\chi(u') + \gamma\frac{d\chi}{du'}\right] e^{-\int_{u'}^{u} \frac{\varkappa^2 D + \Sigma_c}{\xi\Sigma_s + \gamma\Sigma_c + \gamma\varkappa^2 D} du''} du' + \\ &+ \left. \frac{\Sigma_{fT}}{\varkappa^2 D_T + \Sigma_{cT}}\left[\chi(u) + \gamma\frac{d\chi}{du}\right] e^{-\int_{u}^{u_T} \frac{\varkappa^2 D + \Sigma_c}{\xi\Sigma_s + \gamma\Sigma_c + \gamma\varkappa^2 D} du'} \right\}. \end{aligned} \tag{22.5}$$

Similarly, K_{eff} can be obtained from the adjoint equations.

23. Proof of Convergence of the Method of Successive Approximations

According to the definition in Chapter I, in the diffusion-age approximation we can use the method of successive approximations to find K_{eff}, the space and energy distribution of the neutrons, and the neutron importance in the following way.

Let

$$Q(\vec{r}) = \nu_f\left(\int_{-\infty}^{u_T} \Sigma_f \varphi\, du + \Sigma_{fT}\Phi\right)$$

be the neutron and fission sources, and let

$$Q^{*}(\vec{r}) = \nu_f \int_{-\infty}^{u_T} \chi(u)\, \varphi^{*}\, du$$

be the neutron importance sources in the reactor. It then follows directly from the definition of K_{eff} that

$$K_{eff}^{(l)} = \frac{Q^{(l+1)}(\vec{r})}{Q^{(l)}(\vec{r})} \tag{23.1}$$

for the basic equations, and

$$K_{eff}^{(l)} = \frac{Q^{*(l+1)}(\vec{r})}{Q^{*(l)}(\vec{r})} \tag{23.2}$$

for the adjoint equations, where l is the order to which the successive approximations have been carried.

We shall show that the method of successive approximations converges for a nonreflected reactor.

Consider a nonreflected reactor of arbitrary shape and volume G bounded by an extrapolated surface S_e independent of energy.

We shall write the set of basic reactor equations in the form*

$$\left.\begin{aligned} &D\nabla^2\varphi - \Sigma_c\varphi = \frac{\partial\xi\Sigma_s\varphi}{\partial u} - \chi(u)\, Q, \\ &D_T\nabla^2\Phi - \Sigma_{cT}\Phi = -\xi\Sigma_s\varphi(\vec{r}, u_T), \\ &Q(\vec{r}) = \nu_f\left(\int_{-\infty}^{u_T} \Sigma_f\varphi\, du + \Sigma_{fT}\Phi\right). \end{aligned}\right\} \tag{23.3}$$

We shall attempt to find a solution which belongs to the class $\{f\}$.

Then, as we have already established, the eigenvalue for a critical reactor is given by the set of homogeneous reactor equations (23.3). We shall assume that a solution exists. According to the method of successive approximations, we give some function

$$Q^{(l)}(\vec{r}) = \nu_f\left(\int_{-\infty}^{u_T} \Sigma_f\varphi\, du + \Sigma_{fT}\Phi\right), \tag{23.4}$$

which is sufficiently smooth. Further, the problem

$$\left.\begin{aligned} &\nabla^2 Y + \varkappa^2 Y = 0, \\ &Y = 0 \quad \text{on } S_e \end{aligned}\right\} \tag{23.5}$$

defines a complete orthonormal set of functions (Y_n) in **G**, the region bounded by S_e. Here as before, the κ_n^2 are the eigenvalues of (23.5), and they depend on S_e.**

Since the (Y_n) are a complete set of functions, we can write $Q^{(l)}(\vec{r})$ in the form***

$$Q^{(l)}(\vec{r}) = \sum_{n=1}^{\infty} f_n^{(l)} Y_n(\vec{r}), \tag{23.5'}$$

* Recalling the definition of the method of successive approximations and that of K_{eff} as given by (23.1) and (23.2), we may drop the parameter K_{eff} in the basic and adjoint equations.

** It is sufficient to assume, for our purposes, that S_e is a piecewise continuous surface.

*** We assume that $Q^{(2)}(\vec{r})$ satisfies the Dirichlet conditions [42].

where

$$f_n^{(l)} = \int\limits_G Q^{(l)}(\vec{r})\, Y_n(\vec{r})\, d\vec{r}. \tag{23.6}$$

It is easily shown that with (23.5') the solution to the problem can be written

$$\left.\begin{aligned}
\varphi^{(l)} &= \frac{1}{\xi\Sigma_s} \sum_{n=1}^{\infty} f_n^{(l)} \int\limits_{-\infty}^{u} \chi(u')\, e^{-\int\limits_{u'}^{u} \frac{\varkappa_n^2 \cdot D + \Sigma_c}{\xi\Sigma_s} du''} du' \cdot Y_n(\vec{r}),\\
\Phi^{(l)} &= \sum_{n=1}^{\infty} \frac{f_n^{(l)}}{\varkappa_n^2 D_T + \Sigma_{cT}} \int\limits_{-\infty}^{u_T} \chi(u')\, e^{-\int\limits_{u'}^{u_T} \frac{\varkappa_n^2 D + \Sigma_c}{\xi\Sigma_s} du''} du'\, Y_n(\vec{r}).
\end{aligned}\right\} \tag{23.7}$$

Let us now insert (23.7) into (23.4). We then obtain the new approximation $Q^{(l+1)}$, given by

$$Q^{(l\pm 1)}(\vec{r}) = \sum_{n=1}^{\infty} f_n^{(l)} I_n Y_n(\vec{r}), \tag{23.8}$$

where

$$I_n = \nu_f \int\limits_{-\infty}^{u_T} du \left[\frac{\Sigma_f}{\xi\Sigma_s} \int\limits_{-\infty}^{u} \chi(u')\, e^{-\int\limits_{u'}^{u} \frac{\varkappa_n^2 D + \Sigma_c}{\xi\Sigma_s} du''} du' + \right.$$

$$\left. + \frac{\Sigma_{fT}}{\varkappa_n^2 D_T + \Sigma_{cT}} \chi(u) \cdot e^{-\int\limits_{u}^{u_T} \frac{\varkappa_n^2 D + \Sigma_c}{\xi\Sigma_s} du''} \right]. \tag{23.9}$$

We now expand the left side of (23.8), writing

$$Q^{(l+1)}(\vec{r}) = \sum_{n=1}^{\infty} f_n^{(l+1)} Y_n(\vec{r}) \tag{23.10}$$

and then compare coefficients in (23.8) and (23.10). This gives

$$f_n^{(l+1)} = I_n f_n^{(l)}. \tag{23.11}$$

Using this, we can set up the ratio

$$\frac{f_n^{(l+1)}}{f_1^{(l+1)}} = \frac{I_n}{I_1} \frac{f_n^{(l)}}{f_1^{(l)}}.$$

Now the fact that

$$\frac{I_n}{I_1} < 1 \quad \text{for} \quad n > 1, \tag{23.12}$$

implies that the sequence

$$\frac{f_n^{(0)}}{f_1^{(0)}},\ \frac{f_n^{(1)}}{f_1^{(1)}},\ \ldots,\ \frac{f_n^{(l)}}{f_1^{(l)}},\ \ldots \tag{23.13}$$

approaches zero uniformly for $l \to \infty$ as the ratio $(I_n/I_1)^l$. Therefore, in the limit as $l \to \infty$ we obtain a solution of the problem, and this solution is the first eigenfunction $Y_1(\vec{r})$.

Let us now consider the expression for K_{eff}

$$K_{\text{eff}}^{(l)} = \frac{Q^{(l+1)}(\vec{r})}{Q^{(l)}(\vec{r})}. \tag{23.14}$$

If $l > l_0$ is chosen appropriately in (23.5') and (23.8), we need maintain only the first two terms, assuming that the sum of the terms we have dropped is small. We then have

$$\left.\begin{aligned} Q^{(l)}(\vec{r}) &= f_1^{(0)} I_1^l \; Y_1(\vec{r}) \left[1 + A\left(\frac{I_2}{I_1}\right)^l + \dots\right], \\ K_{\text{eff}}^{(l)} &= K_{\text{eff}} \left[1 + B\left(\frac{I_2}{I_1}\right)^l + \dots\right], \end{aligned}\right\} \tag{23.15}$$

where

$$I_1 = K_{\text{eff}}\,, \quad A = \frac{f_2^{(0)} Y_2(\vec{r})}{f_1^{(0)} Y_1(\vec{r})}, \quad B = A\frac{I_2 - I_1}{I_1}.$$

Here it is assumed in addition that A is bounded for all values of $\vec{r}$.

We then arrive at formulas which can be used to estimate $Q^{(l)}$ and $K_{\text{eff}}^{(l)}$, namely:

$$\left.\begin{aligned} Q^{(l)}(\vec{r}) &= f_1 Y_1(\vec{r}) \left\{1 + 0\left[\left(\frac{I_2}{I_1}\right)^l\right]\right\}, \\ K_{\text{eff}}^{(l)} &= K_{\text{eff}} \left\{1 + 0\left[\left(\frac{I_2}{I_1}\right)^l\right]\right\}. \end{aligned}\right\} \tag{23.16}$$

Hence with any choice of constants the successive approximation process converges and can be used to find the eigenvalue K_{eff} and the eigenfunction. This establishes, in particular, the existence of a solution to the problem, which we had previously merely assumed.

In the same way, the convergence of the successive approximation method can be demonstrated for the adjoint reactor equations. We note that in the case of the adjoint equations, the estimates of (23.16) remain valid. All we need do is replace Q by Q^*.

Chapter VI

THE METHOD OF GROUPS. WEAK ABSORPTION OF SLOWING-DOWN NEUTRONS

Let us now consider nuclear reactors with reflectors.

In view of the extreme difficulty in obtaining exact solutions of the basic and adjoint reactor equations for this case, it becomes necessary to use various approximation methods to simplify the problem to the greatest possible extent without significant loss of accuracy.

We shall direct our attention first to the so-called method of groups, which consists essentially of further simplification of the form of the energy operators of the problem by making additional assumptions on the behavior of the solutions within certain energy intervals. In the final analysis these assumptions are related to the use of some type of interpolation formula in the solution. As a result, both the basic and adjoint reactor equations are written approximately as sets of equations of the diffusion type for the desired functions at suitable points of interpolation.

The solutions at the other points in the lethargy intervals are obtained by using the same interpolation formulas as those which have already been used to obtain the sets of diffusion equations [23, 53, 101].

As for the solution for the diffusion equations, this is not particularly difficult and can be done numerically by various methods which will be described below.

We shall consider two separate cases in the method of groups. These are weak and strong absorption of slowing-down neutrons. Weak absorption takes place primarily in thermal reactors, while strong absorption is associated with intermediate ones.

Bearing in mind the specific properties of the processes taking place in these reactors, it is useful to treat each of these types separately.

In this chapter we shall consider the method of groups in the case of weak absorption.

24. The Set of Basic Reactor Equations

Consider a reactor of several zones bounded by the surface S_e. As was shown above, the slowing-down equations in this case* are

$$\nabla D \nabla \varphi - \Sigma_c \varphi = \frac{\partial \xi \Sigma_s \varphi}{\partial u} - \chi(u) Q(\vec{r}) . \tag{24.1}$$

We shall make use of the notation

$$\nabla D \nabla \varphi - \Sigma_c \varphi = \mathbf{M}(\varphi) . \tag{24.2}$$

Then (24.1) can be written

$$\frac{\partial \xi \Sigma_s \varphi}{\partial u} = \mathbf{M}(\varphi) + \chi(u) Q(\vec{r}) . \tag{24.3}$$

*Here and in what follows when attacking the problem by Kellog's method of successive approximations we shall define K_{eff} by Eqs. (7.6) and (7.9), so that in the reactor equations we may drop the factor $1/K_{eff}$ as irrelevant.

Let us now integrate (24.3) over the interval (u_{j-1}, u), where u_{j-1} is a basic interpolation point of the lethargy interval. We then obtain

$$\xi\Sigma_s\varphi = \xi\Sigma_s^{j-1}\varphi^{j-1} + \int_{u_{j-1}}^{u} \mathbf{M}(\varphi)\,du + \chi^j(u)\,Q, \tag{24.4}$$

where

$$\chi^j(u) = \int_{u_{j-1}}^{u} \chi(u)\,du.$$

Further, setting $u = u_j$ in (24.4), we arrive at

$$\xi\Sigma_s^j\varphi^j = \xi\Sigma_s^{j-1}\varphi^{j-1} + \int_{u_{j-1}}^{u_j} \mathbf{M}(\varphi)\,du + \chi^j Q, \tag{24.5}$$

where

$$\chi^j = \chi^j(u^j).$$

We shall use (24.4) and (24.5) as the basis for obtaining the various multigroup representations of the slowing-down equation.

Let us now suppose that it is possible to approximate $\mathbf{M}(\varphi)$ suitably on the interval (u_{j-1}, u_j) using interpolation formulas which relate the value of this function at internal points of the interval with the values at the end points. By using these formulas in (24.4) and (24.5) we obtain both interpolation formulas for the desired function φ, and the required equations for φ at the basic points of interpolation, which are the end points of our intervals. Finally, these latter equations are the set of multigroup slowing-down equations.

Thus the problem is first of all to choose the proper interpolation formulas for $\mathbf{M}(\varphi)$. In an attempt to clarify the principles on which this choice is made, we make the following considerations.

Let us assume that there are no external neutron sources in the interval (u_{j-1}, u_j), which means that $\chi^j(u) = 0$. If, in addition, there is no absorption or leakage of neutrons, (24.4) implies

$$\xi\Sigma_s\varphi(u) = \text{const}. \tag{24.6}$$

Accordingly,

$$\varphi(u) = \frac{\text{const}}{\xi\Sigma_s(u)}. \tag{24.7}$$

Thus, even if there is no absorption or leakage, φ is a rapidly varying function of the lethargy. The variation is due to the behavior of $\xi\Sigma_s$. In its essentials, of course, the situation will not be very different if there is a small amount of absorption and leakage, although $\mathbf{M}(\varphi)$ no longer vanishes. This means that $\mathbf{M}(\varphi)$ also varies rapidly in the interval (u_{j-1}, u_j), and therefore it cannot be approximated sufficiently well by smooth functions such as polynomials of the zeroth and first degrees.

All of this means that in order to obtain the interpolation formulas for $\mathbf{M}(\varphi)$ we must start not with φ, but with other functions whose variation over the interval (u_{j-1}, u_j) is very small. As is known, one such function is the slowing-down density, which in the diffusion-age approximation is

$$q = \xi\Sigma_s\varphi. \tag{24.8}$$

Equation (24.6) shows, among other things, that in the absence of external sources, absorption, and leakage, q is a constant for all points of the lethargy interval. Further, when the absorption and leakage are weak, this function usually varies very slowly, so that it can be approximated by means of the most simple interpolation formulas.

Bearing this in mind, let us use Equation (24.8) to go over from φ to q in the expression for $M(\varphi)$. Then in a fixed zone of the reactor we have

$$M(\varphi)=\frac{D}{\xi\Sigma_s}\nabla^2 q-\frac{\Sigma_c}{\xi\Sigma_s}q. \tag{24.9}$$

Let us now integrate this over the group, obtaining

$$\int_{u_{j-1}}^{u_j} M(\varphi)\,du=\int_{u_{j-1}}^{u_j}\left(\frac{D}{\xi\Sigma_s}\nabla^2 q-\frac{\Sigma_c}{\xi\Sigma_s}q\right)du. \tag{24.10}$$

Since q and $\nabla^2 q$ are assumed to be slowly varying over the interval (u_{j-1}, u_j), we may, according to (24.10), use the approximation *

$$\int_{u_{j-1}}^{u_j} M(\varphi)\,du=\frac{1}{\Delta u_j}\int_{u_{j-1}}^{u_j}\frac{du}{\xi\Sigma_s}\int_{u_{j-1}}^{u_j}(D^j\nabla^2 q-\Sigma_c^j q)\,du, \tag{24.11}$$

where

$$D^j=\frac{\int_{u_{j-1}}^{u_j}\frac{D}{\xi\Sigma_s}du}{\int_{u_{j-1}}^{u_j}\frac{1}{\xi\Sigma_s}du},\qquad \Sigma_c^j=\frac{\int_{u_{j-1}}^{u_j}\frac{\Sigma_c}{\xi\Sigma_s}du}{\int_{u_{j-1}}^{u_j}\frac{du}{\xi\Sigma_s}}. \tag{24.12}$$

Let us now consider several interpolation formulas for $q(\vec{r}, u)$. The simplest of these are

$$\left.\begin{aligned}&\text{procedure I: } q=q^j,\\ &\text{procedure II: } q=\frac{u-u_{j-1}}{\Delta u_j}q^j+\frac{u_j-u}{\Delta u_j}q^{j-1},\\ &\text{procedure III: } q=q^{j-1},\end{aligned}\right\} \tag{24.13}$$

Procedure I assumes that on the interval (u_{j-1}, u_j) the function $q(\vec{r}, u)$ is roughly equal to its value at the point u_j. Procedure II takes account of the linear variation of $q(\vec{r}, u)$ over the interval (u_{j-1}, u_j). Finally, according to procedure III the value of q on this interval is taken as that at the point $u = u_{j-1}$.

These three possible procedures lead to three different sets of multigroup slowing-down equations.

Let us analyze these three procedures. We do this by inserting (24.13) into (24.11). This gives

$$\int_{u_{j-1}}^{u_j} M(\varphi)\,du=A^j(D^j\nabla^2\varphi^j-\Sigma_c^j\varphi^j)+B^j(D^{j-1}\nabla^2\varphi^{j-1}-\Gamma^j\Sigma_c^{j-1}\varphi^{j-1}), \tag{24.14}$$

* In averaging the constants D^j and Σ_c^j for the first groups, it is expedient to set $q = \text{const}\int_{-\infty}^{u}\chi(u)\,du$

where

$$\Gamma^j = \frac{\Sigma_c^j D^{j-1}}{\Sigma_c^{j-1} D^j},$$

and A^j and B^j are constants given in Table 1.

TABLE 1

Definition of A^j and B^j for Different Forms of the Multigroup Slowing-Down Equations of a Reactor for Weak Absorption of the Slowing-Down Neutrons

	A^j	B^j
procedure I	$\xi\Sigma_s^j \int_{u_{j-1}}^{u_j} \frac{du}{\xi\Sigma_s}$	0
procedure II	$\frac{\xi\Sigma_s^j}{2} \int_{u_{j-1}}^{u_j} \frac{du}{\xi\Sigma_s}$	$\frac{D^j}{D^{j-1}} \frac{\xi\Sigma_s^{j-1}}{2} \int_{u_{j-1}}^{u_j} \frac{du}{\xi\Sigma_s}$
procedure III	0	$\frac{D^j}{D^{j-1}} \xi\Sigma_s^{j-1} \int_{u_{j-1}}^{u_j} \frac{du}{\xi\Sigma_s}$

Let us assume further that on the boundaries between the different zones of the reactor we have

$$D^j \nabla \varphi^j = D'^j \nabla \varphi'^j, \tag{24.15}$$

where the primes denote quantities belonging to the neighboring zones. Then (24.14) can be written in a form valid for all points of the reactor, namely:

$$\int_{u_{j-1}}^{u_j} \mathrm{M}(\varphi)\, du = A^j (\nabla D^j \nabla \varphi^j - \Sigma_c^j \varphi^j) + B^j (\nabla D^{j-1} \nabla \varphi^{j-1} - \Gamma^j \Sigma_c^{j-1} \varphi^{j-1}). \tag{24.16}$$

Let us now insert (24.16) into (24.5). Then for procedures I and II we obtain a diffusion equation of the form

$$\nabla D^j \nabla \varphi^j - \Sigma^j \varphi^j = -f^j \quad (j = 1,\ 2,\ \ldots,\ m), \tag{24.17}$$

where, for procedure I

$$f^j = \frac{\xi\Sigma_s^{j-1}}{A^j} \varphi^{j-1} + \frac{\chi^j}{A^j} Q, \quad \Sigma^j = \Sigma_c^j + \frac{\xi\Sigma_s^j}{A^j}, \tag{24.18}$$

while for procedure II

$$f^j = \frac{B^j}{A^j} (\nabla D^{j-1} \nabla \varphi^{j-1} - \Gamma^j \Sigma_c^{j-1} \varphi^{j-1}) + \frac{\xi\Sigma_s^{j-1}}{A^j} \varphi^{j-1} + \frac{\chi^j}{A^j} Q, \tag{24.19}$$

$$\Sigma^j = \Sigma_c^j + \frac{\xi\Sigma_s^j}{A^j}.$$

Let us now use (24.17) to eliminate $\nabla D^{j-1} \nabla \varphi^{j-1}$ from (24.19). Then for procedure II we get the recursion relation

$$f^j = \frac{\xi\Sigma_s^{j-1}}{A^j}\varphi^{j-1} + \frac{B^j}{A^j}\left[-f^{j-1} + (\Sigma^{j-1} - \Gamma^j \Sigma_c^{j-1})\varphi^{j-1}\right] + \frac{\chi^j}{A^j}Q. \tag{24.20}$$

If to (24.17) we now add the diffusion equation for the thermal group*

$$\nabla D_T \nabla \Phi - \Sigma_{cT}\Phi = -\xi\Sigma_s^m \varphi^m, \tag{24.21}$$

we finally arrive at the multigroup set of equations in the form

$$\nabla D^j \nabla \varphi^j - \Sigma^j \varphi^j = -f^j \quad (j = 1, 2, \ldots, m, m+1), \tag{24.22}$$

where we have set

$$\varphi^{m+1} = \Phi, \tag{24.23}$$

and the quantities D^j, Σ^j and f^j are given in Table 2.

Finally, let us consider procedure III. To do this, we insert (24.16) into (24.5). Bearing in mind that $A^j = 0$, we obtain

$$\nabla D^j \nabla \varphi^{j-1} + \Sigma^j \varphi^{j-1} = f^j \quad (j = 1, 2, \ldots, m), \tag{24.24}$$

where

$$f^j = \frac{\xi\Sigma_s^j}{B^j}\varphi^j - \frac{\chi^j}{B^j}Q,$$
$$\Sigma^j = \frac{\xi\Sigma_s^{j-1}}{B^j} - \Gamma^j \Sigma_c^{j-1}. \tag{24.25}$$

TABLE 2

Functions and Constants in the Multigroup Reactor Equations for Weak Absorption of Slowing-Down Neutrons

j	φ^j	D^j	Σ^j	f^j
(1)—(m)	φ^j	D^j	$\Sigma_c^j + \frac{\xi\Sigma_s^j}{A^j}$	$\frac{\xi\Sigma_s^{j-1}}{A^j}\varphi^{j-1} + \frac{B^j}{A^j}[-f^{j-1} + (\Sigma^{j-1} - \Gamma^j\Sigma_c^{j-1})\varphi^{j-1}] + \frac{\chi^j}{A^j}Q$
(m + 1)	Φ	D_T	Σ_{cT}	$\xi\Sigma_s^m \varphi^m$

Obviously the left side of (24.24) contains the given function φ^{j-1}, while the right side contains the unknown φ^j. Let us solve this equation for φ^j. We then obtain

$$\varphi^j = \left[\frac{B^j}{\xi\Sigma_s^j}\left(\nabla D^j \nabla \varphi^{j-1} + \Sigma^j \varphi^{j-1} + \frac{\chi^j}{B^j}Q\right)\right] \quad (j = 1, 2, \ldots, m). \tag{24.26}$$

As before, we add to (24.24) and (24.26), Equation (24.21) for the thermal group. Then the multigroup reactor equations for procedure III are completely defined.

---

*If there are U^{238} nuclei in the core of the reactor, the thermal neutron sources $q = \xi\Sigma^m\varphi^m$ in (24.21) must be multiplied by $\langle\varphi\rangle$, the probability of avoiding resonance capture.

Having obtained the multigroup reactor equations, we must now consider the fission integrals, which are used in computing the neutron sources $Q(\vec{r})$.

Thus, let us consider the expression for the fission sources

$$Q(\vec{r}) = \nu_f \left(\int_{-\infty}^{u_T} \Sigma_f \varphi \, du + \Sigma_{fT} \Phi \right), \tag{24.27}$$

which we may write

$$Q(\vec{r}) = \nu_f \left(\sum_{j=1}^{m} \int_{u_{j-1}}^{u_j} \frac{\Sigma_f}{\xi \Sigma_s} q \, du + \Sigma_{fT} \Phi \right), \tag{24.28}$$

where

$$q = \xi \Sigma_s \varphi.$$

Since by assumption q varies very weakly over (u_{j-1}, u_j), we may approximate it by writing

$$q = \frac{u - u_{j-1}}{\Delta u_j} q^j + \frac{u_j - u}{\Delta u_j} q^{j-1}. \tag{24.29}$$

Now inserting (24.29) into (24.28), we obtain

$$Q(\vec{r}) = \nu_f \sum_{j=1}^{m+1} \alpha^j \varphi^j, \tag{24.30}$$

where

$$\varphi^{m+1} = \Phi,$$

$$\alpha^j = \begin{cases} \alpha_1^j + \alpha_2^{j+1} & (j = 1, 2, \ldots, m-1), \\ \alpha_1^j & (j = m), \\ \Sigma_{fT} & (j = m+1), \end{cases}$$

and

$$\alpha_1^j = \xi \Sigma_s^j \int_{u_{j-1}}^{u_j} \frac{\Sigma_f}{\xi \Sigma_s} \frac{u - u_{j-1}}{\Delta u_j} du, \quad \alpha_2^j = \xi \Sigma_s^{j-1} \int_{u_{j-1}}^{u_j} \frac{\Sigma_f}{\xi \Sigma_s} \frac{u_j - u}{\Delta u_j} du.$$

25. The Set of Adjoint Reactor Equations

Let us now go on to the multigroup representation of the set of adjoint reactor equations.

As was shown in Section 13, the set of adjoint reactor equations for this case can be written

$$\left. \begin{aligned} \nabla D \nabla \varphi^* - \Sigma_c \varphi^* &= -\xi \Sigma_s \frac{\partial \varphi^*}{\partial u} - \Sigma_f Q^*, \\ \nabla D_T \nabla \Phi^* - \Sigma_{cT} \Phi^* &= -\Sigma_{fT} Q^*, \end{aligned} \right\} \tag{25.1}$$

where

$$Q^* = \nu_f \int_{-\infty}^{u_T} \chi(u) \varphi^* du. \tag{25.2}$$

Let us start with the first of (25.1), writing it temporarily in the form

$$\frac{\partial \varphi^*}{\partial u} = -\frac{1}{\xi\Sigma_s} M(\varphi^*) - \frac{\Sigma_f}{\xi\Sigma_s} Q^*. \tag{25.3}$$

Now let us integrate over the interval (u, u_j). This gives

$$\left.\begin{aligned} \varphi^* &= \varphi^{*j} + \int_u^{u_j} \frac{1}{\xi\Sigma_s} \mathbf{M}(\varphi^*)\, du + \eta^j(u)\, Q^*, \\ \eta^j(u) &= \int_u^{u_j} \frac{\Sigma_f}{\xi\Sigma_s}\, du. \end{aligned}\right| \tag{25.4}$$

Setting $u = u_{j-1}$, we have

$$\varphi^{*j-1} = \varphi^{*j} + \int_{u_{j-1}}^{u_j} \frac{1}{\xi\Sigma_s} \mathbf{M}(\varphi^*)\, du + \eta^j Q^*, \tag{25.5}$$

where

$$\eta^j = \eta^j(u_{j-1}).$$

Let us now consider the question of the best way to approximate $\mathbf{M}(\varphi^*)$ within (u_{j-1}, u_j). For this purpose we shall first discuss some properties of φ^*. Assume first that there is no neutron absorption or leakage. Then it follows immediately from (25.4) that

$$\varphi^*(\vec{r}, u) = \varphi^{*j}(\vec{r}).$$

Thus we find that $\varphi^*(\vec{r}, u)$ is in this case independent of u.

It is quite obvious that if there is weak absorption and some leakage in the group, both $\varphi^*(\vec{r}, u)$ and $\mathbf{M}(\varphi^*)$ will be slowly varying functions of u.

Let us now consider some fixed zone of the reactor and integrate $\frac{1}{\xi\Sigma_s}\mathbf{M}(\varphi^*)$ over the group (u_{j-1}, u_j), obtaining

$$\int_{u_{j-1}}^{u_j} \mathbf{M}(\varphi^*) \frac{du}{\xi\Sigma_s} = \int_{u_{j-1}}^{u_j} \left(\frac{D}{\xi\Sigma_s} \nabla^2\varphi - \frac{\Sigma_c}{\xi\Sigma_s}\varphi^*\right) du.$$

Since by assumption φ^* and $\nabla^2\varphi^*$ are slowly varying functions, we may write approximately

$$\int_{u_{j-1}}^{u_j} \mathbf{M}(\varphi^*) \frac{du}{\xi\Sigma_s} = \frac{1}{\Delta u_j} \int_{u_{j-1}}^{u_j} \frac{du}{\xi\Sigma_s} \int_{u_{j-1}}^{u_j} (D^j \nabla^2 \varphi^* - \Sigma_c^j \varphi^*)\, du, \tag{25.6}$$

where D^j and Σ_c^j are again given by (24.12).

Let us consider the following three interpolation formulas for φ^*:

$$\begin{aligned} &\text{procedure I:} \quad \varphi^* = \varphi^{*j-1}, \\ &\text{procedure II:} \quad \varphi^* = \frac{u_j - u}{\Delta u_j}\varphi^{*j-1} + \frac{u - u_{j-1}}{\Delta u_j}\varphi^{*j}, \\ &\text{procedure III:} \quad \varphi^* = \varphi^{*j}. \end{aligned} \tag{25.7}$$

We now insert these interpolation polynomials into (25.6). This gives

$$\int_{u_{j-1}}^{u_j} M(\varphi^*)\frac{du}{\xi\Sigma_s} = A^{*j}(D^j\nabla^2\varphi^{*j-1} - \Sigma_c^j\varphi^{*j-1}) + B^{*j}(D^{j+1}\nabla^2\varphi^{*j} - \Gamma^{*j}\Sigma_c^{j+1}\varphi^{*j}), \tag{25.8}$$

where

$$\Gamma^{*j} = \frac{D^{j+1}}{D^j}\cdot\frac{\Sigma_c^j}{\Sigma_c^{j+1}},$$

and A^{*j} and B^{*j} are given by Table 3.

TABLE 3

Definition of A^{*j} and B^{*j} for Different Forms of the Multigroup Representation of the Adjoint Reactor Equations in the Case of Weak Absorption of Slowing-Down Neutrons

	A^{*j}	B^{*j}
procedure I	$\int_{u_{j-1}}^{u_j}\frac{du}{\xi\Sigma_s}$	0
procedure II	$\frac{1}{2}\int_{u_{j-1}}^{u_j}\frac{du}{\xi\Sigma_s}$	$\frac{1}{2}\frac{D^j}{D^{j+1}}\int_{u_{j-1}}^{u_j}\frac{du}{\xi\Sigma_s}$
procedure III	0	$\frac{D^j}{D^{j+1}}\int_{u_{j-1}}^{u_j}\frac{du}{\xi\Sigma_s}$

Let us assume that on the boundaries separating the zones of the reactor, in addition to φ^{*j-1} being continuous we have

$$D^j\nabla\varphi^{*j-1} = D'^j\nabla\varphi'^{*j-1}. \tag{25.9}$$

Then (25.8) can be written

$$\int_{u_{j-1}}^{u_j} \mathrm{M}(\varphi^*)\frac{du}{\xi\Sigma_s} = A^{*j}(\nabla D^j\nabla\varphi^{*j-1} - \Sigma_c^j\varphi^{*j-1}) + B^{*j}(\nabla D^{j+1}\nabla\varphi^{*j} - \Gamma^{*j}\Sigma_c^{j+1}\varphi^{*j}), \tag{25.10}$$

which is valid for all points in the reactor. We now insert (25.10) into (25.5). This gives

$$\nabla D^j\nabla\varphi^{*j-1} - \Sigma^j\varphi^{*j-1} = -f^{*j} \quad (j = m,\ m-1,\ \ldots,\ 1), \tag{25.11}$$

where

$$\left.\begin{aligned} \Sigma^j &= \Sigma_c^j + \frac{1}{A^{*j}} = \Sigma_c^j + \frac{\xi\Sigma_s^j}{A^j}, \\ f^{*j} &= \frac{1}{A^{*j}}\varphi^{*j} + \frac{B^{*j}}{A^{*j}}(\nabla D^{j+1}\nabla\varphi^{*j} - \Gamma^{*j}\Sigma_c^{j+1}\varphi^{*j}) + \frac{\eta^j}{A^{*j}}Q^*. \end{aligned}\right\} \qquad (25.12)$$

Using (25.11) to eliminate $\nabla D^{j+1}\nabla\varphi^{*j}$ from the last equation, we arrive at

$$f^{*j} = \frac{1}{A^{*j}}\varphi^{*j} + \frac{B^{*j}}{A^{*j}}[(\Sigma^{j+1} - \Gamma^{*j}\Sigma_c^{j+1})\varphi^{*j} - f^{*j+1}] + \frac{\eta^j}{A^{*j}}Q^*. \qquad (25.13)$$

Finally, adding to (25.11) the diffusion equation for the thermal group, namely

$$\nabla D_T\nabla\Phi^* - \Sigma_{cT}\Phi^* = -\Sigma_{fT}Q^*, \qquad (25.14)$$

and making use of the condition

$$\varphi^*(\vec{r}, u_T) = \Phi^*(\vec{r}),$$

we arrive at the multigroup set of adjoint reactor equations

$$\nabla D^j\nabla\varphi^{*j-1} - \Sigma^j\varphi^{*j-1} = -f^{*j} \quad (j = m+1,\ m,\ \ldots,\ 1), \qquad (25.15)$$

where

$$\varphi^{*m} = \Phi^*,$$

and D^j, f^j and Σ^j are given by Table 4. This covers the situation for procedures I and II.

For procedure III one easily obtains the multigroup set of equations in the form

$$\nabla D^{j+1}\nabla\varphi^{*j} + \Sigma^{j+1}\varphi^{*j} = f^{*j}, \qquad (25.16)$$

where

$$\left.\begin{aligned} \Sigma^{j+1} &= \frac{1}{B^{*j}} - \Gamma^{*j}\Sigma_c^{j+1}, \\ f^{*j} &= \frac{1}{B^{*j}}\varphi^{*j-1} - \frac{\eta^j}{B^{*j}}Q^*. \end{aligned}\right\} \qquad (25.17)$$

In accordance with the method of successive approximations, we shall assume that the adjoint multigroup equation is extended successively from u_T towards lower lethargy values.

This means that the left side of Equation (25.16) is known, while the right side contains the unknown. Solving, therefore, for the unknown function φ^{*j-1}, we obtain

$$\varphi^{*j-1} = B^{*j}(\nabla D^{j+1}\nabla\varphi^{*j} + \Sigma^{j+1}\varphi^{*j}) + \eta^j Q^* \quad (j = m,\ m-1,\ \ldots,\ 2,\ 1). \qquad (25.18)$$

To this we must also add the thermal group diffusion equation (25.14).

We now have to obtain the corresponding formulas for $Q^*(\vec{r})$. To do this, consider the expression

$$Q^* = \nu_f \int_{-\infty}^{u_T} \chi(u)\varphi^*\,du,$$

which we shall write in the form

TABLE 4

Functions and Constants in the Adjoint Multigroup Equations for the Case of Weak Absorption of Slowing-Down Neutrons

j	φ^{*j-1}	D^j	Σ^j	f^{*j}
$(m+1)$	Φ^*	D_T	Σ_{cT}	$\Sigma_{fT} Q^*$
$(m)-(1)$	φ^{*j-1}	D^j	$\Sigma_c^j + \frac{\xi \Sigma_s^j}{A^j}$	$\frac{1}{A^{*j}} \varphi^{*j} + \frac{B^{*j}}{A^{*j}} [(\Sigma^{j+1} - \Gamma^{*j} \Sigma_c^{j+1}) \varphi^{*j} - f^{*j+1}] + \frac{\eta^j}{A^{*j}} Q^*$

$$Q^* = \nu_f \sum_{j=1}^{m} \int_{u_{j-1}}^{u_j} \chi(u) \varphi^* du. \tag{25.19}$$

To perform the integrations on the right side we shall use the linear interpolation formula

$$\varphi^* = \frac{u_j - u}{\Delta u_j} \varphi^{*j-1} + \frac{u - u_{j-1}}{\Delta u_j} \varphi^{*j}. \tag{25.20}$$

We then arrive at

$$Q^* = \nu_f \sum_{j=1}^{m} \alpha^{*j} \varphi^{*j}, \tag{25.21}$$

where

$$\alpha^{*j} = \begin{cases} \alpha_1^{*j} & (j = m), \\ \alpha_1^{*j} + \alpha_2^{*j+1} & (j = m-1,\ m-2, \ldots, 1), \end{cases}$$

$$\alpha^{*j} = \int_{u_{j-1}}^{u_j} \chi(u) \frac{u - u_{j-1}}{\Delta u_j} du, \quad \alpha_2^{*j} = \int_{u_{j-1}}^{u_j} \chi(u) \frac{u_j - u}{\Delta u_j} du.$$

26. Inclusion of the Second Moment of Energy in the Multigroup Equations

In the present section we shall derive the basic and adjoint multigroup reactor equations with the second moment of energy included.

It should be mentioned that when one includes the second moment of energy in nuclear reactor calculations, one sometimes gets very much better results. Ordinarily this is true for light elements (such as deuterium, beryllium, etc.), particularly in intermediate reactors with a large gradient of the neutron field due to the "rise" of the neutron flux at the reflector. This improvement of the calculations for these cases leads to a change in K_{eff}. At the same time, the fission source function $Q(\vec{r})$ also changes. It now has a much larger rise on the boundary than in uncorrected age theory, and this gives better agreement with experiment. It is also important to note that the calculated value of $Q(\vec{r})$ on the boundary between the core and the reflector may be changed by several percent, a change which cannot be ignored.

All this implies that it is very important to include the second moment of energy in nuclear reactor theory; in most cases it cannot be ignored.

Let us first consider the basic reactor equations (19.10), which are

$$\left.\begin{aligned}\frac{\partial q}{\partial u} - \nabla D \nabla \varphi + \Sigma_c \varphi = S.\\ \nabla D_T \nabla \Phi - \Sigma_{cT} \Phi = -q(\vec{r}, u_T),\end{aligned}\right\} \quad (26.1)$$

where

$$q = (\xi \Sigma_s + \gamma \Sigma_c)\, \varphi - \gamma \nabla D \nabla \varphi - \gamma S,$$
$$S = \chi(u)\, Q(\vec{r}).$$

As before, let us integrate the first of (26.1) over (u_{j-1}, u), obtaining

$$q = q^{j-1} + \int\limits_{u_{j-1}}^{u} \mathbf{M}(\varphi)\, du + \chi^j(u)\, Q, \quad (26.2)$$

where we have again written

$$\mathbf{M}(\varphi) = \nabla D \nabla \varphi - \Sigma_c \varphi, \text{and } \chi^j(u) = \int\limits_{u_{j-1}}^{u} \chi(u)\, du.$$

If we set $u = u_j$ in (26.2), we arrive at

$$q^j = q^{j-1} + \int\limits_{u_{j-1}}^{u_j} \mathbf{M}(\varphi)\, du + \chi^j Q. \quad (26.3)$$

It is clear that in the absence of external sources, absorption, and neutron leakage, q is independent of u, and we may write

$$q = q^{j-1},$$

where in this case

$$q = (\xi \Sigma_s + \gamma \Sigma_c)\, \varphi.$$

It is therefore natural to suppose that if there is a small amount of absorption and leakage $(\xi \Sigma_s + \gamma \Sigma_c)\varphi$ will be a slowly varying function in the interval (u_{j-1}, u_j), and therefore we can approximate it by one of the following two simple interpolation procedures:

$$\left.\begin{aligned}&\text{procedure I:}\ (\xi\Sigma_s + \gamma\Sigma_c)\,\varphi = (\xi\Sigma_s^j + \gamma\Sigma_c^j)\,\varphi^j,\\ &\text{procedure II:}\ (\xi\Sigma_s + \gamma\Sigma_c)\,\varphi = \frac{u - u_{j-1}}{\Delta u_j}(\xi\Sigma_s^j + \gamma\Sigma_c^j)\,\varphi^j +\\ &+ \frac{u_j - u}{\Delta u_j}(\xi\Sigma_s^{j-1} + \gamma\Sigma_c^{j-1})\,\varphi^{j-1}.\end{aligned}\right\} \quad 26.4)$$

For simplicity, we shall not go into procedure III.

Let us use the approximate relation

$$\int\limits_{u_{j-1}}^{u_j} \mathbf{M}(\varphi)\, du = \frac{1}{\Delta u_j} \int\limits_{u_{j-1}}^{u_j} \frac{du}{\xi\Sigma_s + \gamma\Sigma_c} \int\limits_{u_{j-1}}^{u} (\xi\Sigma_s + \gamma\Sigma_c)(D^j \nabla^2 \varphi - \Sigma_c^j \varphi)\, du', \quad (26.5)$$

where

$$D^j = \frac{\int_{u_{j-1}}^{u_j} \frac{D\,du}{\xi\Sigma_s + \gamma\Sigma_c}}{\int_{u_{j-1}}^{u_j} \frac{du}{\xi\Sigma_s + \gamma\Sigma_c}}, \quad \Sigma_c^j = \frac{\int_{u_{j-1}}^{u_j} \frac{\Sigma_c}{\xi\Sigma_s + \gamma\Sigma_c}\,du}{\int_{u_{j-1}}^{u_j} \frac{1}{\xi\Sigma_s + \gamma\Sigma_c}\,du},$$

with the interpolation procedures of (26.4). We then obtain

$$\int_{u_{j-1}}^{u_j} \mathrm{M}(\varphi)\,du = A^j(\nabla D^j\nabla\varphi^j - \Sigma_c^j\varphi^j) + B^j(\nabla D^{j-1}\nabla\varphi^{j-1} - \Gamma^j\Sigma_c^{j-1}\varphi^{j-1}), \tag{26.6}$$

where

$$\Gamma^j = \frac{D^{j-1}}{D^j}\frac{\Sigma_c^j}{\Sigma_c^{j-1}}$$

and A^j and B^j are again given by Table 1, except that $\xi\Sigma_s$ should be replaced by $\xi\Sigma_s + \gamma\Sigma_c$.

Let us now consider the third expression for $\underline{q}$ in (26.1), setting $u = u_j$. We then have

$$q^j = (\xi\Sigma_s^j + \gamma\Sigma_c^j)\varphi^j - \gamma\nabla D^j\nabla\varphi^j - \gamma S^j. \tag{26.7}$$

We shall assume here that $D(u_j) = D^j$.

Proceeding, we insert (26.6) and (26.7) into (26.3). Then after some simple operations we obtain

$$\nabla D^j\nabla\varphi^j - \Sigma^j\varphi^j = -f^j \quad (j = 1,\ 2,\ \ldots,\ m) \tag{26.8}$$

for interpolation procedures I and II, where

$$\Sigma^j = \frac{A^j\Sigma_c^j + \xi\Sigma_s^j + \gamma\Sigma_c^j}{A^j + \gamma},$$

$$f^j = \frac{1}{A^j + \gamma}\{[\xi\Sigma_s^{j-1} + \gamma\Sigma_c^{j-1} - \gamma\Sigma^{j-1} + B^j(\Sigma^{j-1} - \Gamma^j\Sigma_c^{j-1})]\varphi^{j-1} -$$

$$-(B^j - \gamma)f^{j-1} + \chi^j Q + \gamma(S^j - S^{j-1})\}.$$

Let us now consider the thermal neutron group

$$\nabla D_T\nabla\Phi - \Sigma_{cT}\Phi = -q^m, \tag{26.10}$$

where

$$q^m = (\xi\Sigma_s^m + \gamma\Sigma_c^m)\varphi^m - \gamma\nabla D^m\nabla\varphi^m - \gamma S^m. \tag{26.11}$$

From this relation we can eliminate $\nabla D^m\nabla\varphi^m$ using (26.8). We then have

$$q^m = (\xi\Sigma_s^m + \gamma\Sigma_c^m - \gamma\Sigma^m)\varphi^m - \gamma f^m - \gamma S^m. \tag{26.12}$$

We have thus completely defined the set of multigroup reactor equations.

Finally, let us unite (26.8) and (26.10) into the general system

$$\nabla D^j\nabla\varphi^j - \Sigma^j\varphi^j = -f^j \quad (j = 1,\ 2,\ \ldots,\ m+1), \tag{26.13}$$

where $\varphi^{m+1} = \Phi$, and D^j, Σ^j and f^j are given by Table 5.

As for $Q(\vec{r})$, it is again computed according to (24.30), where $\xi\Sigma_s$ must everywhere be replaced by $\xi\Sigma_s + \gamma\Sigma_c$.

TABLE 5

Functions and Constants in the Multigroup Reactor Equations with the Second Moment of Energy Included and Weak Absorption of Slowing-Down Neutrons

j	φ^j	D^j	Σ^j	f^j
(1) — (m)	φ^j	D^j	$\dfrac{A^j\Sigma_c^j + \xi\Sigma_s^j + \gamma\Sigma_c^j}{A^j + \gamma}$	$\dfrac{1}{A^j+\gamma}\{[\xi\Sigma_s^{j-1} + \gamma\Sigma_c^{j-1} - \gamma\Sigma^{j-1} + B^j(\Sigma^{j-1} - \Gamma^j\Sigma_c^{j-1})]\varphi^{j-1} - (B^j - \gamma)f^{j-1} + \chi^j Q + \gamma(S^j - S^{j-1})\}$
(m + 1)	Φ	D_T	Σ_{cT}	$(\xi\Sigma_s^{j-1} + \gamma\Sigma_c^{j-1} - \gamma\Sigma^{j-1})\varphi^{j-1} - \gamma f^{j-1} - \gamma S^{j-1}$

Let us now go on to a consideration of the set of adjoint reactor equations (19.11), namely

$$\left.\begin{aligned} \xi\Sigma_s \frac{\partial}{\partial u}\frac{q^*}{\xi\Sigma_s} + \nabla D\nabla\varphi^* - \Sigma_c\varphi^* &= -\Sigma_f Q^*, \\ \nabla D_T\nabla\Phi^* - \Sigma_{cT}\Phi &= -\Sigma_{fT}Q^* \\ q^* = (\xi\Sigma_s + \gamma\Sigma_c)\varphi^* - \gamma\nabla D\nabla\varphi^* &- \gamma\Sigma_f Q^*, \\ Q^* = \nu_f \int_{-\infty}^{u_T} \varphi^*\chi(u)\,du. \end{aligned}\right\} \tag{26.14}$$

We multiply the first of these by $1/\xi\Sigma_s$ and integrate over $\underline{u}$ in the interval (u, u_j). We then have

$$\frac{q^*}{\xi\Sigma_s} = \frac{q^{*j}}{\xi\Sigma_s^j} + \int_u^{u_j} M(\varphi)\frac{du}{\xi\Sigma_s} + Q^*\int_u^{u_j}\frac{\Sigma_f}{\xi\Sigma_s}du. \tag{26.15}$$

Setting $u = u_{j-1}$, this becomes

$$\frac{q^{*j-1}}{\xi\Sigma_s^{j-1}} = \frac{q^{*j}}{\xi\Sigma_s^j} + \int_{u_{j-1}}^{u_j} M(\varphi)\frac{du}{\xi\Sigma_s} + \eta^j Q^*, \tag{26.16}$$

where

$$\eta^j = \int_{u_{j-1}}^{u_j}\frac{\Sigma_f}{\xi\Sigma_s}du.$$

Again we shall consider two possible interpolation procedures, namely

$$\left.\begin{aligned} &\text{procedure I:} \quad \varphi^* = \varphi^{*j-1}, \\ &\text{procedure II:} \quad \varphi^* = \frac{u_j - u}{\Delta u_j}\varphi^{*j-1} + \frac{u - u_{j-1}}{\Delta u_j}\varphi^{*j}. \end{aligned}\right\} \tag{26.17}$$

We now insert (26.17) into (26.16) and use the approximation

$$\int_{u_{j-1}}^{u_j} \mathrm{M}(\varphi^*) \frac{du}{\xi\Sigma_s} = \frac{1}{\Delta u_j} \int_{u_{j-1}}^{u_j} \frac{du}{\xi\Sigma_s} \int_{u_{j-1}}^{u_j} (\nabla D^j \nabla\varphi^* - \Sigma_c^j \varphi^*)\, du. \tag{26.18}$$

Proceeding in the same way as in the previous section, we obtain the set of diffusion equations

$$\nabla D^j \nabla \varphi^{*j-1} - \Sigma^j \varphi^{*j-1} = -f^{*j} \quad (j = m,\ m-1,\ \ldots,\ 1), \tag{26.19}$$

where

$$\Sigma^j = \frac{A^{*j}\Sigma_c^j + \dfrac{\xi\Sigma_s^{j-1} + \gamma\Sigma_c^{j-1}}{\xi\Sigma_s^{j-1}}}{A^{*j} + \dfrac{\gamma}{\xi\Sigma_s^{j-1}}}$$

$$f^{*j} = \frac{1}{A^{*j} + \dfrac{\gamma}{\xi\Sigma_s^{j-1}}} \Bigg[\left(\frac{\xi\Sigma_s^j + \gamma\Sigma_c^j}{\xi\Sigma_s^j} - \Gamma^j B^{*j} \Sigma_c^{j+1} \right) \varphi^{*j} +$$

$$+ \left(B^{*j} - \frac{\gamma}{\xi\Sigma_s^j} \right) (\Sigma^{j+1}\varphi^{*j} - f^{*j+1}) + \gamma \left(\frac{S^{*j-1}}{\xi\Sigma_s^{j-1}} - \frac{S^{*j}}{\xi\Sigma_s^j} \right) + \eta^j Q \Bigg],$$

and A^{*j} and B^{*j} are given in Table 3.*

We now add the equation for the thermal group to (26.19). This gives the multigroup set of adjoint reactor equations

$$\nabla D^j \nabla \varphi^{*j-1} - \Sigma^j \varphi^{*j-1} = -f^{*j} \ (j = m+1,\ m,\ \ldots,\ 1),$$

where

$$\varphi^{*m} = \Phi^*,$$

and D^j, Σ^j and f^{*j} are given by Table 6.

TABLE 6

Functions and Constants in Adjoint Multigroup Equations With the Second Moment of Energy and Weak Absorption of Slowing-Down Neutrons

j	φ^{*j-1}	D^j	Σ^j	f^{*j}
$(m+1)$	Φ^*	D_T	Σ_{cT}	$\Sigma_{fT}Q^*$
$(m)-(1)$	φ^{*j-1}	D^j	$\dfrac{A^{*j}\Sigma_c^j}{A^{j*} + \dfrac{\gamma}{\xi\Sigma_s^{j-1}}} + \dfrac{\dfrac{\xi\Sigma_s^{j-1} + \gamma\Sigma_c^{j-1}}{\xi\Sigma_s^{j-1}}}{A^{*j} + \dfrac{\gamma}{\xi\Sigma_s^{j-1}}}$	$\dfrac{1}{A^{*j} + \dfrac{\gamma}{\xi\Sigma_s^{j-1}}} \times \Big[\left(\dfrac{\xi\Sigma_s^j + \gamma\Sigma_c^j}{\xi\Sigma_s^j} - \Gamma^j B^{*j}\Sigma_c^{j+1} \right) \varphi^{*j} + \left(B^{*j} - \dfrac{\gamma}{\xi\Sigma_s^j} \right) (\Sigma^{j+1}\varphi^{*j} - f^{*j+1}) + \gamma \left(\dfrac{S^{*j-1}}{\xi\Sigma_s^{j-1}} - \dfrac{S^{*j}}{\xi\Sigma_s^j} \right) + \eta^j Q^* \Big]$

*In deriving (26.20) we have used the fact that on the interval Δu there is practically no difference between D^{j-1} and D^j. [The Russian text fails to identify any equation as (26.20).]

Chapter VII

THE METHOD OF GROUPS. STRONG ABSORPTION OF SLOWING-DOWN NEUTRONS

In the previous chapter we reduced the basic and adjoint reactor equations to a set of diffusion equations for the case of weak absorption of slowing-down neutrons. Then the slowing-down density $q = \xi\Sigma_s \varphi$ is a slowly varying function of the lethargy, and to solve these problems it is sufficient to treat relatively few energy groups.

The situation is quite different when the reactor contains zones in which the slowing-down neutrons are strongly absorbed. This occurs in intermediate reactors, in whose core q may undergo large changes over a unit lethargy interval due to neutron absorption. If at the same time we recall the fact that the absorption probability increases greatly, in general, at thermal diffusion energies, it becomes clearly necessary to solve the problem with a high degree of accuracy for all groups up to the thermal diffusion energy.

It is possible to increase the accuracy by increasing the number of energy groups. In terms of the necessary operations, however, this way is not the best, since it greatly increases the amount of work necessary. It is thus important to solve the problem to the required accuracy with only a small number of diffusion groups. The present chapter deals with such a method as applied to the set of basic and adjoint reactor equations.

27. The Set of Basic Reactor Equations

Consider the neutron slowing-down equations in the diffusion-age approximation, which we shall write temporarily in the form

$$\frac{\partial q}{\partial u} + \frac{\Sigma_c}{\xi\Sigma_s} q = \nabla D \nabla \varphi + \chi(u) Q(\vec{r}), \tag{27.1}$$

where

$$q = \xi\Sigma_s \varphi.$$

Analysis of (27.1) shows that for energies at which the neutrons are strongly absorbed in the reactor each term on the left side is separately almost everywhere much larger than those terms on the right side, so that to a large extent they compensate each other. This leads one to believe that $\nabla D \nabla \varphi$ might be replaced by some approximate expression, which would give both a result of the required accuracy and a very simple and convenient expression to compute.

Let us therefore assume that $\nabla D \nabla \varphi$ is a given function; then by integrating the linear inhomogeneous first order equation (27.1), we obtain

$$q = p^j(u) q^{j-1} + \int\limits_{u_{j-1}}^{u} \nabla D \nabla \varphi e^{-\int\limits_{u'}^{u} \frac{\Sigma_c}{\xi\Sigma_s} du''} du' + \chi^j(u) Q(\vec{r}), \tag{27.2}$$

where

$$q^j = q(\vec{r}, u_j), \quad p^j(u) = e^{-\int\limits_{u_{j-1}}^{u} \frac{\Sigma_c}{\xi\Sigma_s} du},$$

$$\chi^j(u) = \int_{u_{j-1}}^{u} \chi(u') e^{-\int_{u'}^{u} \frac{\Sigma_c}{\xi \Sigma_s} du''} du'.$$

Further, setting $u = u_j$ in (27.2), we have

$$q^j = p^j q^{j-1} + \int_{u_{j-1}}^{u_j} \nabla D \nabla \varphi \cdot \frac{p^j}{p^j(u')} du' + \chi^j Q(\vec{r}), \tag{27.3}$$

where

$$p^j = p^j(u_j), \quad \chi^j = \chi^j(u_j).$$

We shall now use (27.2) and (27.3) as a basis for obtaining our approximate set of multigroup slowing-down equations.

As was indicated above, we can approximate $\nabla D \nabla \varphi$ in (27.2) and (27.3) by an expression which would take account of the basic feature of its behavior. In this connection we first attempt to represent $q(\vec{r}, u)$ in the form of the product

$$q(\vec{r}, u) = \psi(\vec{r}, u) p^j(u), \tag{27.4}$$

where $\psi(\vec{r}, u)$ is a slowly varying function depending primarily on neutron leakage out of the reactor. It should be noted that in a nonreflected reactor $\psi(\vec{r}, u) = \psi(r)$ will depend only on the leakage, and for this reason the representation of (27.4) is both natural and convenient. It is just as convenient for reactors with weakly moderating or heavy screens.

The situation is very different for intermediate reactors with strongly moderating reflectors, as well as for those in which there is little absorption of the slowing-down neutrons. Then, because of diffusion, a part of the fission neutrons enters the reflector. Most of the core neutrons, however, are strongly absorbed when they are slowed down to intermediate energies, so that there is practically a lethargy cutoff in the neutron flux in the reflector. A return current from the reflector to the core is then established under the influence of the rise in the neutron flux that occurs in the reflector. Inside the core close to its boundary there will then be established a neutron spectrum which is entirely different from that in the center, where it is described, as before, by Equation (27.4). Inside the core, close to the boundary between it and the reflector, q will vary very slowly as a function of lethargy, since it is essentially determined by the neutron flux out of the reflector. As for the slowing-down sources in the reflector, they vary slowly with energy, and only because of leakage both into the core and out of the external boundary of the reactor.

For a distance of roughly $L = \sqrt{D/\Sigma}$ (where D and $\Sigma = \Sigma_s + \Sigma_c$ are averages over the (u_{j-1}, u_j) group in the core) into the core from its boundary with the reflector, the gradient in the neutron flux towards the core establishes a spectrum which varies slowly with u.

Because the core neutrons below a certain energy are replenished by neutrons which flow in from the reflector, it is clear that we must have a very accurate description of neutron leakage particularly for those points which lie close to the boundary between the core and reflector.

The above implies that for a fixed zone q can be described by one of the following interpolation formulas:

$$\left.\begin{array}{ll} \text{procedure I:} & q = q^j, \\ \text{procedure II:} & q = \frac{u - u_{j-1}}{\Delta u_j} q^j + \frac{u_j - u}{\Delta u_j} q^{j-1}, \\ \text{procedure III:} & q = q^{j-1}. \end{array}\right\} \tag{27.5}$$

One may ask whether these interpolation formulas will give a satisfactory description of the slowing-down density in the core far from the reflector boundary. It is clear that these formulas will not do as they stand. They can, however, be used for an approximate calculation of the integral on the right side of (27.3), since for these points

$$\int_{u_{j-1}}^{u_j} \nabla D \nabla \varphi \frac{p^j}{p^j(u)}\, du \tag{27.6}$$

is small compared to the other terms. This means that even a significant error in the calculation of this term will not lead to large errors in the final result.

It is thus seen that interpolation procedures I to III can be used quite satisfactorily for an approximation of (27.6) at all points of the reactor.

Thus consider (27.6), which for a fixed zone of the reactor can be written approximately in the form:*

$$\int_{u_{j-1}}^{u_j} \nabla D \nabla \varphi \frac{p^j}{p^j(u)}\, du = \int_{u_{j-1}}^{u_j} \frac{D}{\xi \Sigma_s} \frac{p^j}{p^j(u)} \nabla^2 q\, du = \frac{1}{\Delta u_j} \int_{u_{j-1}}^{u_j} \frac{p^j}{\xi \Sigma_s} \frac{du}{p^j(u)} \int_{u_{j-1}}^{u_j} D^j \nabla^2 q\, du, \tag{27.7}$$

$$D^j = \frac{\displaystyle\int_{u_{j-1}}^{u_j} \frac{D p^j}{\xi \Sigma_s} \frac{du}{p^j(u)}}{\displaystyle\int_{u_{j-1}}^{u_j} \frac{p^j}{\xi \Sigma_s} \frac{du}{p^j(u)}}. \tag{27.8}$$

Let us calculate (27.7) using (27.5). Noting that on the boundary between zones we have $D^j \nabla \varphi^j = D'^j \nabla \varphi'^j$, we rewrite (27.7) in the form

$$\int_{u_{j-1}}^{u_j} \nabla D \nabla \varphi \frac{p^j}{p^j(u)}\, du = A^j \nabla D^j \nabla \varphi^j + B^j \nabla D^{j-1} \nabla \varphi^{j-1}, \tag{27.9}$$

where A^j and B^j are given by Table 7.

Let us now insert (27.9) into (27.3). Then for procedures I and II we obtain the set of diffusion equations

$$\nabla D^j \nabla \varphi^j - \Sigma^j \varphi^j = -f^j, \qquad (j = 1, 2, \ldots, m), \tag{27.10}$$

where

$$\Sigma^j = \frac{\xi \Sigma_s^j}{A^j}, \qquad f^j = \frac{1}{A^j}\left(p^j \xi \Sigma_s^{j-1} \varphi^{j-1} + B^j \nabla D^{j-1} \nabla \varphi^{j-1} + \chi^j Q\right). \tag{27.11}$$

We now use (27.10) to eliminate $\nabla D^{j-1} \nabla \varphi^{j-1}$ from the expression for f^j. We then obtain:

$$f^j = \frac{1}{A^j}\left[\left(p^j \xi \Sigma_s^{j-1} + B^j \Sigma^{j-1}\right) \varphi^{j-1} - B^j f^{j-1} + \chi^j Q\right]. \tag{27.12}$$

Finally, adding the thermal group diffusion equations to (27.10), we arrive at the multigroup equations

$$\nabla D^j \nabla \varphi^j - \Sigma^j \varphi^j = -f^j \qquad (j = 1, 2, \ldots, m, m+1), \tag{27.13}$$

---

* See also footnote on p. 75.

TABLE 7

Definition of A^j and B^j for the Different Multigroup Reactor Equations for the Case of Strong Absorption of Slowing-Down Neutrons

Procedure	A^j	B^j
I	$\xi\Sigma_s^j \int\limits_{u_{j-1}}^{u_j} \frac{1}{\xi\Sigma_s} \cdot \frac{p^j}{p^j(u)}\, du$	0
II	$\frac{\xi\Sigma_s^j}{2} \int\limits_{u_{j-1}}^{u_j} \frac{1}{\xi\Sigma_s} \cdot \frac{p^j}{p^j(u)}\, du$	$\frac{D^j}{D^{j-1}} \cdot \frac{\xi\Sigma_s^{j-1}}{2} \int\limits_{u_{j-1}}^{u_j} \frac{1}{\xi\Sigma_s} \cdot \frac{p^j}{p^j(u)}\, du$
III	0	$\frac{D^j}{D^{j-1}} \xi\Sigma_s^{j-1} \int\limits_{u_{j-1}}^{u_j} \frac{1}{\xi\Sigma_s} \cdot \frac{p^j}{p^j(u)}\, du$

TABLE 8

Functions and Constants in the Multigroup Reactor Equations for Strong Absorption of Slowing-Down Neutrons

i	φ^j	D^j	Σ^j	f^j
$(1)-(m)$	φ^j	D^j	$\frac{\xi\Sigma_s^j}{A^j}$	$\frac{1}{A^j}\{(p^j\xi\Sigma_s^{j-1} + B^j\Sigma^{j-1})\varphi^{j-1} - B^j f^{j-1} + \chi^j Q\}$
$(m+1)$	Φ	D_T	Σ_{cT}	$\xi\Sigma_s^m \varphi^m$

where $\varphi^{m+1} = \Phi$, and D^j, Σ^j and f^j are given in Table 8.

Finally, in the same way we obtain

$$\nabla D^j \nabla \varphi^{j-1} + \Sigma^j \varphi^{j-1} = f^j, \tag{27.14}$$

for procedure III, where

$$\left.\begin{aligned} D^j &= \frac{\int\limits_{u_{j-1}}^{u_j} \frac{D}{\xi\Sigma_s} \frac{p^j}{p^j(u)}\, du}{\int\limits_{u_{j-1}}^{u_j} \frac{1}{\xi\Sigma_s} \frac{p^j}{p^j(u)}\, du}, \qquad \Sigma^j = p^j \frac{\xi\Sigma_s^{j-1}}{B^j}, \\ f^j &= \frac{\xi\Sigma_s^j}{B^j} \varphi^j - \frac{\chi^j}{B^j} Q. \end{aligned}\right\} \tag{27.15}$$

and B^j are given by Table 7.

Solving (27.14) for φ^j, we have

$$\varphi^j = \frac{B^j}{\xi \Sigma_s^j}\left(\Sigma^j \varphi^{j-1} + \nabla D^j \nabla \varphi^{j-1} + \frac{\chi^j}{B^j} Q \right). \qquad (27.16)$$

Let us now go on to calculate

$$Q(\vec{r}) = \nu_f \left(\int_{-\infty}^{u_T} \Sigma_f \varphi \, du + \Sigma_{fT} \Phi \right). \qquad (27.17)$$

In accordance with the concept of the multigroup computation, we rewrite (27.17) in the form

$$Q(\vec{r}) = \nu_f \left(\sum_{j=1}^{m} \int_{u_{j-1}}^{u_j} \Sigma_f \varphi \, du + \Sigma_{fT} \Phi \right). \qquad (27.18)$$

Thus the problem is how best to approximate φ over the interval (u_{j-1}, u_j) knowing its values at the end points. In the case of weak absorption this can be done by interpolation formulas as simple as (27.5), but now, when $q = \xi \Sigma_s \varphi$ varies greatly over the interval under consideration, such an approximation is invalid. We must now use the exact solutions within a given group as defined by (27.2). However, the integrand in (27.2) contains the unknown function $\nabla D \nabla \varphi$. It is therefore necessary first to express this function in terms of its known values at u_{j-1} and u_j. To do this we must use the same hypotheses as we did in deriving the multigroup reactor equations. If this is possible, we arrive at interpolation formulas which, except for the errors already inherent in φ^j due to the multigroup calculation, will introduce no new errors in calculating the integrals in the expression for $Q(\vec{r})$.

Let us therefore now use (27.2) to attempt to find suitable interpolation formulas. For this purpose we first consider the integral

$$\int_{u_{j-1}}^{u} \nabla D \nabla \varphi e^{-\int_{u'}^{u} \frac{\Sigma_c}{\xi \Sigma_s} du''} du' = \int_{u_{j-1}}^{u} \nabla D \nabla \varphi \frac{p^j(u)}{p^j(u')} du'. \qquad (27.19)$$

For a fixed zone of the reactor we can write

$$\int_{u_{j-1}}^{u} \nabla D \nabla \varphi \frac{p^j(u)}{p^j(u')} du' = \int_{u_{j-1}}^{u} \frac{D}{\xi \Sigma_s} \frac{p^j(u)}{p^j(u')} \nabla^2 q \, du'. \qquad (27.20)$$

The function $\underline{q}$ in this equation can be expressed by means of the interpolation formulas of (27.5), which were already used in obtaining the multigroup equations. We then obtain

$$\int_{u_{j-1}}^{u} \nabla D \nabla \varphi \frac{p^j(u)}{p^j(u')} du' = A^j(u) \nabla D^j \nabla \varphi^j + B^j(u) \nabla D^{j-1} \nabla \varphi^{j-1}, \qquad (27.21)$$

where $A^j(u)$ and $B^j(u)$ are given in Table 9.

Let us now insert (27.21) into (27.2). We then obtain

$$q = p^j(u) q^{j-1} + A^j(u) \nabla D^j \nabla \varphi^j + B^j(u) \nabla D^{j-1} \nabla \varphi^{j-1} + \chi^j(u) Q. \qquad (27.22)$$

Let us eliminate $\nabla D^j \nabla \varphi^j$ and $\nabla D^{j-1} \nabla \varphi^{j-1}$ from this equation using (27.13).

We then obtain the interpolation formulas

TABLE 9

Definition of $A^j(u)$ and $B^j(u)$ for the Different Multigroup Representations for the Case of Strong Absorption of Slowing-Down Neutrons

Procedure	$A^j(u)$	$B^j(u)$
I	$\frac{p^j(u)}{D^j}\cdot\xi\Sigma_s^j\int\limits_{u_{j-1}}^{u}\frac{D}{\xi\Sigma_s}\cdot\frac{du'}{p^j(u')}$	0
II	$\frac{p^j(u)}{D^j}\xi\Sigma_s^j\int\limits_{u_{j-1}}^{u}\frac{D}{\xi\Sigma_s}\times \times\frac{u'-u_{j-1}}{\Delta u_j}\cdot\frac{du'}{p^j(u')}$	$\frac{p^j(u)}{D^{j-1}}\xi\Sigma_s^{j-1}\int\limits_{u_{j-1}}^{u}\frac{D}{\xi\Sigma_s}\times \times\frac{u_j-u'}{\Delta u_j}\cdot\frac{du'}{p^j(u')}$
III	0	$\frac{p^j(u)}{D^{j-1}}\xi\Sigma_s^{j-1}\int\limits_{u_{j-1}}^{u}\frac{D}{\xi\Sigma_s}\cdot\frac{du'}{p^j(u')}$

$$\varphi=\frac{1}{\xi\Sigma_s}[p^j(u)\,\xi\Sigma_s^{j-1}\varphi^{j-1}+A^j(u)(\Sigma^j\varphi^j-f^j)+ + B^j(u)(\Sigma^{j-1}\varphi^{j-1}-f^{j-1})+\chi^j(u)\,Q] \tag{27.23}$$

for procedures I and II, and

$$\varphi=\frac{1}{\xi\Sigma_s}[p^j(u)\,\xi\Sigma_s^{j-1}\varphi^{j-1}+B^j(u)(f^j-\Sigma^j\varphi^{j-1})+\chi^j(u)\,Q] \tag{27.23'}$$

for procedure III.

We now insert (27.23) into (27.18). This gives

$$Q=\nu_f\sum_{j=1}^{m+1}(\alpha^j\varphi^j-\beta^j f^j)+\nu_f\varepsilon Q, \tag{27.24}$$

where

$$\varepsilon=\sum_{j=1}^{m}\int\limits_{u_{j-1}}^{u_j}\frac{\Sigma_f}{\xi\Sigma_s}\chi^j(u)\,du, \tag{27.25}$$

and α^j and β^j are given in Table 10.

$$\left.\begin{aligned}\alpha_1^j&=\int\limits_{u_{j-1}}^{u_j}\frac{\Sigma_f}{\xi\Sigma_s}[\xi\Sigma_s^{j-1}p^j(u)+\Sigma^{j-1}B^j(u)]\,du,\\ \alpha_2&=\Sigma^j\int\limits_{u_{j-1}}^{u_j}\frac{\Sigma_f}{\xi\Sigma_s}A^j(u)\,du,\\ \beta_2^j&=\int\limits_{u_{j-1}}^{u_j}\frac{\Sigma_f}{\xi\Sigma_s}A^j(u)\,du,\quad \beta_1^j=\int\limits_{u_{j-1}}^{u_j}\frac{\Sigma_f}{\xi\Sigma_s}B^j(u)\,du.\end{aligned}\right\} \tag{27.26}$$

TABLE 10

The Constants α^j and β^j for Calculating the Slowing-Down Sources for the Case of Strong Absorption of Slowing-Down Neutrons

j	α^j	β^j
$(1)-(m-1)$	$\alpha_1^{j+1}+\alpha_2^j$	$\beta_1^{j+1}+\beta_2^j$
(m)	α_2^j	β_2^j
$(m+1)$	Σ_{fT}	0

Finally, solving (27.24) for $Q(\vec{r})$, we have

$$Q(\vec{r})=\frac{\nu_f}{1-\nu_{fs}}\sum_{j=1}^{m+1}(\alpha^j\varphi^j-\beta^j f^j). \tag{27.27}$$

An analogous equation is obtained for procedure III.

In conclusion it should be noted that in calculating resonance capture by U^{238} nuclei and structural elements, one must take into account the spectrum of the slowing-down neutrons. If we know that there are several resonance levels in a (u_{j-1}, u_j) group, and we have calculated the probability $\langle\varphi\rangle_l$ for avoiding resonance capture on each of these levels, we can take into account resonance capture in this group according to the multigroup method by replacing

$$p^j(u) \text{ by } \kappa(u)\cdot p^j(u),$$

in all of the formulas in this section, where

$$\varkappa(u)=\prod_l \vartheta_l(u-u_l)$$

and

$$\vartheta_l(u-u_l)=\begin{cases}1 & \text{for} \quad u<u_l,\\ \langle\varphi\rangle_l & \text{for} \quad u>u_l,\end{cases}$$

where the index l denotes the ordinal number of the resonance level in the given group.

28. The Set of Adjoint Reactor Equations

Let us now consider the set of adjoint equations for the case of strong absorption of slowing-down neutrons. As in the case of weak absorption, we shall start with the equation

$$\nabla D\nabla\varphi^*-\Sigma_c\varphi^*=-\xi\Sigma_s\frac{\partial\varphi^*}{\partial u}-\Sigma_f Q^*(\vec{r}), \tag{28.1}$$

where

$$Q^*(\vec{r})=\nu_f\int_{-\infty}^{u_T}\chi(u)\varphi^*\,du. \tag{28.2}$$

Let us write (28.1) in the form

$$\frac{\partial \varphi^*}{\partial u} - \frac{\Sigma_c}{\xi \Sigma_s} \varphi^* = -\frac{1}{\xi \Sigma_s} \nabla D \nabla \varphi^* - \frac{\Sigma_f}{\xi \Sigma_s} Q^* (\vec{r}). \tag{28.3}$$

Assuming the right side of this equation to be known, let us treat it as a homogeneous first-order differential equation in the interval (u, u_j). We then have

$$\varphi^* = p^{*j}(u)\varphi^{*j} + \int_u^{u_j} \frac{1}{\xi \Sigma_s} \nabla D \nabla \varphi^* e^{-\int_u^{u'} \frac{\Sigma_c}{\xi \Sigma_s} du''} du' + \eta^j(u) Q^* (\vec{r}), \tag{28.4}$$

where

$$p^{*j}(u) = e^{-\int_u^{u_j} \frac{\Sigma_c}{\xi \Sigma_s} du}, \qquad \eta^j(u) = \int_u^{u_j} \frac{\Sigma_f}{\xi \Sigma_s} e^{-\int_u^{u'} \frac{\Sigma_c}{\xi \Sigma_s} du''} du'.$$

Setting $u = u_{j-1}$ in (28.4), we arrive at

$$\varphi^{*j-1} = p^j \varphi^{*j} + \int_{u_{j-1}}^{u_j} \frac{1}{\xi \Sigma_s} \nabla D \nabla \varphi^* e^{-\int_{u_{j-1}}^{u'} \frac{\Sigma_c}{\xi \Sigma_s} du''} du' + \eta^j Q^* (\vec{r}), \tag{28.5}$$

where

$$\eta^j = \eta^j(u_{j-1}), \qquad p^j = p^{*j}(u_{j-1}) = p^j(u_j).$$

In order to obtain the multigroup diffusion equations we must assume a form for the function approximating $\nabla D \nabla \varphi^*$ in (28.5) on the interval (u_{j-1}, u_j). To do this, we consider the integral

$$\int_{u_{j-1}}^{u_j} \nabla D \nabla \varphi^* \frac{e^{-\int_{u_{j-1}}^{u} \frac{\Sigma_c}{\xi \Sigma_s} du'}}{\xi \Sigma_s} du = \int_{u_{j-1}}^{u_j} \nabla D \nabla \varphi^* \frac{p^j(u)}{\xi \Sigma_s} du,$$

which, in a fixed zone of the reactor, we shall write in the form

$$\int_{u_{j-1}}^{u_j} \nabla D \nabla \varphi^* \frac{p^j(u)}{\Sigma \xi_s} du = \int_{u_{j-1}}^{u_j} \frac{D}{\xi \Sigma_s} p^j(u) \nabla^2 \varphi^* du. \tag{28.6}$$

Finally, introducing the notation

$$D^j = \frac{\int_{u_{j-1}}^{u_j} \frac{D}{\xi \Sigma_s} p^j(u)\, du}{\int_{u_{j-1}}^{u_j} \frac{1}{\xi \Sigma_s} p^j(u)\, du}, \tag{28.7}$$

we rewrite (28.6) in the form*

$$\int_{u_{j-1}}^{u_j} \nabla D \nabla \varphi^* \frac{p^j(u)}{\xi\Sigma_s} du = \int_{u_{j-1}}^{u_j} \frac{1}{\xi\Sigma_s} p^j(u)\, du \int_{u_{j-1}}^{u_j} D^j \nabla^2 \varphi^* du. \tag{28.8}$$

Again we perform the integration in (28.8) by using very simple interpolation formulas, namely:

$$\left.\begin{aligned} &\text{procedure I:} && \varphi^* = \varphi^{*j-1}, \\ &\text{procedure II:} && \varphi^* = \frac{u_j - u}{\Delta u_j} \varphi^{*j-1} + \frac{u - u_{j-1}}{\Delta u_j} \varphi^{*j}, \\ &\text{procedure III:} && \varphi^* = \varphi^{*j}. \end{aligned}\right\} \tag{28.9}$$

We then obtain

$$\int_{u_{j-1}}^{u_j} \nabla D \nabla \varphi^* \frac{p^j(u)}{\xi\Sigma_s} du = A^{*j} \nabla D^j \nabla \varphi^{*j-1} + B^{*j} \nabla D^{j+1} \nabla \varphi^{*j}, \tag{28.10}$$

where A^{*j} and B^{*j} are given by the formulas in Table 11.

TABLE 11

Definition of A^{*j} and B^{*j} for the Different Multigroup Representations of the Adjoint Equations for the Case of Strong Absorption of Slowing-Down Neutrons

Procedure	A^{*j}	B^{*j}
I	$\int_{u_{j-1}}^{u_j} \frac{p^j(u)}{\xi\Sigma_s} du$	0
II	$\int_{u_{j-1}}^{u_j} \frac{p^j(u)}{\xi\Sigma_s} \cdot \frac{u_j - u}{\Delta u_j} du$	$\frac{D^j}{D^{j+1}} \int_{u_{j-1}}^{u_j} \frac{p^j(u)}{\xi\Sigma_s} \cdot \frac{u - u_{j-1}}{\Delta u_j} du$
III	0	$\frac{D^j}{D^{j+1}} \cdot \int_{u_{j-1}}^{u_j} \frac{p^j(u)}{\xi\Sigma_s} du$

*It should be mentioned that both the basic and adjoint equations in the expression for D^j can sometimes be considerably simplified by approximating it as

$$D^j = \frac{\Delta\tau^j}{\Lambda^j},$$

where

$$\Delta\tau^j = \int_{u_{-1}}^{u_j} \frac{D}{\xi\Sigma_s} du, \qquad \Lambda^j = \int_{u_{j-1}}^{u_j} \frac{du}{\xi\Sigma_s}.$$

Let us now insert (28.10) into (28.5). For procedures I and II we then obtain

$$\nabla D^j \nabla \varphi^{*j-1} - \Sigma^j \varphi^{*j-1} = -f^{*j} \qquad (j = m,\ m-1,\ \ldots,\ 1), \tag{28.11}$$

where

$$\Sigma^j = \frac{1}{A^{*j}}, \qquad f^{*j} = \frac{1}{A^{*j}} (p^j \varphi^{*j} + B^{*j} \nabla D^{j+1} \nabla \varphi^{*j} + \eta^j Q^*). \tag{28.12}$$

Let us now use (28.11) to eliminate $\nabla D^{j+1} \nabla \varphi^{*j}$ from the last equation. We then obtain the recursion relation

$$f^{*j} = \frac{1}{A^{*j}} [p^j \varphi^{*j} + B^{*j} (\Sigma^{j+1} \varphi^{*j} - f^{*j+1}) + \eta^j Q^*]. \tag{28.13}$$

Finally, adding the diffusion equation for the thermal group to (28.11), we arrive at the multigroup set of adjoint equations, namely:

$$\nabla D^j \nabla \varphi^{*j-1} - \Sigma^j \varphi^{*j-1} = -f^{*j} \qquad (j = m+1,\ m,\ \ldots,\ 1), \tag{28.14}$$

where $\varphi^{*m+1} = \Phi^*$, and D^j, Σ^j and f^j are given by Table 12.

TABLE 12

Functions and Constants for the Adjoint Multigroup Reactor Equations for the Case of Strong Absorption of Slowing Down Neutrons

j	φ^{*j-1}	D^j	Σ^j	f^{*j}
$(m+1)$	Φ^*	D_T	Σ_{cT}	$\Sigma_{fT} Q^*$
$(m)-(1)$	φ^{*j-1}	D^j	$\frac{1}{A^{*j}}$	$\frac{1}{A^{*j}} [p^j \varphi^{*j} + B^{*j} (\Sigma^{j+1} \varphi^{*j} - f^{*j+1}) + \eta^j Q^*]$

For procedure III we obtain

$$\nabla D^{j+1} \nabla \varphi^{*j} + \Sigma^{j+1} \varphi^{*j} = f^{*j}, \tag{28.15}$$

where

$$\Sigma^{j+1} = \frac{p^j}{B^{*j}}, \qquad f^{*j} = \frac{1}{B^{*j}} \varphi^{*j-1} - \frac{\eta^j}{B^{*j}} Q^*. \tag{28.16}$$

Solving (28.15) for φ^{*j-1}, we obtain

$$\left.\begin{array}{c} \varphi^{*j-1} = B^{*j} (\nabla D^{j+1} \nabla \varphi^{*j} + \Sigma^{j+1} \varphi^{*j}) + \eta^j Q^* \\ (j = m,\ m-1,\ \ldots,\ 1). \end{array}\right\} \tag{28.17}$$

To these equations we must add the equation for Φ^*.

Let us now consider $Q^*(\vec{r})$. We consider the expression

$$Q^*(\vec{r}) = \nu_f \int_{-\infty}^{u_T} \chi(u) \varphi^* \, du, \tag{28.18}$$

which we shall write in the form

$$Q^* = \nu_f \sum_{j=1}^{m} \int_{u_{j-1}}^{u_j} \chi(u)\, \varphi^*\, du. \tag{28.19}$$

To calculate the necessary integrals we must have interpolation formulas for φ^* in the (u_{j-1}, u_j) intervals. Such formulas can be obtained in the same way as we did in the previous section when calculating $Q(\vec{r})$. In this way, starting with (28.4) and using procedures I-III, we obtain

$$\varphi^* = p^{*j}(u)\, \varphi^{*j} + A^{*j}(u)\, \nabla D^j \nabla \varphi^{*j-1} + B^{*j}(u)\, \nabla D^{j+1} \nabla \varphi^{*j} + \eta^j(u)\, Q^*, \tag{28.20}$$

where $A^{*j}(u)$ and $B^{*j}(u)$ are given in Table 13.

TABLE 13

Definition of $A^{*j}(u)$ and $B^{*j}(u)$ for the Different Multigroup Representations of the Adjoint Reactor Equations for the Case of Strong Absorption of Slowing-Down Neutrons

Procedure	$A^{*j}(u)$	$B^{*j}(u)$
I	$\frac{1}{D^j} \int_u^{u_j} \frac{D}{\xi\Sigma_s} \cdot \frac{p^{*j}(u)}{p^{*j}(u')}\, du'$	0
II	$\frac{1}{D^j} \int_u^{u_j} \frac{D}{\xi\Sigma_s} \cdot \frac{p^{*j}(u)}{p^{*j}(u')} \cdot \frac{u_j - u'}{\Delta u_j}\, du'$	$\frac{1}{D^{j+1}} \int_u^{u_j} \frac{D}{\xi\Sigma_s} \cdot \frac{p^{*j}(u)}{p^{*j}(u')} \cdot \frac{u' - u_{j-1}}{\Delta u_j}\, du'$
III	0	$\frac{1}{D^{j+1}} \int_u^{u_j} \frac{D}{\xi\Sigma_s} \cdot \frac{p^{*j}(u)}{p^{*j}(u')}\, du'$

Further, eliminating the expressions $\nabla D^j \nabla \varphi^{*j-1}$ and $\nabla D^{j+1} \nabla \varphi^{*j}$ from (28.20) we arrive at the interpolation formulas

$$\varphi^* = p^{*j}(u)\, \varphi^{*j} + A^{*j}(u)\, (\Sigma^j \varphi^{*j-1} - f^{*j}) + B^{*j}(u)\, (\Sigma^{j+1} \varphi^{*j} - f^{*j+1}) + \eta^j(u)\, Q^* \tag{28.21}$$

for procedures I and II, and

$$\varphi^* = p^{*j}(u)\, \varphi^{*j} + B^{*j}(u)\, (f^{*j} - \Sigma^{j+1} \varphi^{*j}) + \eta^j(u)\, Q^* \tag{28.22}$$

for procedure III.

Let us now insert (28.21) into (28.19). This gives

$$Q^* = \frac{\nu_f}{1 - \nu_f \varepsilon^*} \sum_{j=1}^{m} (\alpha^{*j} \varphi^{*j-1} - \beta^{*j} f^{*j}), \tag{28.23}$$

where

$$\varepsilon^* = \sum_{j=1}^{m} \int_{u_{j-1}}^{u_j} \chi(u)\, \eta^j(u)\, du, \tag{28.24}$$

and

$$\alpha^{*j} = \Sigma^j \int_{u_{j-1}}^{u_j} \chi(u)\, A^{*j}(u)\, du + \int_{u_{j-2}}^{u_{j-1}} \chi(u)\, [p^{*j-1}(u) + \Sigma^j B^{*j-1}(u)]\, du,$$

$$\beta^{*j} = \int_{u_{j-1}}^{u_j} \chi(u)\, A^{*j}(u)\, du + \int_{u_{j-2}}^{u_{j-1}} \chi(u)\, B^{*j-1}(u)\, du. \tag{28.25}$$

Similarly, Equation (28.23) can be obtained for procedure III.

In conclusion, it should be noted that in calculating resonance capture by U^{238} nuclei and structural elements, one must take into account the spectrum of the slowing-down neutrons. If we know that there are several resonance levels in a (u_{j-1}, u_j) group, and we have calculated the probability $\langle\varphi\rangle_l$ for avoiding resonance capture on each of these levels, we can take into account resonance capture in this group according to the multigroup method by replacing $p_*^j(u)$ by $[1 - \kappa(u)]p_*^j(u)$, in all of the formulas of this section, where $\kappa(u) = \prod_l \vartheta_l(u - u_l)$ and

$$\vartheta_l(u - u_l) = \begin{cases} 1, & \text{for} \quad u < u_l, \\ \langle\varphi\rangle_l, & \text{for} \quad u > u_l, \end{cases}$$

where the index l denotes the ordinal number of the resonance level in the given group, and $p_*^j(u)$ should be replaced by $\frac{\kappa(u_j)}{\kappa(u)} p_*^j(u)$.

29. Inclusion of the Second Moment of Energy in the Multigroup Equations

As in the case of weak absorption which we treated in Section 26, we shall start with the slowing-down equations which we write temporarily in the form

$$\left.\begin{aligned} \frac{\partial q}{\partial u} + \frac{\Sigma_c}{\xi\Sigma_s + \gamma\Sigma_c} q &= \mu\nabla D\nabla\varphi + \mu S, \\ q &= (\xi\Sigma_s + \gamma\Sigma_c)\,\varphi - \gamma\nabla D\nabla\varphi - \gamma S, \end{aligned}\right\} \tag{29.1}$$

where

$$\mu = 1 - \gamma \frac{\Sigma_c}{\xi\Sigma_s + \gamma\Sigma_c}.$$

The first of Equations (29.1) will be treated on the interval (u_{j-1}, u), as a linear inhomogeneous first-order equation for q, and we solve it, obtaining

$$q = p^j(u)\, q^{j-1} + \int_{u_{j-1}}^{u} \nabla D\nabla\varphi\mu \frac{p^j(u)}{p^j(u')} du' + \chi^j(u)\, Q(\vec{r}), \tag{29.2}$$

where

$$p^j(u) = e^{-\int\limits_{u_{j-1}}^{u} \frac{\Sigma_c}{\xi\Sigma_s+\gamma\Sigma_c} du'}, \quad \chi^j(u) = \int\limits_{u_{j-1}}^{u} \chi(u')\,\mu \frac{p^j(u)}{p^j(u')} du'.$$

Setting $u = u_j$ in (29.2), we arrive at

$$q^j = p^j q^{j-1} + \int\limits_{u_{j-1}}^{u_j} \nabla D \nabla \varphi \mu \frac{p^j}{p^j(u')} du' + \chi^j Q(\vec{r}). \tag{29.3}$$

Further, using the approximate expression

$$\int\limits_{u_{j-1}}^{u_j} \nabla D \nabla \varphi\, \mu \frac{p^j}{p^j(u')} du' = A^j \nabla D^j \nabla \varphi^j + B^j \nabla D^{j-1} \nabla \varphi^{j-1}, \tag{29.4}$$

we rewrite (29.3) in the form

$$q^j = p^j q^{j-1} + A^j \nabla D^j \nabla \varphi^j + B^j \nabla D^{j-1} \nabla \varphi^{j-1} + \chi^j Q, \tag{29.5}$$

where A^j and B^j are given by Table 7, in which

$$\int\limits_{u_{j-1}}^{u_j} \frac{1}{\xi\Sigma_s} \frac{p^j}{p^j(u)} du$$

should everywhere be replaced by

$$\int\limits_{u_{j-1}}^{u_j} \frac{\mu}{\xi\Sigma_s+\gamma\Sigma_c} \frac{p^j}{p^j(u)} du,$$

and D^j by

$$D^j = \frac{\int\limits_{u_{j-1}}^{u_j} \frac{D \cdot \mu}{\xi\Sigma_s+\gamma\Sigma_c} \frac{p^j}{p^j(u)} du}{\int\limits_{u_{j-1}}^{u_j} \frac{\mu}{\xi\Sigma_s+\gamma\Sigma_c} \frac{p^j}{p^j(u)} du}.$$

Since μ hardly changes within a group, we can make the approximation

$$D^j = \frac{\int\limits_{u_{j-1}}^{u_j} \frac{D}{\xi\Sigma_s} \frac{p^j}{p^j(u)} du}{\int\limits_{u_{j-1}}^{u_j} \frac{1}{\xi\Sigma_s} \frac{p^j}{p^j(u)} du}.$$

Consider the second of Equations (29.1), setting $u = u_{j-1}$ and $u = u_j$. We insert the expression so obtained into (29.5). Then for procedures I and II we obtain

$$\nabla D^j \nabla \varphi^j - \Sigma^j \varphi^j = -f^j \quad (j=1,\ 2,\ \ldots,\ m), \tag{29.6}$$

where

$$\Sigma^j = \frac{\xi\Sigma_s^j + \gamma\Sigma_c^j}{A^j + \gamma},$$

$$\left.\begin{aligned} f^j &= \frac{1}{A^j+\gamma}\{p^j(\xi\Sigma_s^{j-1} + \gamma\Sigma_c^{j-1})\varphi^{j-1} + (B^j - p^j\gamma)\nabla D^{j-1}\nabla\varphi^{j-1} + \\ &\quad + \omega^j Q\}, \\ \omega^j &= \chi^j + \gamma(\chi_j - p^j\chi_{j-1}),\ \chi^j = \int_{u_{j-1}}^{u_j} \chi(u)\,\mu\,\frac{p^j}{p^j(u)}\,du,\ \ \chi_j = \chi(u_j). \end{aligned}\right\} \tag{29.7}$$

Using (29.6) we can obtain the recursion relation

$$f^j = \frac{1}{A^j+\gamma}\{p^j(\xi\Sigma_s^{j-1} + \gamma\Sigma_c^{j-1})\varphi^{j-1} + (B^j - \gamma p^j)(\Sigma^{j-1}\varphi^{j-1} - f^{j-1}) + \omega^j Q\}. \tag{29.8}$$

Adding the diffusion equation for the thermal neutrons to (29.6), we arrive at the multigroup equations

$$\nabla D^j \nabla \varphi^j - \Sigma^j \varphi^j = -f^j \quad (j=1,\ 2,\ \ldots,\ m,\ m+1), \tag{29.9}$$

where $\varphi^{m+1} = \Phi$, and D^j, Σ^j and f^j are given by Table 14.

TABLE 14

Functions and Constants in the Multigroup Reactor Equations With the Second Moment of Energy and Strong Absorption of Slowing-Down Neutrons

j	φ^j	D^j	Σ^j	f^j
(1) — (m)	φ^j	D^j	$\dfrac{\xi\Sigma_s^j + \gamma\Sigma_c^j}{A^j+\gamma}$	$\dfrac{1}{A^j+\gamma}\{p^j(\xi\Sigma_s^{j-1}+\gamma\Sigma_c^{j-1})\varphi^{j-1} + (B^j - \gamma p^j)(\Sigma^{j-1}\varphi^{j-1} - f^{j-1}) + \omega^j Q\}$
(m + 1)	Φ	D_T	Σ_{cT}	$(\xi\Sigma_s^m + \gamma\Sigma_c^m - \gamma\Sigma^m)\varphi^m + \gamma f^m$

We now go on to a consideration of

$$Q = \nu_f\left(\sum_{j=1}^{m}\int_{u_{j-1}}^{u_j} \Sigma_f\varphi\,du + \Sigma_{fT}\Phi\right). \tag{29.10}$$

To calculate the integrals we must use suitable interpolation formulas for φ, which we can obtain from (29.2).

Bearing in mind the approximate relation (27.21), that is,

$$\int_{u_{j-1}}^{u} \nabla D\nabla\varphi\,\frac{p^j(u)}{p^j(u')}\,du' = A^j(u)\nabla D^j\nabla\varphi^j + B^j(u)\nabla D^{j-1}\nabla\varphi^{j-1},$$

we can use (29.2) to obtain

$$q = p^j(u)\,q^{j-1} + A^j(u)\,\nabla D^j \nabla \varphi^j + B^j(u)\,\nabla D^{j-1}\nabla \varphi^{j-1} + \chi^j(u)\,Q, \tag{29.11}$$

where $A^j(u)$ and $B^j(u)$ are given by Table 9 in which again

$$\int\limits_{u_{j-1}}^{u} \frac{1}{\xi\Sigma_s}\,\frac{p^j(u)}{p^j(u')}\,du'$$

is replaced by

$$\int\limits_{u_{j-1}}^{u} \frac{\mu}{\xi\Sigma_s + \gamma\Sigma_c}\,\frac{p^j(u)}{p^j(u')}\,du'.$$

Let us now use the second of (29.1) to eliminate $\underline{q}$ from (29.11). We then have

$$\begin{aligned}\varphi = \frac{1}{\xi\Sigma_s + \gamma\Sigma_c}\{p^j(u)\,(\xi\Sigma_s^{j-1} + \gamma\Sigma_c^{j-1})\,\varphi^{j-1} + [A^j(u) + \gamma]\,\nabla D^j \nabla\varphi^j + \\ + [B^j(u) - p^j(u)\,\gamma]\,\nabla D^{j-1}\nabla\varphi^{j-1} + \omega^j(u)\,Q\},\end{aligned} \tag{29.12}$$

where

$$\omega^j(u) = \chi^j(u) + \gamma\,[\chi(u) - p^j(u)\,\chi_{j-1}].$$

It should be noted that because $\gamma\nabla D\nabla\varphi$ is small we have replaced it by $\gamma\nabla D^j\nabla\varphi^j$ in (29.12).

Finally let us eliminate $\nabla D^j\nabla\varphi^j$ and $\nabla D^{j-1}\nabla\varphi^{j-1}$ from (29.12); using the diffusion equations (29.9) we obtain

$$\begin{aligned}\varphi = \frac{1}{\xi\Sigma_s + \gamma\Sigma_c}\{p^j(u)\,(\xi\Sigma_s^{j-1} + \gamma\Sigma_c^{j-1})\,\varphi^{j-1} + [A^j(u) + \gamma]\,(\Sigma^j\varphi^j - f^j) + \\ + [B^j(u) - p^j(u)\,\gamma]\,(\Sigma^{j-1}\varphi^{j-1} - f^{j-1}) + \omega^j(u)\,Q\}.\end{aligned} \tag{29.13}$$

Inserting the interpolation formula (29.13) into (29.10) and integrating, we arrive at

$$Q = \frac{\nu_f}{1 - \nu_f\varepsilon}\sum_{j=1}^{m+1}(\alpha^j\varphi^j - \beta^j f^j), \tag{29.14}$$

where α^j and β^j are taken from Table 10, except that in (27.26) $\xi\Sigma_s$ should be replaced by $\xi\Sigma_s + \gamma\Sigma_c$, $A^j(u)$ by $A^j(u) + \gamma$, and $B^j(u)$ by $B^j(u) - p^j(u)\,\gamma$, while Σ^j is given by Table 14. As for ϵ, it is given by (27.25), where $\chi^j(u)$ is replaced by $\omega^j(u)$ and $\xi\Sigma_s$ by $\xi\Sigma_s + \gamma\Sigma_c$.

Let us now consider the adjoint equations, which we shall first write in the form

$$\left.\begin{aligned}&\frac{\partial}{\partial u}\frac{q^*}{\xi\Sigma_s} - \frac{\Sigma_c}{\xi\Sigma_s + \gamma\Sigma_c}\,\frac{q^*}{\xi\Sigma_s} = -\frac{\mu}{\xi\Sigma_s}(\nabla D\nabla\varphi^* + \Sigma_f Q^*),\\ &q^* = (\xi\Sigma_s + \gamma\Sigma_c)\,\varphi^* - \gamma\,(\nabla D\nabla\varphi^* + \Sigma_f Q^*),\end{aligned}\right\} \tag{29.15}$$

where

$$Q^* = \nu_f\int\limits_{-\infty}^{u_T}\varphi^*\chi(u)\,du.$$

We now integrate the first of (29.15) over (u, u_j). This gives

$$\frac{q^*}{\xi\Sigma_s} = p^{*j}(u)\frac{q^{*j}}{\xi\Sigma_s^j} + \int\limits_u^{u_j} \frac{\mu}{\xi\Sigma_s} \nabla D \nabla \varphi^* e^{-\int\limits_u^{u'} \frac{\Sigma_c}{\xi\Sigma_s+\gamma\Sigma_c} du''} du' + \eta^j(u) Q^*, \tag{29.16}$$

where

$$p^{*j}(u) = e^{-\int\limits_u^{u_j} \frac{\Sigma_c}{\xi\Sigma_s+\gamma\Sigma_c} du}, \quad \eta^j(u) = \int\limits_u^{u_j} \mu \frac{\Sigma_f}{\xi\Sigma_s} e^{-\int\limits_u^{u'} \frac{\Sigma_c}{\xi\Sigma_s+\gamma\Sigma_c} du''} du'. \tag{29.16'}$$

Further, we set $u = u_{j-1}$ in (29.16). We then obtain

$$\frac{q^{*j-1}}{\xi\Sigma_s^{j-1}} = p^j \frac{q^{*j}}{\xi\Sigma_s^j} + \int\limits_{u_{j-1}}^{u_j} \frac{\mu}{\xi\Sigma_s} \nabla D \nabla \varphi^* e^{-\int\limits_{u_{j-1}}^{u'} \frac{\Sigma_c}{\xi\Sigma_s+\gamma\Sigma_c} du''} du' + \eta^j Q^*, \tag{29.17}$$

where

$$p^j = p^{*j}(u_{j-1}), \quad \eta^j = \eta^j(u_{j-1}).$$

Using (28.9) as before, we arrive at

$$\int\limits_{u_{j-1}}^{u_j} \frac{\mu}{\xi\Sigma_s} \nabla D \nabla \varphi^* e^{-\int\limits_{u_{j-1}}^{u} \frac{\Sigma_c}{\xi\Sigma_s+\gamma\Sigma_c} du'} du = A^{*j} \nabla D^j \nabla \varphi^{*j-1} + B^{*j} \nabla D^{j+1} \nabla \varphi^{*j}, \tag{29.18}$$

where A^{*j} and B^{*j} are given by Table 11, where in the integrands $p^j(u)/\xi\Sigma_s$ must be replaced by $\mu\frac{p^j(u)}{\xi\Sigma_s}$. Inserting (29.18) and the second of (29.15) into (29.17), performing some simple operations, and dropping some terms in (29.15) because of the smallness of $\gamma\nabla D\nabla\varphi^*$, we arrive at the set of diffusion equations

$$\nabla D^j \nabla \varphi^{*j-1} - \Sigma^j \varphi^{*j-1} = -f^{*j} \quad (j = m, m-1, \ldots, 1), \tag{29.19}$$

where

$$\left.\begin{aligned}
D^j &= \frac{\int\limits_{u_{j-1}}^{u_j} \frac{D}{\xi\Sigma_s+\gamma\Sigma_c} \mu \frac{p^j}{p^j(u)} du}{\int\limits_{u_{j-1}}^{u_j} \frac{1}{\xi\Sigma_s+\gamma\Sigma_c} \mu \frac{p^j}{p^j(u)} du}, \\
\Sigma^j &= \frac{\xi\Sigma_s^{j-1} + \gamma\Sigma_c^{j-1}}{\xi\Sigma_s^{j-1}} \cdot \frac{1}{A^{*j} + \frac{\gamma}{\xi\Sigma_s^{j-1}}}, \\
f^{*j} &= \frac{1}{A^{*j} + \frac{\gamma}{\xi\Sigma_s^{j-1}}} \Big[p^j \frac{\xi\Sigma_s^j + \gamma\Sigma_c^j}{\xi\Sigma_s^j} \varphi^{*j} + \\
&\quad + \left(B^{*j} - \frac{\gamma p^j}{\xi\Sigma_s^j} \right) (\Sigma^{j+1}\varphi^{*j} - f^{*j+1}) + \omega^{*j} Q^* \Big], \\
\omega^{*j} &= \eta^j + \gamma \left(\frac{\Sigma_f^{j-1}}{\xi\Sigma_s^{j-1}} - \frac{p^j}{\xi\Sigma_s^j} \Sigma_f^j \right).
\end{aligned}\right\} \tag{29.20}$$

Finally, adding the equation for the thermal group to (29.19), we arrive at the adjoint multigroup reactor equations

$$\nabla D^j \nabla \varphi^{*j-1} - \Sigma^j \varphi^{*j-1} = -f^{*j} \quad (j = m+1,\ m,\ \ldots,\ 1), \tag{29.21}$$

where $\varphi^{*m} = \Phi^*$, and D^j, Σ^j and f^j are given by Table 15.

TABLE 15

Functions and Constants in the Adjoint Multigroup Reactor Equations With the Second Moment of Energy and Strong Absorption of Slowing-Down Neutrons

j	φ^{*j-1}	D^j	Σ^j	f^{*j}
$(m+1)$	Φ^*	D_T	Σ_{cT}	$\Sigma_{fT} Q^*$
$(m)-(1)$	φ^{*j-1}	D^i	$\frac{\xi\Sigma_s^{j-1} + \gamma\Sigma_c^{j-1}}{\xi\Sigma_s^{j-1}} \times \frac{1}{A^{*j} + \frac{\gamma}{\xi\Sigma_s^{j-1}}}$	$\frac{1}{A^{*j} + \frac{\gamma}{\xi\Sigma_s^{j-1}}} \left[p^j \frac{\xi\Sigma_s^j + \gamma\Sigma_c^j}{\xi\Sigma_s^j} \varphi^{*j} + \left(B^{*j} - \frac{\gamma p^j}{\xi\Sigma_s^j} \right) (\Sigma^{i+1}\varphi^{*j} - f^{*j+1}) + \omega^{*j} Q^* \right]$

In conclusion, we consider the computation of

$$Q^* = \nu_f \int_{-\infty}^{u_T} \chi(u)\, \varphi^*\, du.$$

Repeating the considerations of Section 28, we again arrive at (28.23), where ϵ^*, α^{*j} and β^{*j} are given by (28.24) and (28.25), except that we must first replace

$$p^{*j}(u) \quad \text{by} \quad \frac{\xi\Sigma_s}{\xi\Sigma_s + \gamma\Sigma_c} \cdot \frac{\xi\Sigma_s^j + \gamma\Sigma_c^j}{\xi\Sigma_s^j}\, p^{*j}(u),$$

$$A^{*j}(u) \quad \text{»} \quad \frac{\xi\Sigma_s}{\xi\Sigma_s + \gamma\Sigma_c} \cdot \left[A^{*j}(u) + \frac{\gamma}{\xi\Sigma_s} \right],$$

$$B^{*j}(u) \quad \text{»} \quad \frac{\xi\Sigma_s}{\xi\Sigma_s + \gamma\Sigma_c} \left[B^{*j}(u) - p^{*j}(u) \frac{\gamma}{\xi\Sigma_s} \right],$$

$$\eta^j(u) \quad \text{»} \quad \omega^{*j}(u),$$

and $A^{*j}(u)$ and $B^{*j}(u)$ are given by Table 13, where $1/\xi\Sigma_s$ in the integrand must be replaced by $\mu/\xi\Sigma_s$.

Chapter VIII

FINITE-DIFFERENCE DIFFUSION EQUATIONS

The calculation involved in an exact solution of the differential diffusion equations, even for very simple regions, is of extreme difficulty. It is therefore necessary to use various approximation methods for solving these problems [22, 58, 66, 78].

The most convenient of these is the method of finite differences, which consists essentially of replacing a differential operator by a finite difference operator. In this way a problem formulated in differential terms is reduced to a simple problem in linear algebra.

It is well known that there are many finite difference approximations to a given differential equation. We shall use the most effective method, namely the "continuous calculation" method which was first developed for diffusion equations by A.N. Tikhonov and A.A. Samarskii [78, 79]. This is a simple method for obtaining unique finite difference equations for a wide class of differential equations with piecewise continuous coefficients.

The present chapter is a discussion of ways to construct finite difference diffusion equations both for one-dimensional and two-dimensional problems.

30. A General Method for Constructing Finite Difference Diffusion Equations in One-Dimensional Regions

Consider the diffusion equation in a one-dimensional region with plane, cylindrical and spherical geometry. This equation is

$$\frac{1}{r^\alpha}\frac{d}{dr}r^\alpha D\frac{d\varphi}{dr}-\Sigma\varphi=-f(r), \tag{30.1}$$

where D is the diffusion coefficient, Σ is the macroscopic capture cross section, and

$$\alpha=\begin{cases}0 \text{ for plane geometry}\\ 1 \text{ for cylindrical geometry}\\ 2 \text{ for spherical geometry}\end{cases}$$

Let us assume that D and Σ are piecewise constant, and that $f(r)$ is piecewise continuous in the region of the reactor.

We wish to find a solution of (30.1) which has a continuous flux

$$I=r^\alpha D\frac{d\varphi}{dr}. \tag{30.2}$$

We shall consider the interval of variation of r (that is, $0\le r\le r_n$, where r_n is the radius of the extrapolated reactor boundary), to be divided up by two sets of interpolation points, the basic set r_k and the subsidiary set $r_{k+1/2}$. These two sets intermesh so that $r_{k-1/2}<r_k<r_{k+1/2}$ and $r_k<r_{k+1/2}<r_{k+1}$. Later we shall have more to say about how r_k and $r_{k+1/2}$ are related.

Let us multiply (30.1) by r^α and integrate over $r_{k-1/2}\le r\le r_{k+1/2}$. We then obtain

$$I_{k+1/2} - I_{k-1/2} = \int_{r_{k-1/2}}^{r_{k+1/2}} (\Sigma\varphi - f)\, r^{\alpha}\, dr, \tag{30.3}$$

where

$$I_{k+1/2} = I\,(r_{k+1/2}).$$

We find $I_{k-1/2}$ by proceeding as follows. We multiply (30.1) by r^{α} and integrate over $(r_{k-1/2}, r)$. This gives

$$r^{\alpha} D \frac{d\varphi}{dr} = I_{k-1/2} + \int_{r_{k-1/2}}^{r} (\Sigma\varphi - f)\, r^{\alpha}\, dr. \tag{30.4}$$

We now divide (30.4) by $r^{\alpha}D$ and again integrate over (r_{k-1}, r_k). As a result we have

$$\varphi_k - \varphi_{k-1} = I_{k-1/2} \int_{r_{k-1}}^{r_k} \frac{dr}{r^{\alpha} D} + \int_{r_{k-1}}^{r_k} \frac{dr}{r^{\alpha} D} \int_{r_{k-1/2}}^{r} (\Sigma\varphi - f)\, r^{\alpha}\, dr. \tag{30.5}$$

Solving this for $I_{k-1/2}$, we arrive at

$$I_{k-1/2} = \frac{\varphi_k - \varphi_{k-1}}{\int_{r_{k-1}}^{r_k} \frac{dr}{r^{\alpha} D}} - \frac{1}{\int_{r_{k-1}}^{r_k} \frac{dr}{r^{\alpha} D}} \int_{r_{k-1}}^{r_k} \frac{dr}{r^{\alpha} D} \int_{r_{k-1/2}}^{r} (\Sigma\varphi - f)\, r^{\alpha}\, dr. \tag{30.6}$$

We have thus been able to express the flux and the desired solution in terms of known functions. We note that (30.6) is not an approximate solution. Let us now insert (30.6) into (30.3). We have obtained

$$\begin{aligned} &\frac{\varphi_{k+1} - \varphi_k}{\int_{r_k}^{r_{k+1}} \frac{dr}{r^{\alpha} D}} - \frac{\varphi_k - \varphi_{k-1}}{\int_{r_{k-1}}^{r_k} \frac{dr}{r^{\alpha} D}} - \int_{r_{k-1/2}}^{r_k} (\Sigma\varphi - f)\, r^{\alpha} dr - \\ &- \frac{1}{\int_{r_k}^{r_{k+1}} \frac{dr}{r^{\alpha} D}} \int_{r_k}^{r_{k+1}} \frac{dr}{r^{\alpha} D} \int_{r_{k+1/2}}^{r} (\Sigma\varphi - f)\, r^{\alpha}\, dr + \\ &+ \frac{1}{\int_{r_{k-1}}^{r_k} \frac{dr}{r^{\alpha} D}} \int_{r_{k-1}}^{r_k} \frac{dr}{r^{\alpha} D} \int_{r_{k-1/2}}^{r} (\Sigma\varphi - f)\, r^{\alpha}\, dr = 0. \end{aligned} \tag{30.7}$$

Now (30.7) will be the basis on which we will obtain the different finite difference systems, which we shall call the finite-difference diffusion equations. In accordance, what will follow will be a replacement of φ and f within the intervals of integration by suitable interpolation polynomials.

A very accurate finite difference representation of the diffusion equation (30.1) can be obtained by using interpolation formulas of high order. This, however, usually leads to very complicated sets of linear equations which cannot be solved simply. This is the decisive factor in the choice of the interpolation formulas. Further, let us direct our attention to the following fact. In obtaining the necessary relation between the flux at an interpolation point and the solution of the problem, we arrived at Equation (30.6), which was used later to obtain the basic equation (30.7). In principle (30.7) can be used to obtain a finite-difference equation of any given accuracy.

This method for obtaining the finite-difference diffusion equation is not, however, the only one. In practice it is sometimes even more convenient to use a different method based on more understandable assumptions. Essentially, this involves deriving the finite-difference equations by using the set of equations

$$I_{k+1/2} - I_{k-1/2} = \int_{r_{k-1/2}}^{r_{k+1/2}} (\Sigma\varphi - f)\, r^{\alpha}\, dr, \tag{30.8}$$

$$\varphi_{k+1} - \varphi_k = \int_{r_k}^{r_{k+1}} \frac{I}{Dr^{\alpha}}\, dr, \tag{30.9}$$

where I(r) satisfies, as before, Equation (30.1), which we shall now write

$$\frac{1}{r^{\alpha}} \frac{dI}{dr} = \Sigma\varphi - f. \tag{30.10}$$

We next replace the integrands in (30.8) and (30.9) by suitable interpolation polynomials. By eliminating $I_{k+1/2}$ and $I_{k-1/2}$ from the set of equations so obtained, we arrive at the desired finite-difference system.*

Let us now go into more detail concerning the choice of the r_k and $r_{k+1/2}$. Consider two different cases, the first in which a point of discontinuity of D, Σ_c and f lies on r_k, and the second in which it lies on $r_{k+1/2}$. In the first case it is convenient to define the $r_{k+1/2}$ by $r_{k+1/2} = \frac{1}{2}(r_{k+1} + r_k)$. In the second case, on the other hand, we define r_k by $r_k = \frac{1}{2}(r_{k-1/2} + r_{k+1/2})$.

31. The Simplest Finite-Difference Diffusion Equations

In obtaining the simplest finite-difference diffusion equations, we shall start from (30.8) and (30.9), and make use only of the fundamental properties of the solution.

First consider the case in which a point of discontinuity of D, Σ and f lies on an interpolation point r_k. Analysis of the integrands in (30.8) shows that in the interval $(r_{k-1/2}, r_{k+1/2})$ the function $\Sigma\varphi - f$ has a discontinuity at r_k. With this in mind we can break up the integral into the two parts

$$\int_{r_{k-1/2}}^{r_{k+1/2}} (\Sigma\varphi - f)\, r^{\alpha}\, dr = \int_{r_{k-1/2}}^{r_k} (\Sigma\varphi - f)\, r^{\alpha}\, dr + \int_{r_k}^{r_{k+1/2}} (\Sigma\varphi - f)\, r^{\alpha}\, dr. \tag{31.1}$$

The integrands in the resulting two terms are now continuous, and they can therefore be expanded in a Taylor's series about r_k. Then, keeping only the first terms of the expansion

$$\left.\begin{aligned} (\Sigma\varphi - f)\, r^{\alpha} &= (\Sigma_k^{-}\varphi_k - f_k^{-})\, r_k^{\alpha} + \ldots, \quad r_{k-1/2} \leqslant r < r_k, \\ (\Sigma\varphi - f)\, r^{\alpha} &= (\Sigma_k^{+}\varphi_k - f_k^{+})\, r_k^{\alpha} + \ldots, \quad r_k < r \leqslant r_{k+1/2}, \end{aligned}\right\} \tag{31.2}$$

where we have written

$$\Phi_k^{+} = \lim_{\varepsilon\to 0} \Phi(r_k + \varepsilon), \qquad \Phi_k^{-} = \lim_{\varepsilon\to 0} \Phi(r_k - \varepsilon), \tag{31.3}$$

---

* This method for constructing finite-difference systems for diffusion equations with constant coefficients was first proposed by Tikhonov and Samarskii [70].

we arrive at an approximation of the form

$$\int_{r_{k-1/2}}^{r_{k+1/2}} (\Sigma\varphi - f)\, r^{\alpha}\, dr = [(\Sigma\Delta r)_k \varphi_k - (f\Delta r)_k]\, r_k^{\alpha}. \tag{31.4}$$

We have here introduced the notation

$$(\Phi)_k = \frac{1}{2}(\Phi_k^+ + \Phi_k^-), \quad \Delta r_k^+ = r_{k+1} - r_k, \quad \Delta r_k^- = r_k - r_{k-1}. \tag{31.5}$$

Inserting (31.4) into (30.8), we obtain

$$I_{k+1/2} - I_{k-1/2} - (\Sigma\Delta r)_k \varphi_k r_k^{\alpha} = -(f\Delta r)_k r_k^{\alpha}. \tag{31.6}$$

Let us now consider (30.9). The integrand $I/r^{\alpha}D$ is continuous on the interval (r_k, r_{k+1}), so that we can expand it in a Taylor's series about $r_{k+1/2}$. Maintaining only the first term of the expansion, we have

$$\frac{I}{r^{\alpha}D} = \frac{I_{k+1/2}}{r_{k+1/2}^{\alpha} D_{k+1/2}} + \dots \tag{31.7}$$

therefore,

$$\int_{r_k}^{r_{k+1}} \frac{I}{r^{\alpha}D}\, dr = \Delta r_{k+1/2} \frac{I_{k+1/2}}{r_{k+1/2}^{\alpha} D_{k+1/2}}, \tag{31.8}$$

where

$$\Delta r_{k+1/2} = r_{k+1} - r_k. \tag{31.9}$$

Inserting (31.8) into (30.9), we arrive at

$$\varphi_{k+1} - \varphi_k = \Delta r_{k+1/2} \frac{I_{k+1/2}}{r_{k+1/2}^{\alpha} D_{k+1/2}}. \tag{31.10}$$

Solving for $I_{k+1/2}$, we have

$$I_{k+1/2} = \mu_k (\varphi_{k+1} - \varphi_k), \tag{31.11}$$

where

$$\mu_k = \frac{r_{k+1/2}^{\alpha} D_{k+1/2}}{\Delta r_{k+1/2}}.$$

Let us use (31.6) to eliminate $I_{k+1/2}$, $I_{k-1/2}$ from (31.11). We then arrive at the finite-difference equation

$$\mu_k (\varphi_{k+1} - \varphi_k) - \mu_{k-1} (\varphi_k - \varphi_{k-1}) - r_k^{\alpha} (\Sigma\Delta r)_k \varphi_k = -r_k^{\alpha} (f\Delta r)_k. \tag{31.12}$$

Finally, we transform (31.12) to the form*

$$a_k \varphi_{k+1} - b_k \varphi_k + c_k \varphi_{k-1} = -F_k, \tag{31.13}$$

*It should be recalled that if $f = 0$ in the reflector, we must set $f_k^+ = 0$, but $f_k^- \neq 0$ at the point of discontinuity.

where

$$a_k = \left(1 + \alpha \frac{\Delta r_{k+1/2}}{2r_k}\right) \frac{D_{k+1/2}}{\Delta r_{k+1/2}}, \qquad b_k = a_k + c_k + (\Sigma \Delta r)_k,$$
$$c_k = \left(1 - \alpha \frac{\Delta r_{k-1/2}}{2r_k}\right) \frac{D_{k-1/2}}{\Delta r_{k-1/2}}, \qquad F_k = (f \Delta r)_k. \tag{31.14}$$

In these equations we have written

$$\left(1 + \frac{\Delta r_k}{2r_k}\right)^\alpha \simeq 1 + \alpha \frac{\Delta r_k}{2r_k}, \quad \left(1 - \frac{\Delta r_k}{2r_k}\right)^\alpha \simeq 1 - \alpha \frac{\Delta r_k}{2r_k},$$

which is an approximation only if $\alpha = 2$.

Because spherically symmetric regions are of most interest, let us consider this problem in somewhat more detail. It turns out that in this case the problem is greatly simplified if $r_k \varphi_k$ is chosen as the unknown function in (31.12). If $r_k \varphi_k$ and $r_k f_k$ are denoted by φ_k and f_k as before, we arrive again at the finite difference equation (31.13), whose coefficients a_k, b_k and c_k are now given by the computationally simpler expressions

$$a_k = \frac{D_{k+1/2}}{\Delta r_{k+1/2}}, \qquad c_k = \frac{D_{k-1/2}}{\Delta r_{k-1/2}}.$$
$$b_k = a_k + c_k + \frac{\Delta r_{k+1/2}}{r_k}\left(a_k - \frac{\Delta r_{k-1/2}}{\Delta r_{k+1/2}} c_k\right) + (\Sigma \Delta r)_k. \tag{31.15}$$

We have here made the approximation

$$r^2_{k+1/2} = r_k r_{k+1} + 0\,(\Delta r^2).$$

In the new variables the coefficients a_k, b_k and c_k no longer depend on the radius, except for b_k at points of discontinuity.

Let us now go on to a consideration of the case in which a point of discontinuity lies on an internal $r_{k+1/2}$ point. We again consider Equations (30.8) and (30.9). With a given choice of the interpolation points, the integrand of (30.8) will be continuous, and therefore it can be expanded in a Taylor's series* about r_k:

$$(\Sigma\varphi - f)\, r^\alpha = (\Sigma_k \varphi_k - f_k)\, r_k^\alpha + \ldots \tag{31.16}$$

Keeping only the first term of the expansion, we insert (31.16) into (30.8). We then obtain

$$I_{k+1/2} - I_{k-1/2} = (\Sigma_k \varphi_k - f_k)\, r_k^\alpha \Delta r_k, \tag{31.17}$$

where

$$\Delta r_k = r_{k+1/2} - r_{k-1/2}.$$

Let us now write $I_{k+1/2}$ and $I_{k-1/2}$ in (31.17) in terms of values of φ_k. To do this consider Equation (30.9).

The integrand in (30.9) contains the product of the discontinuous function D by the continuous one I/r^α. Let us expand I/r^α in a series about $r_{k+1/2}$ and keep only the first term of the expansion:

$$\frac{I}{r^\alpha} = \frac{I_{k+1/2}}{r^\alpha_{k+1/2}} + \ldots \tag{31.18}$$

---

*Here and in what follows we shall assume that this is possible, since in the neighborhood of the point under consideration the principal parts of both Taylor's series expansions for $r_k + 0$ and $r_k - 0$ are the same.

We now insert this into (30.9) and integrate. This gives

$$\varphi_{k+1}-\varphi_k=\frac{I_{k+1/2}}{r^{\alpha}_{k+1/2}}\left(\frac{\Delta r}{D}\right)_{k+1/2}, \tag{31.19}$$

where

$$(\Phi)_{k+1/2}=\frac{1}{2}(\Phi_k+\Phi_{k+1}), \tag{31.20}$$

Solving (31.19) for $I_{k+1/2}$, we have

$$I_{k+1/2}=\mu_k(\varphi_{k+1}-\varphi_k), \tag{31.21}$$

where

$$\mu_k=\frac{r^{\alpha}_{k+1/2}}{\left(\frac{\Delta r}{D}\right)_{k+1/2}}. \tag{31.22}$$

Let us now insert (31.21) into (31.17). This gives the finite-difference equation

$$\mu_k(\varphi_{k+1}-\varphi_k)-\mu_{k-1}(\varphi_k-\varphi_{k-1})-\Sigma_k\Delta r_k\varphi_k r^{\alpha}_k=-\Delta r_k f_k r^{\alpha}_k, \tag{31.23}$$

which can be written

$$a_k\varphi_{k+1}-b_k\varphi_k+c_k\varphi_{k-1}=-F_k, \tag{31.24}$$

where

$$a_k=\frac{1+\alpha\frac{\Delta r_k}{2r_k}}{\left(\frac{\Delta r}{D}\right)_{k+1/2}},\qquad b_k=a_k+c_k+\Sigma_k\Delta r_k,$$

$$c_k=\frac{1-\alpha\frac{\Delta r_k}{2r_k}}{\left(\frac{\Delta r}{D}\right)_{k-1/2}},\qquad F_k=f_k\Delta r_k.$$

Just as in the previous case, we shall treat the spherical case by going to the new functions $\varphi_k r_k$ and $f_k r_k$. Then, writing φ_k and f_k as before, we arrive at an equation of the same type as (31.24), in which

$$\left.\begin{aligned} a_k&=\frac{1}{\left(\frac{\Delta r}{D}\right)_{k+1/2}},\qquad c_k=\frac{1}{\left(\frac{\Delta r}{D}\right)_{k-1/2}},\\ b_k&=a_k+c_k+\frac{\Delta r_k}{r_k}(a_k-c_k)+\Sigma_k\Delta r_k.\end{aligned}\right\} \tag{31.25}$$

The finite-difference diffusion equation (31.23) was first obtained by Tikhonov and Samarskii [79].

Let us now consider the boundary conditions which must be given at the center of the reactor and on the external extrapolated boundary. For any region the latter condition reduces to

$$\varphi_m=0, \tag{31.26}$$

where m is the index of the interpolation point at the boundary of the reactor. Two cases are possible for $r=0$. These are $\left.\frac{d\varphi}{dr}\right|_{r=0}=0$ and $\varphi|_{r=0}=0$. The latter of these holds for spherical systems, in which the solution is $r\varphi$.

This can be written

$$\varphi_0 = 0. \tag{31.27}$$

In order to satisfy the symmetry condition $\left.\frac{d\varphi}{dr}\right|_{r=0} = 0$ with the greatest accuracy, it is desirable to choose the points of interpolation so that $r = 0$ lies at the center of the first Δr interval. Then the condition at the center of the reactor can be written

$$\varphi_0 = \varphi_1, \tag{31.28}$$

where $k = 0$ and $k = 1$ are the indices of the interpolation points at a distance $\frac{1}{2}\Delta r$ on both sides of $r = 0$.

32. Improved Finite-Difference Diffusion Equations

The finite-difference diffusion equations obtained in the previous section are simple and very convenient for practical use. With a sufficiently large number of interpolation points they can give highly accurate results.

The largest errors involved in using finite-difference equations arise at points close to discontinuities. If the different zones of the reactor have very different properties, these errors may be quite large. In solving such problems, therefore, it becomes necessary to decrease the dimensions of the interpolation intervals and to increase the number of interpolation points, which may lead to an extreme amount of work. For this reason such a procedure is often without advantage.

It is possible, however, to use more accurate interpolation polynomials in deriving the finite-difference diffusion equations, polynomials which will account in more detail for the fundamental properties of the solution in the neighborhood of a discontinuity. It is found that at the expense of an insignificant increase in the amount of calculation at an interpolation point, one can introduce an improved finite-difference system which will give the desired accuracy with less work.

The present section will give a derivation of some refinements of the finite-difference equations.

First let us consider what occurs when a discontinuity point lies on an $r = r_k$ interpolation point. We again use (30.8) and (30.9) to obtain the finite-difference equations.

Because the integrand of (30.8) has a discontinuity at r_k, we can break up the integral, writing

$$\int_{r_{k-1/2}}^{r_{k+1/2}} (\Sigma\varphi - f)\, r^\alpha\, dr = \int_{r_{k-1/2}}^{r_k} (\Sigma\varphi - f)\, r^\alpha\, dr + \int_{r_k}^{r_{k+1/2}} (\Sigma\varphi - f)\, r^\alpha\, dr. \tag{32.1}$$

The integrands in the new integrals are continuous, and we expand them in power series

$$\left.\begin{aligned} (\Sigma\varphi - f)\, r^\alpha &= (\Sigma_k^{-}\varphi_k - f_k^{-})\, r_k^\alpha + \frac{d}{dr}(\Sigma\varphi - f)\, r^\alpha \Big|_{r_k-0} (r - r_k) + \dots \\ &\qquad r < r_k, \\ (\Sigma\varphi - f)\, r^\alpha &= (\Sigma_k^{+}\varphi_k - f_k^{+})\, r_k^\alpha + \frac{d}{dr}(\Sigma\varphi - f)\, r^\alpha \Big|_{r_k+0} (r - r_k) + \dots \\ &\qquad r > r_k. \end{aligned}\right\} \tag{32.2}$$

Inserting (32.2) into (32.1), we obtain

$$\begin{aligned} \int_{r_{k-1/2}}^{r_{k+1/2}} (\Sigma\varphi - f)\, r^\alpha\, dr &= (\Sigma\Delta r)_k \varphi_k r_k^\alpha - \frac{1}{8}\left(\Sigma\Delta r^2 \frac{d}{dr}\varphi r^\alpha \Big|_{r_k-0} - \Sigma\Delta r^2 \frac{d}{dr}\varphi r^\alpha \Big|_{r_k+0} \right) - \\ &- (f\Delta r)_k\, r_k^\alpha + \frac{1}{8}\left(\Delta r^2 \frac{d}{dr} f r^\alpha \Big|_{r_k-0} - \Delta r^2 \frac{d}{dr} f r^\alpha \Big|_{r_k+0} \right). \end{aligned} \tag{32.3}$$

We note that for

$$\left.\begin{aligned}\Sigma\Delta r^2 \frac{d}{dr}\varphi r^\alpha \Big|_{r_k-0} &= \frac{\Sigma_k^- \Delta r_k^{2-} I_k}{D_k^-} + \alpha\Sigma_k^- \Delta r_k^{2-} \varphi_k r_k^{\alpha-1}, \\ \Sigma\Delta r^2 \frac{d}{dr}\varphi r^\alpha \Big|_{r_k+0} &= \frac{\Sigma_k^+ \Delta r_k^{2+} I_k}{D_k^+} + \alpha\Sigma_k^+ \Delta r_k^{2+} \varphi_k r_k^{\alpha-1},\end{aligned}\right\} \tag{32.4}$$

We now insert (32.4) into (32.3). This gives

$$\int_{r_{k-1/2}}^{r_{k+1/2}} (\Sigma\varphi - f)\, r^\alpha\, dr = \left\{(\Sigma\Delta r)_k + \frac{\alpha}{8r_k}[\Sigma\Delta r^2]_k\right\} \varphi_k r_k^\alpha + \frac{1}{8}\left[\frac{1}{D}\Sigma\Delta r^2\right]_k I_k -$$

$$- \left\{(f\Delta r)_k + \frac{\alpha}{8r_k}[f\Delta r^2]_k + \frac{1}{8}[f'\Delta r^2]_k\right\} r_k^\alpha, \tag{32.5}$$

where

$$[\Phi]_k = \Phi_k^+ - \Phi_k^-, \quad f' = \frac{df}{dr}. \tag{32.6}$$

The second terms in the braces of (32.5) are small compared to the first. Using (32.5), we can rewrite (30.8) in the form

$$I_{k+1/2} - I_{k-1/2} - \frac{1}{8}\left[\frac{1}{D}\Sigma\Delta r^2\right]_k I_k - \left\{(\Sigma\Delta r^2)_k + \frac{\alpha}{8r_k}[\Sigma\Delta r^2]_k\right\} \varphi_k r_k^\alpha =$$

$$= - \left\{(f\Delta r)_k + \frac{1}{8}[f'\Delta r^2]_k + \frac{\alpha}{8r_k}[f\Delta r^2]_k\right\} r_k^\alpha. \tag{32.7}$$

We eliminate I_k from (32.7) using the relation

$$I_k = \frac{1}{2}(I_{k+1/2} + I_{k-1/2}). \tag{32.8}$$

Equation (32.8) is an assumption which is sufficiently valid because $\frac{1}{8}[\Sigma\Delta r^2/D]_k I_k$ is usually small compared to the other terms of (32.7).

Thus, with (32.8) we transform (32.7) to

$$\frac{1}{r_k^\alpha}[(1-\varkappa_k) I_{k+1/2} - (1+\varkappa_k) I_{k-1/2}] - \left\{(\Sigma\Delta r)_k + \frac{\alpha}{8r_k}[\Sigma\Delta r^2]_k\right\} \varphi_k =$$

$$= - \left\{(f\Delta r)_k + \frac{\alpha}{8r_k}[f\Delta r^2]_k + \frac{1}{8}[f'\Delta r^2]_k\right\}, \tag{32.9}$$

where

$$\varkappa_k = \frac{1}{16}\left[\frac{\Sigma\Delta r^2}{D}\right]_k.$$

We now have to eliminate $I_{k+1/2}$ and $I_{k-1/2}$ from (32.9). For this purpose, consider (30.9). Because the integrand in (30.9) is continuous, we can expand it in a series, and keeping just the first two terms of the expansion we have

$$\frac{I}{Dr^{\alpha}} = \frac{I_{k+1/2}}{D_{k+1/2}\, r^{\alpha}_{k+1/2}} + \frac{d}{dr}\frac{I}{Dr^{\alpha}}\Big|_{r_{k+1/2}} (r - r_{k+1/2}) + \ldots \tag{32.10}$$

We now insert (32.10) into (30.9). This gives

$$I_{k+1/2} = \mu_k\,(\varphi_{k+1} - \varphi_k), \tag{32.11}$$

where μ_k is given by (31.11). Eliminating $I_{k+1/2}$ and $I_{k-1/2}$ from (32.9) we arrive at

$$a_k\varphi_{k+1} - b_k\varphi_k + c_k\varphi_{k-1} = -F_k, \tag{32.12}$$

where

$$\left.\begin{aligned} a_k &= (1-\varkappa_k)\left(1+\alpha\frac{\Delta r_{k+1/2}}{2r_k}\right)\frac{D_{k+1/2}}{\Delta r_{k+1/2}}, \\ b_k &= a_k + c_k + (\Sigma\Delta r)_k + \frac{\alpha}{8r_k}[\Sigma\Delta r^2]_k, \\ c_k &= (1+\varkappa_k)\left(1-\alpha\frac{\Delta r_{k-1/2}}{2r_k}\right)\frac{D_{k-1/2}}{\Delta r_{k-1/2}}, \\ F_k &= (f\Delta r)_k + \frac{\alpha}{8r_k}[f\Delta r^2]_k + \frac{1}{8}[f'\Delta r^2]_k. \end{aligned}\right\} \tag{32.13}$$

From well-known assumptions concerning the form of the solution and F in a spherically symmetric region, we again arrive at (32.12), where

$$\left.\begin{aligned} a_k &= \frac{(1-\varkappa_k)\,D_{k+1/2}}{\Delta r_{k+1/2}}, \qquad c_k = \frac{(1+\varkappa_k)\,D_{k-1/2}}{\Delta r_{k-1/2}}, \\ b_k &= a_k + c_k + \frac{\Delta r_{k+1/2}}{r_k}\left(a_k - \frac{\Delta r_{k-1/2}}{\Delta r_{k+1/2}}c_k\right) + (\Sigma\Delta r)_k + \frac{1}{4r_k}[\Sigma\Delta r^2]_k, \\ F_k &= (f\Delta r)_k + \frac{1}{4r_k}[f\Delta r^2]_k + \frac{1}{8}[f'\Delta r^2]_k. \end{aligned}\right\} \tag{32.14}$$

It is easily seen that, except at discontinuities, (32.13) and (32.14) are everywhere identical with the analogous equations (31.14) and (31.15) for the simplest finite-difference diffusion equation.

We now go on to a derivation of the improved finite-difference diffusion equation for the case in which the discontinuity is at an internal $r_{k+1/2}$ point of the (r_k, r_{k+1}) interval. To do this, we first consider (30.8).

The integrand of (30.8) is continuous, and we expand it, writing

$$(\Sigma\varphi - f)\,r^{\alpha} = (\Sigma_k\varphi_k - f_k)\,r_k^{\alpha} + \frac{d}{dr}(\Sigma\varphi - f)\,r^{\alpha}\Big|_{r=r_k}(r - r_k) + \ldots \tag{32.15}$$

Inserting (32.15) into (30.8) and integrating, we have

$$I_{k+1/2} - I_{k-1/2} - \Sigma_k\,\Delta r_k\,\varphi_k\,r_k^{\alpha} = -f_k\,\Delta r_k\,r_k^{\alpha}. \tag{32.16}$$

We now use (30.9) to obtain expressions for $I_{k+1/2}$ and $I_{k-1/2}$. As before, we break up the integral (30.9) into two parts because the integrand contains the discontinuous function D. We then have

$$\int\limits_{r_k}^{r_{k+1}} \frac{I}{r^{\alpha}D}\,dr = \int\limits_{r_k}^{r_{k+1/2}} \frac{I}{r^{\alpha}D}\,dr + \int\limits_{r_{k+1/2}}^{r_{k+1}} \frac{I}{r^{\alpha}D}\,dr. \tag{32.17}$$

We expand $I/r^{\alpha}D$, writing

$$\left.\begin{aligned} \frac{I}{r^{\alpha}D} &= \frac{I_{k+1/2}}{r^{\alpha}_{k+1/2}D^{-}_{k+1/2}} + \frac{d}{dr}\left(\frac{I}{r^{\alpha}D}\right)\Bigg|_{r_{k+1/2}-0}(r - r_{k+1/2}) + \ldots, \\ &\qquad (r_k \leqslant r < r_{k+1/2}), \\ \frac{I}{r^{\alpha}D} &= \frac{I_{k+1/2}}{r^{\alpha}_{k+1/2}D^{+}_{k+1/2}} + \frac{d}{dr}\left(\frac{I}{r^{\alpha}D}\right)\Bigg|_{r_{k+1/2}+0}(r - r_{k+1/2}) + \ldots, \\ &\qquad (r_{k+1/2} < r \leqslant r_{k+1}). \end{aligned}\right\} \tag{32.18}$$

We now insert (32.18) into the right side of (32.17) and integrate. The result is

$$\int\limits_{r_k}^{r_{k+1}} \frac{I}{r^{\alpha}D}\,dr = \left(\frac{\Delta r}{D}\right)_{k+1/2}\frac{I_{k+1/2}}{r^{\alpha}_{k+1/2}} + \frac{1}{8}\left[\frac{\Delta r^{2+}_{k+1/2}}{D^{+}_{k+1/2}}\frac{d}{dr}\left(\frac{I}{r^{\alpha}}\right)\Bigg|_{r_{k+1/2}+0} - \right.$$
$$\left. - \frac{\Delta r^{2-}_{k+1/2}}{D^{-}_{k+1/2}}\frac{d}{dr}\left(\frac{I}{r^{\alpha}}\right)\Bigg|_{r_{k+1/2}-0}\right]. \tag{32.19}$$

We recall that

$$\frac{d}{dr}\frac{I}{r^{\alpha}} = \frac{1}{r^{\alpha}}\frac{dI}{dr} - \frac{\alpha}{r^{\alpha+1}}I. \tag{32.20}$$

We use (30.10) to eliminate $\frac{1}{r^{\alpha}}\frac{dI}{dr}$ from this equation. We then have

$$\left.\begin{aligned} \frac{d}{dr}\frac{I}{r^{\alpha}}\Bigg|_{r_{k+1/2}+0} &= \Sigma^{+}_{k+1/2}\varphi_{k+1/2} - f^{+}_{k+1/2} - \frac{\alpha}{r^{\alpha+1}_{k+1/2}}I_{k+1/2}, \\ \frac{d}{dr}\frac{I}{r^{\alpha}}\Bigg|_{r_{k+1/2}-0} &= \Sigma^{-}_{k+1/2}\varphi_{k+1/2} - f^{-}_{k+1/2} - \frac{\alpha}{r^{\alpha+1}_{k+1/2}}I_{k+1/2}. \end{aligned}\right\} \tag{32.21}$$

According to (32.19) and (32.21), we have

$$\int\limits_{r_k}^{r_{k+1}} \frac{I}{r^{\alpha}D}\,dr = \frac{1}{\mu_k}I_{k+1/2} + \frac{1}{8}\left[\frac{\Sigma\Delta r^{2}}{D}\right]_{k+1/2}\varphi_{k+1/2} - \frac{1}{8}\left[\frac{f\Delta r^{2}}{D}\right]_{k+1/2}, \tag{32.22}$$

where

$$\mu_k = \frac{r^{\alpha}_{k+1/2}}{\left(\frac{\Delta r}{D}\right)_{k+1/2} - \frac{1}{8}\frac{\alpha}{r_{k+1/2}}\left[\frac{\Delta r^{2}}{D}\right]_{k+1/2}}. \tag{32.23}$$

We now insert (32.22) into (30.9) and solve for $I_{k+1/2}$, obtaining

$$I_{k+1/2} = \mu_k\left\{\varphi_{k+1} - \varphi_k - \frac{1}{8}\left[\frac{\Sigma\Delta r^{2}}{D}\right]_{k+1/2}\varphi_{k+1/2} + \frac{1}{8}\left[\frac{f\Delta r^{2}}{D}\right]_{k+1/2}\right\} \tag{32.24}$$

and similarly for $I_{k-1/2}$, obtaining

$$I_{k-1/2} = \mu_{k-1}\left\{\varphi_k - \varphi_{k-1} - \frac{1}{8}\left[\frac{\Sigma\Delta r^{2}}{D}\right]_{k-1/2}\varphi_{k-1/2} + \right.$$
$$\left. + \frac{1}{8}\left[\frac{f\Delta r^{2}}{D}\right]_{k-1/2}\right\}. \tag{32.25}$$

The terms $\frac{1}{8}[\Sigma\Delta r^2/D]_{k\pm 1/2}$ in (32.24) and (32.25) are small, so that we may write

$$\varphi_{k+1/2}=\varphi_k \quad \text{and} \quad \varphi_{k-1/2}=\varphi_k. \tag{32.26}$$

Then $I_{k+1/2}$ and $I_{k-1/2}$ become

$$\left.\begin{aligned} I_{k+1/2} &= \mu_k\left\{\varphi_{k+1}-(1+\varkappa_k)\varphi_k+\frac{1}{8}\left[\frac{f\Delta r^2}{D}\right]_{k+1/2}\right\},\\ I_{k-1/2} &= \mu_{k-1}\left\{(1-\varkappa_{k-1})\varphi_k-\varphi_{k-1}+\frac{1}{8}\left[\frac{f\Delta r^2}{D}\right]_{k-1/2}\right\},\end{aligned}\right\} \tag{32.27}$$

where

$$\varkappa_k=\frac{1}{8}\left[\frac{\Sigma\Delta r^2}{D}\right]_{k+1/2}.$$

We now insert (32.27) into (32.16). This gives the finite-difference diffusion equation in the form

$$a_k\varphi_{k+1}-b_k\varphi_k+c_k\varphi_{k-1}=-F_k, \tag{32.28}$$

where

$$\left.\begin{aligned} a_k &= \frac{1+\alpha\cdot\dfrac{\Delta r_k}{2r_k}}{\left(\dfrac{\Delta r}{D}\right)_{k+1/2}-\dfrac{1}{8}\dfrac{\alpha}{r_{k+1/2}}\left[\dfrac{\Delta r^2}{D}\right]_{k+1/2}},\\ b_k &= (1+\varkappa_k)a_k+(1-\varkappa_{k-1})c_k+\Sigma_k\Delta r_k,\\ c_k &= \frac{1-\alpha\cdot\dfrac{\Delta r_k}{2r_k}}{\left(\dfrac{\Delta r}{D}\right)_{k-1/2}-\dfrac{1}{8}\dfrac{\alpha}{r_{k-1/2}}\left[\dfrac{\Delta r^2}{D}\right]_{k-1/2}},\\ F_k &= f_k\Delta r_k+\frac{1}{8}a_k\left[\frac{f\Delta r^2}{D}\right]_{k+1/2}-\frac{1}{8}c_k\left[\frac{f\Delta r^2}{D}\right]_{k-1/2}.\end{aligned}\right\} \tag{32.29}$$

For spherical symmetry, instead of (32.28) we have*

$$\left.\begin{aligned} a_k &= \frac{1}{\left(\dfrac{\Delta r}{D}\right)_{k+1/2}-\dfrac{1}{4}\dfrac{1}{r_{k+1/2}}\left[\dfrac{\Delta r^2}{D}\right]_{k+1/2}}, \qquad c_k=a_{k-1},\\ b_k &= (1+\varkappa_k)a_k+(1-\varkappa_{k-1})c_k+\\ &\quad+\frac{\Delta r_k}{r_k}\left[(1+\varkappa_k)a_k-(1-\varkappa_{k-1})c_k\right]+\Sigma_k\Delta r_k,\\ F_k &= f_k\Delta r_k+\frac{1}{8}a_k\left[\frac{f\Delta r^2}{D}\right]_{k+1/2}-\frac{1}{8}a_{k-1}\left[\frac{f\Delta r^2}{D}\right]_{k-1/2}\end{aligned}\right\} \tag{32.30}$$

Analysis of (32.29) and (32.30) shows that the a_k, b_k, c_k and F_k are the same as those of (31.24) almost everywhere except at points in the neighborhood of discontinuities.

Let us consider in more detail the calculation of F_k in (32.29) and (32.30). To do this, we write f_k in the form

$$f_k=\rho_k\varphi_k^0,$$

*In the formula for F_k, the coefficients a_k and $c_k = a_{k-1}$ are given by (32.29) for $\alpha = 2$.

where φ_k^0 is a solution of the preceding group, continuous on going from one zone to another, and ρ_k is a factor whose meaning becomes clear from the multigroup equations.* Setting, as before,

$$\varphi_{k+1/2}^0 = \frac{1}{2}(\varphi_k^0 + \varphi_{k+1}^0),$$

we obtain

$$F_k = \left(\rho_k \Delta r_k + \frac{1}{8} a_k \left[\frac{\rho \Delta r^2}{D}\right]_{k+1/2} - \frac{1}{8} a_{k-1} \left[\frac{\rho \Delta r^2}{D}\right]_{k-1/2} \right) \varphi_k^0. \tag{32.31}$$

In conclusion, we remark that in order to have a completely specified problem using the finite-difference diffusion equation, we need the boundary conditions. These are the same as (31.26)-(31.28) of the simpler form.

33. Finite-Difference Diffusion Equations for a Two-Dimensional Region

We now go on to a consideration of the diffusion equation for a two-dimensional region. In this case we have

$$\frac{1}{r^\alpha}\frac{\partial}{\partial r} r^\alpha D \frac{\partial \varphi}{\partial r} + \frac{\partial}{\partial z} D \frac{\partial \varphi}{\partial z} - \Sigma\varphi = -f(r, z). \tag{33.1}$$

We write (33.1) in the form of the set of equations

$$\frac{1}{r^\alpha}\frac{\partial J}{\partial r} + \frac{\partial Y}{\partial z} - \Sigma\varphi = -f, \tag{33.2}$$

$$J = r^\alpha D \cdot \frac{\partial \varphi}{\partial r}, \tag{33.3}$$

$$Y = D \cdot \frac{\partial \varphi}{\partial z}, \tag{33.4}$$

where

$$\alpha = \begin{cases} 0 \text{ for rectangular regions} \\ 1 \text{ for cylindrical regions} \end{cases}$$

We shall start our considerations from the simplest finite-difference system. As in the one-dimensional case, let us consider the two possibilities whereby discontinuities in D, Σ and f lie either on basic interpolation points (r_k, z_l) or on some internal points of the intervals. We first consider the former possibility.

We start by multiplying (33.2) by r^α and integrating over $r_{k-1/2} < r < r_{k+1/2}$ and $z_{l-1/2} < z < z_{l+1/2}$. This leads to

$$\int_{z_{l-1/2}}^{z_{l+1/2}} (J_{k+1/2} - J_{k-1/2})\, dz + \int_{r_{k-1/2}}^{r_{k+1/2}} (Y_{l+1/2} - Y_{l-1/2})\, r^\alpha\, dr = \\ = \int_{r_{k-1/2}}^{r_{k+1/2}} \int_{z_{l-1/2}}^{z_{l+1/2}} (\Sigma\varphi - f)\, r^\alpha\, dz\, dr. \tag{33.5}$$

In accordance with the assumptions of the simplified scheme, we write

*For the first group j = 1, and the right side of f_k is nonzero only in the core, so that we must set $\rho_{k+1/2}^{+}$ equal to zero at discontinuity points.

$$\left.\begin{aligned}
&\int\limits_{z_{l-1/2}}^{z_{l+1/2}} (J_{k+1/2} - J_{k-1/2})\, dz = (J_{k+1/2,\, l} - J_{k-1/2,\, l})\, \Delta z_l,\\
&\int\limits_{r_{k-1/2}}^{r_{k+1/2}} (Y_{l+1/2} - Y_{l-1/2})\, r^{\alpha}\, dr = (Y_{k,\, l+1/2} - Y_{k,\, l-1/2})\, \Delta r_k r_k^{\alpha},\\
&\int\limits_{r_{k-1/2}}^{r_{k+1/2}} \int\limits_{z_{l-1/2}}^{z_{l+1/2}} (\Sigma\varphi - f)\, r^{\alpha}\, dz\, dr = (\Sigma \Delta r \Delta z)_{kl}\, \varphi_{kl} r_k^{\alpha} - (f \Delta r \Delta z)_{kl}\, r_k^{\alpha},
\end{aligned}\right\} \tag{33.6}$$

where

$$\begin{aligned}
(\Phi)_{kl} = \frac{1}{4} \Big[&\Phi (r_k + 0,\ z_l + 0) + \Phi (r_k - 0,\ z_l + 0) + \\
&+ \Phi (r_k + 0,\ z_l - 0) + \Phi (r_k - 0,\ z_l - 0) \Big],
\end{aligned} \tag{33.7}$$

and

$$\Delta r\,|_{r_k+0} = \Delta r_{k+1/2} = r_{k+1} - r_k, \quad \Delta z\,|_{z_l+0} = \Delta z_{l+1/2} = z_{l+1} - z_l \tag{33.8}$$

Using (33.6), we transform (33.5) to the form

$$\begin{aligned}
(J_{k+1/2,\, l} - J_{k-1/2,\, l})\, \Delta z_l + (Y_{k,\, l+1/2} - Y_{k,\, l-1/2})\, r_k^{\alpha} \Delta r_k - (\Sigma \Delta r\, \Delta z)_{kl} \varphi_{kl} r_k^{\alpha} = \\
= -(f \Delta r\, \Delta z)_{kl}\, r_k^{\alpha}.
\end{aligned} \tag{33.9}$$

Let us now use (33.3) and (33.4) to eliminate J and Y from (33.9). We do this by first solving for the derivative in the former equations and integrating the first over ($r_k < r < r_{k+1};\ z_{l-1/2} < z < z_{l+1/2}$), and the second over ($r_{k-1/2} < r < r_{k+1/2};\ z_l < z < z_{l+1}$). Then, bearing in mind the behavior of the integrand and the assumptions of the simplified scheme, we arrive at

$$\left.\begin{aligned}
\varphi_{k+1,\, l} - \varphi_{kl} &= \left(\frac{\Delta z}{D}\right)_{k+1/2,\, l} \frac{\Delta r_{k+1/2}}{\Delta z_l} \frac{J_{k+1/2,\, l}}{r_{k+1/2}^{\alpha}},\\
\varphi_{k,\, l+1} - \varphi_{kl} &= \left(\frac{\Delta r}{D}\right)_{k,\, l+1/2} \frac{\Delta z_{l+1/2}}{\Delta r_k} Y_{k,\, l+1/2}.
\end{aligned}\right\} \tag{33.10}$$

It follows from (33.10) that

$$\left.\begin{aligned}
J_{k+1/2,\, l} &= r_{k+1/2}^{\alpha} \frac{\Delta z_l}{\Delta r_{k+1/2}} \frac{\varphi_{k+1,\, l} - \varphi_{kl}}{\left(\dfrac{\Delta z}{D}\right)_{k+1/2,\, l}},\\
Y_{k,\, l+1/2} &= \frac{\Delta r_k}{\Delta z_{l+1/2}} \frac{\varphi_{k,\, l+1} - \varphi_{kl}}{\left(\dfrac{\Delta r}{D}\right)_{k,\, l+1/2}},
\end{aligned}\right\} \tag{33.11}$$

where

$$(\Phi)_{k+1/2,\, l} = \frac{1}{2} [\Phi (r_{k+1/2},\ z_l + 0) + \Phi (r_{k+1/2},\ z_l - 0)],$$

$$(\Phi)_{k,\, l+1/2} = \frac{1}{2} [\Phi (r_k + 0,\ z_{l+1/2}) + \Phi (r_k - 0,\ z_{l+1/2})].$$

Now we insert (33.11) into (33.9). As a result we obtain the finite-difference diffusion equation

$$L(\varphi_{kl}) + \alpha \cdot \frac{\Delta r_{k+1/2}}{2r_k} N(\varphi_{kl}) - \Lambda_{kl}\varphi_{kl} = -F_{kl}, \tag{33.12}$$

where

$$\left.\begin{aligned} L(\varphi_{kl}) &= \mu^l_{k+1/2}\varphi_{k+1,l} + \mu^l_{k-1/2}\varphi_{k-1,l} + \mu^k_{l+1/2}\varphi_{k,l+1} + \mu^k_{l-1/2}\varphi_{k,l-1}, \\ N(\varphi_{kl}) &= \mu^l_{k+1/2}\varphi_{k+1,l} - \frac{\Delta r_{k-1/2}}{\Delta r_{k+1/2}}\mu^l_{k-1/2}\varphi_{k-1,l}, \\ \Lambda_{kl} &= \mu^l_{k+1/2} + \mu^l_{k-1/2} + \mu^k_{l+1/2} + \mu^k_{l-1/2} + \alpha\frac{\Delta r_{k+1/2}}{2r_k}\Big(\mu^l_{k+1/2} - \\ &\quad - \frac{\Delta r_{k-1/2}}{\Delta r_{k+1/2}}\mu^l_{k-1/2}\Big) + (\Sigma\Delta r\,\Delta z)_{kl}, \\ F_{kl} &= (f\Delta r\,\Delta z)_{kl}, \end{aligned}\right\} \tag{33.13}$$

$$\mu^l_{k+1/2} = \frac{\Delta z_l^2}{\Delta r_{k+1/2}\left(\frac{\Delta z}{D}\right)_{k+1/2,\,l}}, \qquad \mu^k_{l+1/2} = \frac{\Delta r_k^2}{\Delta z_{l+1/2}\left(\frac{\Delta r}{D}\right)_{k,\,l+1/2}}. \tag{33.14}$$

It is useful to note that, far from discontinuity points, (33.13) and (33.14) take on the very simple form

$$\left.\begin{aligned} L(\varphi_{kl}) &= D(\varphi_{k+1,l} + \varphi_{k-1,l} + \varphi_{k,l+1} + \varphi_{k,l-1}), \\ N(\varphi_{kl}) &= D(\varphi_{k+1,l} - \varphi_{k-1,l}), \\ \Lambda_{kl} &= D\left[4 + \left(\frac{\Sigma}{D}\right)_{kl} h^2\right], \\ F_{kl} &= f_{kl}h^2, \end{aligned}\right\} \tag{33.15}$$

where

$$h = \Delta r = \Delta z.$$

At discontinuity points, however, the unknowns φ_{kl} must be calculated using (33.12-33.14).

Let us now go on to consider the simplest finite-difference diffusion equation when the discontinuities of D, Σ and $f(r, z)$ lie at internal points of the ($r_k < r < r_{k+1}$; $z_l < z < z_{l+1}$) regions, namely on the line given by $r_{k+1/2}$ and $z_{l+1/2}$.

As before, we will derive the finite-difference equation by using (33.5). Noting that φ in the integrand is continuous, we arrive at the approximation

$$\begin{aligned} (J_{k+1/2,\,l} - J_{k-1/2,\,l})\,\Delta z_l + (Y_{k,\,l+1/2} - Y_{k,\,l-1/2})\,\Delta r_k r_k^{\alpha} - \Sigma_{kl}\varphi_{kl} r_k^{\alpha}\Delta r_k\,\Delta z_l = \\ = -f_{kl} r_k^{\alpha}\Delta r_k \Delta z_l. \end{aligned} \tag{33.16}$$

We must now write J and Y in (33.16) in terms of the values of φ at the lattice points. To do this we again consider (33.3) and (33.4). We solve these for the derivatives and then integrate the first over ($r_k < r < r_{k+1}$; $z_{l-1/2} < z < z_{l+1/2}$), and the second over ($r_{k-1/2} < r < r_{k+1/2}$; $z_l < z < z_{l+1}$), which gives

$$\left.\begin{aligned} \int_{z_{l-1/2}}^{z_{l+1/2}} (\varphi_{k+1} - \varphi_k)\,dz &= \int_{z_{l-1/2}}^{z_{l+1/2}} dz \int_{r_k}^{r_{k+1}} \frac{J}{r^{\alpha}D}\,dr, \\ \int_{r_{k-1/2}}^{r_{k+1/2}} (\varphi_{l+1} - \varphi_l)\,dr &= \int_{r_{k-1/2}}^{r_{k+1/2}} dr \int_{z_l}^{z_{l+1}} \frac{Y}{D}\,dz. \end{aligned}\right\} \tag{33.17}$$

With the assumptions we have been making, we arrive at

$$\left.\begin{aligned} J_{k+1/2,\, l} &= \frac{r^{\alpha}_{k+1/2}}{\left(\frac{\Delta r}{D_l}\right)_{k+1/2}} (\varphi_{k+1,\, l} - \varphi_{k,\, l}), \\ Y_{k,\, l+1/2} &= \frac{1}{\left(\frac{\Delta z}{D_k}\right)_{l+1/2}} (\varphi_{k,\, l+1} - \varphi_{kl}), \end{aligned}\right\} \tag{33.18}$$

where we have again written

$$\left.\begin{aligned} (\Phi)_{k+1/2} &= \frac{1}{2}\left[\Phi(r_{k+1/2}+0) + \Phi(r_{k+1/2}-0)\right], \\ (\Phi)_{l+1/2} &= \frac{1}{2}\left[\Phi(z_{l+1/2}+0) + \Phi(z_{l+1/2}-0)\right], \end{aligned}\right\} \tag{33.19}$$

We now insert (33.18) into (33.16). This leads to the finite-difference equation

$$L(\varphi_{kl}) + \alpha\frac{\Delta r_k}{2r_k} N(\varphi_{kl}) - \Lambda_{kl}\varphi_{kl} = -F_{kl}, \tag{33.20}$$

where

$$\left.\begin{aligned} L(\varphi_{kl}) &= \mu^l_{k+1/2}\varphi_{k+1,\, l} + \mu^l_{k-1/2}\varphi_{k-1,\, l} + \mu^k_{l+1/2}\varphi_{k,\, l+1} + \mu^k_{l-1/2}\varphi_{k,\, l-1}, \\ N(\varphi_{kl}) &= \mu^l_{k+1/2}\varphi_{k+1,\, l} - \mu^l_{k-1/2}\varphi_{k-1,\, l}, \\ \Lambda_{kl} &= \mu^l_{k+1/2} + \mu^l_{k-1/2} + \mu^k_{l+1/2} + \mu^k_{l-1/2} + \\ &\quad + \alpha\frac{\Delta r_k}{2r_k}(\mu^l_{k+1/2} - \mu^l_{k-1/2}) + \Sigma_{kl}\Delta r_k \Delta z_l, \\ F_{kl} &= f_{kl}\Delta r_k \Delta z_l, \end{aligned}\right\} \tag{33.21}$$

$$\mu^l_{k+1/2} = \frac{\Delta z_l}{\left(\frac{\Delta r}{D_l}\right)_{k+1/2}}, \quad \mu^k_{l+1/2} = \frac{\Delta r_k}{\left(\frac{\Delta z}{D_k}\right)_{l+1/2}}. \tag{33.22}$$

If $\Delta z = \Delta r = h$ and the points at which we are making the calculation lie no less than 1.5 interpolation intervals from a line of discontinuity, Equation (33.21) simplifies greatly and takes on the form of (33.15).

In order to obtain a completely specified problem corresponding to Equations (33.12) and (33.20), let us consider the boundary conditions. They are

$$\left.\begin{aligned} J_{1/2,\, l} = 0, \quad Y_{k,\, 1/2} = 0, \\ \varphi(r_k, H_e) = 0, \quad \varphi(R_e\; z_l) = 0. \end{aligned}\right\} \tag{33.23}$$

Here it is assumed that the centers of the first interpolation intervals Δr and Δz are at $z = 0$ and $r = 0$.

In order to obtain a highly accurate solution of the diffusion equation with a small number of interpolation intervals, we must use a more accurate finite-difference system, which accounts in more detail for the behavior of the functions and parameters in the neighborhood of a line of discontinuity. Let us therefore consider the improved finite-difference equation for the case in which the points of discontinuity lie on $r = r_k$ or $z = z_l$. First let us consider (33.5) in the form

$$(J_{k+1/2,\, l} - J_{k-1/2,\, l})\Delta z_l + (Y_{k,\, l+1/2} - Y_{k,\, l-1/2}) r^{\alpha}_k \Delta r_k = \\ = \int\limits_{r_{k-1/2}}^{r_{k+1/2}} \int\limits_{z_{l-1/2}}^{z_{l+1/2}} (\Sigma\varphi - f) r^{\alpha}\, dr\, dz. \tag{33.24}$$

Of the terms in (33.24), let us consider in particular

$$\int_{r_{k-1/2}}^{r_{k+1/2}} r^\alpha\, dr \int_{z_{l-1/2}}^{z_{l+1/2}} \Sigma\varphi\, dz \quad \text{and} \int_{r_{k-1/2}}^{r_{k+1/2}} r^\alpha\, dr \int_{z_{l-1/2}}^{z_{l+1/2}} f\, dz. \tag{33.25}$$

We first treat these expressions for points lying on a line of discontinuity $r = r_k$. We assume that $\Sigma\varphi$ and f are linear in $\underline{z}$. This leads to two new integrals

$$\int_{r_{k-1/2}}^{r_{k+1/2}} \Sigma_l \varphi_l r^\alpha \Delta z_l dr \quad \text{and} \int_{r_{k-1/2}}^{r_{k+1/2}} f_l r^\alpha \Delta z_l\, dr. \tag{33.26}$$

Since by assumption Σ_l and f_l are discontinuous at r_k, we can break up the integrals over $(r_{k-1/2}, r_{k+1/2})$. We then have

$$\int_{r_{k-1/2}}^{r_{k+1/2}} = \int_{r_{k-1/2}}^{r_k} + \int_{r_k}^{r_{k+1/2}}$$

and within the limits of integration the integrands will then be considered linear functions of $\underline{r}$. Then as in the one-dimensional case, we obtain

$$\left.\begin{aligned}
\int_{r_{k-1/2}}^{r_{k+1/2}} r^\alpha\, dr \int_{z_{l-1/2}}^{z_{l+1/2}} \Sigma\varphi\, dz &= \left\{ (\Sigma_l \Delta r)_k + \frac{\alpha}{8r_k} [\Sigma_l \Delta r^2]_k \right\} \varphi_{kl} r_k^\alpha \Delta z_l + \\
&\quad + \frac{1}{8} \left[\frac{\Sigma_l \Delta r^2}{D_l} \right]_k \Delta z_l J_{kl}, \\
\int_{r_{k-1/2}}^{r_{k+1/2}} r^\alpha\, dr \int_{z_{l-1/2}}^{z_{l+1/2}} f\, dz &= \left\{ (f_l \Delta r)_k + \frac{\alpha}{8r_k} [f_l \Delta r^2]_k \right\} r_k^\alpha \Delta z_l + \\
&\quad + \frac{1}{8} \left[\frac{\partial f_l}{\partial r} \Delta r^2 \right]_k r_k^\alpha \Delta z_l.
\end{aligned}\right\} \tag{33.27}$$

We have here used the notation

$$(\Phi_l)_k = \frac{1}{2} [\Phi_l (r_k + 0) + \Phi_l (r_k - 0)], \quad [\Phi_l]_k = \Phi_l (r_k + 0) - \Phi_l (r_k - 0).$$

Similarly, taking $z = z_l$ as a line of discontinuity, we have

$$\left.\begin{aligned}
\int_{z_{l-1/2}}^{z_{l+1/2}} dz \int_{r_{k-1/2}}^{r_{k+1/2}} \Sigma\varphi r^\alpha\, dr &= (\Sigma_k \Delta z)_l \Delta r_k r_k^\alpha \varphi_{kl} + \\
&\quad + \frac{1}{8} \left[\frac{\Sigma_k \Delta z^2}{D_k} \right]_l \Delta r_k r_k^\alpha Y_{kl}, \\
\int_{z_{l-1/2}}^{z_{l+1/2}} dz \int_{r_{k-1/2}}^{r_{k+1/2}} f r^\alpha\, dr &= (f_k \Delta z)_l \Delta r_k r_k^\alpha + \frac{1}{8} \left[\frac{\partial f_k}{\partial z} \Delta z^2 \right]_l r_k^\alpha \Delta r_k
\end{aligned}\right\} \tag{33.28}$$

If we now assume that

$$J_{hl}=\frac{1}{2}(J_{h+1/2,l}+J_{h-1/2,l}),\ Y_{hl}=\frac{1}{2}(Y_{h,l+1/2}+Y_{h,l-1/2}), \tag{33.29}$$

we arrive at the set of finite-difference diffusion equations

$$L(\varphi_{hl})+\alpha\frac{\Delta r_{h+1/2}}{2r_h}N(\varphi_{hl})-\Lambda_{hl}\varphi_{hl}=-F_{hl}, \tag{33.30}$$

where

$$\left.\begin{aligned} L(\varphi_{hl}) &= A_{hl}\varphi_{h+1,l}+B_{hl}\varphi_{h-1,l}+C_{hl}\varphi_{h,l+1}+D_{hl}\varphi_{h,l-1},\\ N(\varphi_{hl}) &= A_{hl}\varphi_{h+1,l}-\frac{\Delta r_{h-1/2}}{\Delta r_{h+1/2}}B_{hl}\varphi_{h-1,l},\\ \Lambda_{hl} &= A_{hl}+B_{hl}+C_{hl}+D_{hl}+(\Sigma\Delta r\Delta z)_{hl}+R_{hl}, \end{aligned}\right\} \tag{33.31}$$

$$\left.\begin{aligned} R_{hl} &= \alpha\left\{\frac{\Delta z_l}{8r_h}[\Sigma_l\Delta r^2]_h+\frac{\Delta r_{h+1/2}}{2r_h}\left(A_{hl}-\frac{\Delta r_{h-1/2}}{\Delta r_{h+1/2}}B_{hl}\right)\right\},\\ A_{hl} &= (1-\lambda_h^l)\,\mu_{h+1/2}^l,\quad B_{hl}=(1+\lambda_h^l)\,\mu_{h-1/2}^l,\\ C_{hl} &= (1-\lambda_l^h)\,\mu_{l+1/2}^h,\quad D_{hl}=(1+\lambda_l^h)\,\mu_{l-1/2}^h,\\ \lambda_h^l &= \frac{1}{16}\left[\frac{\Sigma_l\Delta r^2}{D_l}\right]_h,\quad \lambda_l^h=\frac{1}{16}\left[\frac{\Sigma_h\Delta z^2}{D_h}\right]_l,\\ F_{hl} &= (f\cdot\Delta r\cdot\Delta z)_{hl}+\alpha\frac{\Delta z_l}{8r_h}[f_l\Delta r^2]_h+\\ &\quad+\frac{1}{8}\left\{\left[\frac{\partial f_l}{\partial r}\Delta r^2\right]_h+\left[\frac{\partial f_h}{\partial z}\Delta z^2\right]_l\right\}. \end{aligned}\right\} \tag{33.32}$$

Away from discontinuity points, (33.31) and (33.32) go over into (33.15).

34. The Method of Conditional Separation of Variables

In discussing cylindrical reactor calculations, one cannot fail to mention the method of conditional separation of variables, which has been widely used in critical mass calculations.

Essentially the method consists of reducing the calculations for a cylindrical reactor to the calculation of two cylindrical reactors, one of which has no reflector at the ends, and the other of which has none around the circumference.

When the problem is so stated, one must be careful that the neutron spectrum in the core of the reactor far from the boundary with the reflector is not altered by the absence of the reflector. This can be corrected for by an increase in the effective dimension of the core in the direction of the eliminated reflector. The difference between the effective and true dimensions of the cylinder will be called the radial increment δR and the increment in height δH. Essentially the problem involves finding these quantities such that the critical mass of both imaginary cylindrical reactors of the effective dimensions be approximately equal to the critical mass of the original reactor.

It is not difficult to find the critical mass of the imaginary reactors, since for them the variables can be separated. Indeed, in the reactor without end reflectors we may write*

$$\left.\begin{aligned} \varphi(r,z,u) &= \varphi(r,u)\cos\frac{\pi}{H_e}z,\\ \Phi(r,z) &= \Phi(r)\cos\frac{\pi}{H_e}z, \end{aligned}\right\} \tag{34.1}$$

where $H_e = H + \delta H$, and H is the height of the cylindrical core. For the reactor without a reflector at the circumference, we have

*We have here made the assumption that the extrapolated boundary is independent of u.

$$\left.\begin{aligned}\varphi(r, z, u) &= \varphi(z, u)\, J_0\left(\frac{\varkappa}{R_e} r\right), \\ \Phi(r, z) &= \Phi(z)\, J_0\left(\frac{\varkappa}{R_e} r\right),\end{aligned}\right\} \tag{34.2}$$

where $R_e = R + \delta R$, and R is the true radius of the core of the reactor, with $\kappa = 2.405$. Thus, by separating the variables we obtain one-dimensional problems which are easily solved by the methods described above. It should be borne in mind that in the one-dimensional problems the capture cross section Σ_c should everywhere be replaced by $\Sigma_c + (\pi/H_e)^2 D$ for the one-dimensional cylindrical reactor, and $\Sigma_c + (\kappa/R_e)^2 D$ for the one-dimensional plane reactor.

To find δR and δH we use the following successive approximation method.

Consider a cylindrical reactor without end reflectors whose radial dimensions are the same as those of the original one. Choosing some equivalent height H_e of this reactor and, finding K_{eff} according to (20.9), we can find the extrapolated boundary $R_e = R + \delta R$, for a reactor which has no reflector on the circumference but which has the same value of K_{eff}.

Let us consider in addition another one-dimensional reactor, whose height is that of the real one and whose peripheral reflector is replaced by one with the increment calculated above. If we now find K_{eff} for this reflector, we can use Equation (20.9) to calculate the increment in height, etc.

These calculations should be continued until the same limiting value of K_{eff} is found for both reflectors. Then this value of K_{eff} will be the required approximation of the eigenvalue.

In order to find the critical mass of the reactor, one must perform similar calculations either for cores of different size with a fixed fuel concentration, or for a reactor of fixed dimensions and for different fuel concentrations. The desired critical mass is obtained when the uranium concentration is that which gives $K_{eff} = 1$.

In conclusion, it must be emphasized that the conditional separation of variables gives very good agreement with the exact solution for K_{eff}. The neutron fields, on the other hand, are not so well reproduced. This method leads to particularly large errors in the neighborhood of "corners."

Thus the method of conditional separation of variables can be used to find the critical mass of a reactor.

Chapter IX

SOLUTION OF THE FINITE-DIFFERENCE DIFFUSION EQUATIONS

An important stage in multigroup reactor calculations is the solution of the finite-difference diffusion equations. In fact, the practical effectiveness of the multigroup method is determined by the way in which the finite difference equations are solved.

The direct solution of the second-order finite-difference equations by the methods of linear algebra is complicated by the rapid accumulation of rounding-off errors. Primarily, this is due to the fact that a second-order difference equation has two eigenvalues,one of which is positive. Therefore, the errors are propagated from interpolation point to interpolation point exponentially with positive eigenvalues.

The difficulties in solving one-dimensional finite-difference equations of the diffusion type have been eliminated only by the difference factorization method proposed by Gel'fand and Lokutsievskii [47] and independently by Kronrod (see [20]) and Stark (see [101]). The method is based on factorization of ordinary differential equations as discussed by Ince [2] and later developed for diffusion equations by Thompson [138] and Vladimirov [14].

The present chapter will describe methods for solving finite-difference diffusion equations for both one-dimensional and two-dimensional regions.

35. Solution of Finite-Difference Diffusion Equations in a One-Dimensional Region

In the preceding chapter we obtained several finite-difference diffusion equations, and we will now solve them. This will be done using the method of difference factorization. Essentially, this involves solving a second-order difference equation by successively solving a set of simpler first-order difference equations.

Thus, consider the finite-difference diffusion equation

$$a_k\varphi_{k+1} - b_k\varphi_k + c_k\varphi_{k-1} = -f_k, \tag{35.1}$$

where a_k, b_k, c_k and f_k are given positive numbers. We now divide this equation by a_k. This gives

$$\varphi_{k+1} - B_k\varphi_k + C_k\varphi_{k-1} = -F_k, \tag{35.2}$$

where

$$B_k = \frac{b_k}{a_k}, \quad C_k = \frac{c_k}{a_k}, \quad F_k = \frac{f_k}{a_k}. \tag{35.3}$$

Equation (35.2) can be written in the form*

$$(\nabla + \sigma_{k+1})(I_{k+1} + \lambda_{k+1}\varphi_{k+1}) = -F_k, \tag{35.4}$$

$$\nabla\Phi_{k+1} = \Phi_{k+1} - \Phi_k, \quad I_{k+1} = \varphi_{k+1} - C_{k+1}\varphi_k, \tag{35.5}$$

where σ_k and λ_k are yet to be determined.

We now write out the operator on the left side of (35.4). This gives the finite-difference equation

* The present method for deriving the factored system was suggested by A.A. Samarskii [70].

$$[1+\sigma_{k+1}+\lambda_{k+1}(1+\sigma_{k+1})]\varphi_{k+1}-$$
$$-[1+\lambda_k+C_{k+1}(1+\sigma_{k+1})]\varphi_k+C_k\varphi_{k-1}=-F_k. \tag{35.6}$$

We now choose the parameters λ_k and σ_k so that Equation (35.6) be identically equal to (35.2). This is done, obviously, by requiring that

$$\left.\begin{aligned}\lambda_{k+1}(1+\sigma_{k+1})+\sigma_{k+1}=0,\\ C_{k+1}(1+\sigma_{k+1})+(1+\lambda_k)=B_k.\end{aligned}\right\} \tag{35.7}$$

From the first of (35.7) we have

$$\lambda_{k+1}=\beta_{k+1}-1, \tag{35.8}$$

where

$$\beta_{k+1}=\frac{1}{1+\sigma_{k+1}}. \tag{35.9}$$

Inserting (35.8) and (35.9) into the second of (35.7), we arrive at

$$\beta_{k+1}=\frac{C_{k+1}}{B_k-\beta_k}. \tag{35.10}$$

Let us now consider (35.4), which we can write as a set of two equations, namely

$$\left.\begin{aligned}I_{k+1}+\lambda_{k+1}\varphi_{k+1}=-z_{k+1},\\ z_{k+1}-z_k+\sigma_{k+1}z_{k+1}=F_k.\end{aligned}\right\} \tag{35.11}$$

Bearing in mind (35.5), we can write the first of Equations (35.11) in the form

$$\varphi_{k+1}-C_{k+1}\varphi_k+(\beta_{k+1}-1)\varphi_{k+1}=-z_{k+1},$$

or

$$\varphi_k=\frac{\beta_{k+1}\varphi_{k+1}+z_{k+1}}{C_{k+1}}. \tag{35.12}$$

Further, since

$$\sigma_{k+1}=\frac{1}{\beta_{k+1}}-1,$$

the second of Equations (35.11) can be written in the form

$$z_{k+1}=\beta_{k+1}(z_k+F_k). \tag{35.13}$$

Thus the finite-difference diffusion equation has been represented in the form of a set of three difference equations, namely

$$\left.\begin{aligned}\beta_{k+1}=\frac{C_{k+1}}{B_k-\beta_k},\\ z_{k+1}=\beta_{k+1}(z_k+F_k),\\ \varphi_k=\frac{\beta_{k+1}\varphi_{k+1}+z_{k+1}}{C_{k+1}}.\end{aligned}\right\} \tag{35.14}$$

It can be shown [47] that if one bases the calculations on Equations (35.14), the rounding-off errors will not increase from interpolation point to interpolation point, and therefore the computation will be stable. This is particularly simple to see when the coefficients a_k, b_k and c_k in the difference equation (35.1) are independent of the index k. For simplicity, consider the case of a plane region. In this case Equation (35.1) becomes

$$\varphi_{k+1} - \left(2 + \frac{\Sigma}{D}\Delta x^2\right)\varphi_k + \varphi_{k-1} = -f_k,$$

and therefore,

$$B_k = 2 + \frac{\Sigma}{D}\Delta x^2, \quad C_k = 1 \qquad (k = 1,\ 2,\ \ldots).$$

We then have, using the first of (35.14),

$$\beta_{k+1} = \frac{1}{2 + \frac{\Sigma}{D}\Delta x^2 - \beta_k}.$$

It can be shown that β_k has an upper limit as $k \to \infty$, namely

$$\beta_k \leqslant 1 + \frac{\Sigma}{2D}\Delta x^2 - \sqrt{\frac{\Sigma}{D}\Delta x + \left(\frac{\Sigma}{2D}\Delta x\right)^2} = q < 1.$$

This means that in solving the second and third equations of (35.14) the rounding-off error at an interpolation point will decay according to

$$e_n = q^{n-k} e_k.$$

Lokutsievskii [47] has given a general proof of the stability of the solution based on the factored system.

Let us now define the initial values of β_k, z_k and φ_k so as to satisfy the boundary conditions for the diffusion equation. To do this, we distinguish between two cases. In the first case we require that

$$\varphi_0 = \varphi_1, \ \varphi_m = 0, \tag{35.15}$$

and in the second, that

$$\varphi_0 = 0, \ \varphi_m = 0. \tag{35.16}$$

We recall that the first of these corresponds to the boundary condition

$$\left.\frac{d\varphi}{dr}\right|_{r=0} = 0,$$

and that the second holds for spherical regions in which the unknown function is $r\varphi$.

In order to satisfy Conditions (35.15) we must choose the initial conditions in the form

$$\beta_1 = G_1, \ z_1 = 0, \ \varphi_m = 0. \tag{35.17}$$

When using Conditions (35.16) we must set

$$\beta_1 = 0, \ z_1 = 0, \ \varphi_m = 0. \tag{35.18}$$

Let us now turn our attention to the following situation. As will be shown in Chapter XII, when one attempts to calculate the reactivity equivalent of a control rod in the center of the core of a cylindrical reactor without a reflector around its circumference, one is forced to solve the multigroup set of diffusion equations for which the boundary condition on the rod surface ($r = r_0$) is

$$\frac{1}{\varphi}\frac{d\varphi}{dr}\Big|_{r=r_0} = \frac{1}{\gamma}, \tag{35.19}$$

where $1/\gamma$ is a quantity which is calculated by means of the asymptotic solution of the kinetic equation.

Let us divide up the radius of the reactor core into intervals of length Δr so that the boundary point $r = r_0$ of our rod lies at the center of the first interval. Then the boundary condition (35.19) can be written

$$\varphi_0 = \frac{1 - \frac{\Delta r}{2\gamma}}{1 + \frac{\Delta r}{2\gamma}}\varphi_1. \tag{35.20}$$

Further, let us approximate the diffusion equation by a finite-difference equation which we will attempt to solve by the difference factorization method. We must, however, take care that our solution satisfies Condition (35.20) at $r = r_0$. Assuming that at the outer extrapolated boundary of the reactor $\varphi = 0$, it is easily shown that all the conditions will be satisfied by writing

$$z_1 = 0,\ \beta_1 = \frac{1 - \frac{\Delta r}{2\gamma}}{1 + \frac{\Delta r}{2\gamma}} C_1,\ \varphi_m = 0. \tag{35.21}$$

So far we have always assumed that at the outer extrapolated boundary of the reactor the neutron density vanishes. In many cases, however, it is often convenient to use not the extrapolated boundary, which is a function of the energy, but the true boundary. In this case, a condition analogous to that for the control rod is set up at the boundary, namely

$$\frac{1}{\varphi}\frac{d\varphi}{dr} = -\frac{1}{\gamma}, \tag{35.22}$$

where*

$$\gamma = \frac{2}{3}\lambda_{tr}.$$

As in the case of a control rod, let us write (35.22) in finite-difference form. To do this, we pick the difference net so that the boundary point $r = R$ be at the center of the last interval. Then in terms of central differences, Equation (35.22) can be written in the form

$$\varphi_{m-1} = \frac{1 + \frac{\Delta r}{2\gamma}}{1 - \frac{\Delta r}{2\gamma}}\varphi_m, \tag{35.23}$$

where r_m lies a half of an interval away from the reactor boundary.

Condition (35.23) cannot be used to find the value of the desired function at r_m, a quantity which is needed if we are to calculate φ_k. We therefore add the condition

$$\varphi_{m-1} = \frac{\beta_m\varphi_m + z_m}{C_m}, \tag{35.24}$$

which is obtained with the aid of (35.14).

---

* See Section 14.

In view of the fact that the unknowns in (35.24) are only φ_{m-1} and φ_m, we can combine it with the boundary condition (35.23) to arrive at a set of equations whose solution gives

$$\varphi_m = \frac{z_m}{\dfrac{1+\dfrac{\Delta r}{2\gamma}}{1-\dfrac{\Delta r}{\Delta\gamma}} C_m - \beta_m}. \tag{35.25}$$

We have thus obtained the additional boundary condition necessary to solve the third of Equations (35.14).

In conclusion, it should be mentioned that the difference factorization method is very general and that computers are easily programmed for this method.

36. Solution of Finite-Difference Diffusion Equations in a Two-Dimensional Region

Let us now proceed to a consideration of finite-difference diffusion equations in two-dimensional regions. We will first direct our attention to a method based on matrix factorization.*

Consider Equation (33.20), the finite-difference diffusion equation for a two-dimensional plane or cylindrically symmetric region. We will write it in the form

$$a_{kl}\varphi_{k+1,\,l} + b_{kl}\varphi_{k,\,l-1} + c_{kl}\varphi_{k-1,\,l} + d_{kl}\varphi_{k,\,l+1} - p_{kl}\varphi_{kl} = -f_{kl}, \tag{36.1}$$

where the coefficients a_{kl}, b_{kl}, c_{kl}, d_{kl} and p_{kl} are given positive numbers satisfying the condition

$$a_{kl} + b_{kl} + c_{kl} + d_{kl} \leqslant p_{kl}, \tag{36.2}$$

the inequality being satisfied at least at one interpolation point of the (k, l) net.

To (36.1) we must add the boundary conditions

$$\varphi_{0l} = \varphi_{1l}, \qquad \varphi_{n+1,\,l} = 0 \qquad (l = 1,\ 2,\ \ldots,\ m), \tag{36.3}$$

and

$$\varphi_{k0} = \varphi_{k1}, \qquad \varphi_{k,\,m+1} = 0 \qquad (k = 1,\ 2,\ \ldots,\ n). \tag{36.4}$$

The first of each of (36.3) and (36.4) are symmetry conditions with respect to the r and z axes, and the second are consequences of the fact that the solution vanishes at the external boundary S.

We shall write the finite-difference equation (36.1) with Condition (36.4) in the matrix form:

$$\mathbf{a}_k\varphi_{k+1} - \mathbf{b}_k\varphi_k + \mathbf{c}_k\varphi_{k-1} = -\mathbf{f}_k, \tag{36.5}$$

where

$$\varphi_k = \begin{vmatrix} \varphi_{k1} \\ \varphi_{k2} \\ \cdots \\ \varphi_{km} \end{vmatrix}, \qquad \mathbf{f}_k = \begin{vmatrix} f_{k1} \\ f_{k2} \\ \cdots \\ f_{km} \end{vmatrix}$$

are vectors, and where $\mathbf{a}_k$, $\mathbf{b}_k$ and $\mathbf{c}_k$ are the matrices

* The author is deeply grateful to M.V. Keldysh, I.M. Gel'fand, and O.V. Lokutsievskii, as well as to K.I. Babenko and N.N. Chentsov for allowing him to publish the method of matrix factorization in the present monograph.

$$\mathbf{a}_k = \begin{Vmatrix} a_{k1} & 0 & \ldots & 0 \\ 0 & a_{k2} & \ldots & 0 \\ \cdot & \cdot & \cdot & \cdot \\ 0 & 0 & \ldots & a_{km} \end{Vmatrix}, \qquad \mathbf{c}_k = \begin{Vmatrix} c_{k1} & 0 & \ldots & 0 \\ 0 & c_{k2} & \ldots & 0 \\ \cdot & \cdot & \cdot & \cdot \\ 0 & 0 & \ldots & c_{km} \end{Vmatrix}.$$

$$\mathbf{b}_k = \begin{Vmatrix} b_{k1}-p_{k1} & d_{k1} & 0 & 0 & \ldots 0 & 0 \\ b_{k2} & -p_{k2} & d_{k2} & 0 & \ldots 0 & 0 \\ 0 & b_{k3} & -p_{k3} & d_{k3} & \ldots 0 & 0 \\ \cdot & \cdot & \cdot & \cdot & \cdot & \cdot \\ 0 & 0 & 0 & 0 & b_{km} & -p_{km} \end{Vmatrix}.$$

Let us multiply (36.5) on the left by $\mathbf{a}_k^{-1}$. We then obtain

$$\boldsymbol{\varphi}_{k+1} - \boldsymbol{B}_k\boldsymbol{\varphi}_k + \boldsymbol{C}_k\boldsymbol{\varphi}_{k-1} = -\boldsymbol{F}_k, \tag{36.6}$$

where

$$\boldsymbol{B}_k = \boldsymbol{a}_k^{-1}\boldsymbol{b}_k, \quad \boldsymbol{C}_k = \boldsymbol{a}_k^{-1}\boldsymbol{c}_k, \quad \boldsymbol{F}_k = \boldsymbol{a}_k^{-1}\boldsymbol{f}_k.$$

To (36.6) we add the boundary conditions which, using (36.3), are

$$\left.\begin{aligned} \boldsymbol{\varphi}_0 &= \boldsymbol{\varphi}_1, \\ \boldsymbol{\varphi}_{n+1} &= 0. \end{aligned}\right\} \tag{36.7}$$

Now (36.6) and (36.7) form a complete self-consistent set of equations. Formally it is a set of difference equations analogous to those of the one-dimensional case, except that the solution is now a vector, and the coefficients are matrices. For these reasons (36.6) must be factored in accordance with the rules of matrix calculus. Let us, therefore, write it in the form

$$(\nabla + \boldsymbol{\sigma}_{k+1})(\boldsymbol{I}_{k+1} + \boldsymbol{\mu}_{k+1}\boldsymbol{\varphi}_{k+1}) = -\boldsymbol{F}_k, \tag{36.8}$$

where $\mathbf{I}_{k+1} = \boldsymbol{\varphi}_{k+1} - \mathbf{C}_{k+1}\boldsymbol{\varphi}_k$, and ∇ is the difference operator defined by

$$\nabla\psi_{k+1} = \psi_{k+1} - \psi_k,$$

and $\boldsymbol{\sigma}_{k+1}$ and $\boldsymbol{\mu}_{k+1}$ are matrices whose form will be found from the condition that Equations (36.6) and (36.8) are equivalent.

Writing out the operator on the left side of (36.8), we arrive at

$$\begin{gathered}(\boldsymbol{E} + \boldsymbol{\mu}_{k+1} + \boldsymbol{\sigma}_{k+1} + \boldsymbol{\sigma}_{k+1}\boldsymbol{\mu}_{k+1})\,\boldsymbol{\varphi}_{k+1} - (\boldsymbol{E} + \boldsymbol{C}_{k+1} + \boldsymbol{\mu}_k + \boldsymbol{\sigma}_{k+1}\boldsymbol{C}_{k+1})\boldsymbol{\varphi}_k + \\ + \boldsymbol{C}_k\boldsymbol{\varphi}_{k-1} = -\boldsymbol{F}_k,\end{gathered} \tag{36.9}$$

where $\mathbf{E}$ is the unit matrix.

Comparing coefficients in (36.6) and (36.9), we arrive at the set of matrix equations

$$\left.\begin{aligned} \boldsymbol{E} &= \boldsymbol{E} + \boldsymbol{\mu}_{k+1} + \boldsymbol{\sigma}_{k+1}\boldsymbol{\mu}_{k+1} + \boldsymbol{\sigma}_{k+1}, \\ \boldsymbol{B}_k &= \boldsymbol{E} + \boldsymbol{C}_{k+1} + \boldsymbol{\mu}_k + \boldsymbol{\sigma}_{k+1}\boldsymbol{C}_{k+1}, \end{aligned}\right\} \tag{36.10}$$

which we may write in the form

$$\left.\begin{aligned} \boldsymbol{\sigma}_{k+1} + \boldsymbol{\mu}_{k+1}(\boldsymbol{E} + \boldsymbol{\sigma}_{k+1}) &= 0, \\ \boldsymbol{\mu}_k + \boldsymbol{\sigma}_{k+1}\boldsymbol{C}_{k+1} &= \boldsymbol{B}_k - \boldsymbol{C}_{k+1} - \boldsymbol{E}. \end{aligned}\right\} \tag{36.11}$$

Let us multiply the first of (36.11) on the right by $(\mathbf{E} + \boldsymbol{\sigma}_{k+1})^{-1}$. We then have

$$\mu_{k+1} = -\sigma_{k+1}(E+\sigma_{k+1})^{-1}. \tag{36.12}$$

We make use of the identity transformation

$$\begin{aligned}\sigma_{k+1}(E+\sigma_{k+1})^{-1} &= \sigma_{k+1}(E+\sigma_{k+1})^{-1} + (E+\sigma_{k+1})^{-1} - (E+\sigma_{k+1})^{-1} = \\ &= (E+\sigma_{k+1})^{-1}(\sigma_{k+1}+E) - (E+\sigma_{k+1})^{-1},\end{aligned} \tag{36.13}$$

to rewrite (36.12) in the form

$$\mu_{k+1} = \beta_{k+1} - E, \tag{36.14}$$

where

$$\beta_{k+1} = (E+\sigma_{k+1})^{-1}. \tag{36.15}$$

Let us now write the second of (36.11) in the form

$$E + (E+\sigma_{k+1})\, C_{k+1} + \mu_k = B_k. \tag{36.16}$$

From this equation we eliminate μ_k by (36.14). This gives

$$(E+\sigma_{k+1})\, C_{k+1} = B_k - \beta_k. \tag{36.17}$$

We now multiply on the right by C_{k+1}^{-1}, obtaining

$$E + \sigma_{k+1} = (B_k - \beta_k)\, C_{k+1}^{-1}, \tag{36.18}$$

which leads to

$$(E+\sigma_{k+1})^{-1} = C_{k+1}(B_k - \beta_k)^{-1}. \tag{36.19}$$

We have made use of the well-known relation

$$(AB)^{-1} = B^{-1}A^{-1}.$$

Finally, in accordance with (36.15), we rewrite (36.19) in the form

$$\beta_{k+1} = C_{k+1}(B_k - \beta_k)^{-1}. \tag{36.20}$$

Let us now again consider (36.8), rewriting it in the form of the two equations

$$\left.\begin{aligned}(\nabla + \sigma_{k+1})z_{k+1} &= F_k, \\ I_{k+1} + \mu_{k+1}\varphi_{k+1} &= -z_{k+1}.\end{aligned}\right\} \tag{36.21}$$

The first of these can be written

$$z_{k+1} = \beta_{k+1}(z_k + F_k). \tag{36.22}$$

We have here used the fact that

$$\sigma_{k+1} = \beta_{k+1}^{-1} - E,$$

which is a trivial consequence of (36.15). Similarly, the second of Equations (36.21) can be written

$$\varphi_k = C_{k+1}^{-1}(\beta_{k+1}\varphi_{k+1} + z_{k+1}). \tag{36.23}$$

Thus (36.6) has been reduced to the set of three equations

$$\left.\begin{aligned} \beta_{k+1} &= C_{k+1}(B_k - \beta_k)^{-1}, \\ z_{k+1} &= \beta_{k+1}(z_k + F_k), \\ \varphi_k &= C_{k+1}^{-1}(\beta_{k+1}\varphi_{k+1} + z_{k+1}). \end{aligned}\right\} \tag{36.24}$$

To these we must add the boundary conditions (36.7). In order that the first of these conditions be fulfilled, we must obviously require that

$$z_1 = 0, \quad \beta_1 = C_1, \quad \varphi_{n+1} = 0. \tag{36.25}$$

Thus, matrix factorization methods have reduced the boundary value problem to a succession of three Cauchy problems.

K.I. Babenko and N.N. Chentsov have shown that numerical calculation using Equations (36.24) and (36.25) does not lead to accumulation of rounding-off errors, so that the solution is stable.

It should be borne in mind that in inverting matrices of high order accuracy may be lost due to rounding-off, so that one must be careful to use inversion methods free of such deficiencies [83].*

In conclusion, it should be noted that the matrix factorization method is very easy to program for calculating machines. It is not, however, always the most effective method to use. In many cases it turns out that ordinary iteration procedures analogous to those of Liebmann, Seidel, and others (see [34, 83]) are more convenient. Let us consider the simplest of these.

To do this, we first solve (36.1) for φ_{kl}. We then have

$$\varphi_{kl} = \frac{1}{p_{kl}}(a_{kl}\varphi_{k+1,\,l} + b_{kl}\varphi_{k,\,l-1} + c_{kl}\varphi_{k-1,\,l} + d_{kl}\varphi_{k,\,l+1} + f_{kl}), \tag{36.26}$$

which can be written in the matrix form

$$\varphi = A\varphi + F, \tag{36.27}$$

where φ and F are the vectors whose components are $\{\varphi_{kl}\}$ and $\{f_{kl}\}$, and A is a matrix which transforms φ into the new vector $A\varphi$ whose components are

$$\left\{\frac{a_{kl}}{p_{kl}}\varphi_{k+1,\,l} + \frac{b_{kl}}{p_{kl}}\varphi_{k,\,l-1} + \frac{c_{kl}}{p_{kl}}\varphi_{k-1,\,l} + \frac{d_{kl}}{p_{kl}}\varphi_{k,\,l+1} + \frac{f_{kl}}{p_{kl}}\right\}.$$

The method of successive approximations is defined by

$$\varphi^{(n)} = A\varphi^{(n-1)} + F. \tag{36.28}$$

This process must be continued until the solution attained has the desired degree of accuracy.

It can be shown [21] that this procedure always converges. We note only that the convergence is more rapid when the ratio

$$\frac{a_{kl} + b_{kl} + c_{kl} + d_{kl}}{p_{kl}}$$

is smaller (it is always less than unity).

---

*See Appendix C.

If may happen that the successive approximation method of (36.28) does not converge sufficiently rapidly. Then one can take certain steps to increase the rate of this convergence. We mention the method developed by Liusternik [49].*

In addition to Equations (36.27), consider the set of homogeneous equations

$$(A-\lambda_i E)\psi_i = 0 \qquad (i=1,\ 2,\ \dots,\ N), \tag{36.29}$$

where the ψ_i are a set of linearly independent vectors belonging to the eigenvalues λ_i [21]. Let us expand the solution of (36.27) in a linear combination of the form

$$\varphi = \sum_{i=1}^{N} a_i\psi_i. \tag{36.30}$$

We write the zeroth approximation for φ in the form

$$\varphi^{(0)} = \sum_{i=1}^{N} \alpha_i\psi_i. \tag{36.31}$$

If, further, we write

$$F = \sum_{i=1}^{N} \beta_i\psi_i, \tag{36.32}$$

the iteration process of (36.28) gives

$$\varphi^{(n)} = \sum_{i=1}^{N} \beta_i\psi_i\,(1+\lambda_i+\ \dots\ +\lambda_i^{n-1}) + \sum_{i=1}^{N} \alpha_i\lambda_i^n\psi_i. \tag{36.33}$$

Now let $\lambda_1 = \max\{\lambda_i\}$. It is clear that the series in (36.33) will converge as $N \to \infty$ only if $\lambda_1 < 1$.

Consider the difference

$$\varphi^{(n+1)} - \varphi^{(n)} = \sum_{i=1}^{N} \beta_i\lambda_i^n\psi_i + \sum_{i=1}^{N} \alpha_i\lambda_i^n(\lambda_i-1)\psi_i,$$

which we will write in the form

$$\varphi^{(n+1)} - \varphi^{(n)} = \lambda_1^n\left[\gamma_1\psi_1 + \sum_{i=2}^{N}\gamma_i\left(\frac{\lambda_i}{\lambda_1}\right)^n\psi_i\right],$$

where

$$\gamma_i = \beta_i + (\lambda_i - 1)\alpha_i.$$

If n is sufficiently large,

$$\varphi^{(n+1)} - \varphi^{(n)} = \lambda_1^n\gamma_1\psi_1\left[1 + 0\left(\frac{\lambda_i^n}{\lambda_1^n}\right)\right]. \tag{36.34}$$

*We will not here go into the proofs involved, but will merely describe the principal points. The interested reader is referred to the literature.

Similarly,

$$\varphi^{(n)} - \varphi^{(n-1)} = \lambda_1^{n-1}\gamma_1\psi_1\left[1 + 0\left(\frac{\lambda_i^{n-1}}{\lambda_1^{n-1}}\right)\right]. \tag{36.35}$$

Consider the ratio

$$\frac{\varphi^{(n+1)} - \varphi^{(n)}}{\varphi^{(n)} - \varphi^{(n-1)}} = \lambda_1 + 0\left(\frac{\lambda_2^{n-1}}{\lambda_1^{n-1}}\right).$$

It is seen that the first eigenvalue of (36.29) can be approximated by the expression

$$\lambda_1 = \frac{\varphi^{(n+1)} - \varphi^{(n)}}{\varphi^{(n)} - \varphi^{(n-1)}} + 0\left(\frac{\lambda_2^{n-1}}{\lambda_1^{n-1}}\right). \tag{36.36}$$

If φ is the exact solution of Equations (36.27), then obviously

$$\varphi - \varphi^{(n)} = (\varphi^{(n+1)} - \varphi^{(n)}) + (\varphi^{(n+2)} - \varphi^{(n+1)}) + \ldots \tag{36.37}$$

Consider the difference

$$\varphi^{(n+1)} - \varphi^{(n)} = A\,(\varphi^{(n)} - \varphi^{(n-1)}) = \ldots = A^n\,(\varphi^{(1)} - \varphi^{(0)}).$$

Noting that

$$\varphi^{(1)} - \varphi^{(0)} = \sum_{i=1}^{N} \gamma_i\psi_i,$$

we have

$$\varphi^{(n+1)} - \varphi^{(n)} = \sum_{i=1}^{N} \gamma_i\lambda_i^n\psi_i \tag{36.38}$$

or, for large values of n,

$$\varphi^{(n+1)} - \varphi^{(n)} = \gamma_1\lambda_1^n\psi_1\left[1 + 0\left(\frac{\lambda_2^n}{\lambda_1^n}\right)\right]. \tag{36.38'}$$

Inserting (36.38) into (36.37), we obtain

$$\varphi - \varphi^{(n)} = \sum_{i=1}^{N} \gamma_i\psi_i\lambda_i^n\,(1 + \lambda_i + \lambda_i^2 + \ldots) = \sum_{i=1}^{N} \frac{\gamma_i\lambda_i^n}{1 - \lambda_i}\psi_i \tag{36.39}$$

or, for large values of n,

$$\varphi - \varphi^{(n)} = \frac{\gamma_1\lambda_1^n\psi_1}{1 - \lambda_1}\left[1 + 0\left(\frac{\lambda_2^n}{\lambda_1^n}\right)\right]. \tag{36.40}$$

Recalling (36.38'), we may write approximately

$$\varphi - \varphi^{(n)} = \frac{\varphi^{(n+1)} - \varphi^{(n)}}{1 - \lambda_1}. \tag{36.41}$$

Solving this for φ, we have

$$\varphi = \varphi^{(n)} + \frac{\varphi^{(n+1)} - \varphi^{(n)}}{1-\lambda_1}. \tag{36.42}$$

This equation gives us the desired improvement of the approximate solution when the iteration procedure converges too slowly.

The vector equation (36.42) written in scalar form is

$$\varphi_{kl} = \varphi_{kl}^{(n)} + \frac{\varphi_{kl}^{(n+1)} - \varphi_{kl}^{(n)}}{1-\lambda_1},$$

where λ_1 is some average value of the numbers

$$\frac{\varphi_{kl}^{(n+1)} - \varphi_{kl}^{(n)}}{\varphi_{kl}^{(n)} - \varphi_{kl}^{(n-1)}}.$$

N.I. Buleev has suggested the following method of successive approximations.

Consider the difference equation

$$\begin{aligned} &a_{kl}\varphi_{k+1,\,l} + b_{kl}\varphi_{k,\,l-1} + c_{kl}\varphi_{k-1,\,l} + d_{kl}\varphi_{k,\,l+1} - \\ &- p_{kl}\varphi_{kl} = -F_{kl}, \qquad \begin{pmatrix} k=1, 2, \dots n \\ l=1, 2, \dots m \end{pmatrix} \end{aligned} \tag{36.43}$$

where the coefficients a_{kl}, b_{kl}, c_{kl}, d_{kl} and p_{kl} are given numbers. To both sides of this equation we add

$$-\sigma_{kl}\varphi_{k-1,\,l+1} - \tau_{kl}\varphi_{k+1,\,l-1} + \varkappa(\sigma_{kl} + \tau_{kl})\varphi_{kl}, \tag{36.44}$$

where the coefficients σ_{kl} and τ_{kl} are as yet arbitrary, and the parameter κ is chosen from the interval $0 \leq \kappa \leq 1$. We then have

$$\begin{aligned} &a_{kl}\varphi_{k+1,\,l} + b_{kl}\varphi_{k,\,l-1} + c_{kl}\varphi_{k-1,\,l} + d_{kl}\varphi_{k,\,l+1} - \\ &- (p_{kl} - \varkappa\sigma_{kl} - \varkappa\tau_{kl})\varphi_{kl} - \sigma_{kl}\varphi_{k-1,\,l+1} - \tau_{kl}\varphi_{k+1,\,l-1} = -f_{kl}, \end{aligned} \tag{36.45}$$

where

$$f_{kl} = F_{kl} + \sigma_{kl}(\varphi_{k-1,\,l+1} - \varkappa\varphi_{kl}) + \tau_{kl}(\varphi_{k+1,\,l-1} - \varkappa\varphi_{kl}). \tag{36.46}$$

We can replace (36.45) by the equivalent set of equations

$$z_{kl} = m_{kl}z_{k-1,\,l} + \beta_{kl}z_{k,\,l-1} + \gamma_{kl}f_{kl}, \tag{36.47}$$

$$\varphi_{kl} = \alpha_{kl}\varphi_{k+1,\,l} + n_{kl}\varphi_{k,\,l+1} + z_{kl}, \tag{36.48}$$

where α_{kl}, β_{kl}, m_{kl}, n_{kl}, and γ_{kl} are coefficients yet to be determined.

To find these coefficients we solve (36.48) for z_{kl} and insert this solution into (36.47). We obtain

$$\begin{aligned} &\alpha_{kl}\varphi_{k+1,\,l} + \beta_{kl}\varphi_{k,\,l-1} + m_{kl}\varphi_{k-1,\,l} + n_{kl}\varphi_{k,\,l+1} - \\ &- (1 + m_{kl}\alpha_{k-1,\,l} + \beta_{kl}n_{k,\,l-1})\varphi_{kl} - m_{kl}n_{k-1,\,l}\varphi_{k-1,\,l+1} - \\ &- \beta_{kl}\alpha_{k,\,l-1}\varphi_{k+1,\,l-1} = -\gamma_{kl}f_{kl}. \end{aligned} \tag{36.49}$$

In order for this to be equivalent to (36.45), the coefficients equations must be proportional. Therefore,

$$\left.\begin{aligned} \alpha_{kl} &= a_{kl}\gamma_{kl} \\ \beta_{kl} &= b_{kl}\gamma_{kl} \\ m_{kl} &= c_{kl}\gamma_{kl} \\ n_{kl} &= d_{kl}\gamma_{kl} \end{aligned}\right\} \tag{36.50}$$

$$\left.\begin{aligned} \beta_{kl}\alpha_{k,\,l-1} &= \tau_{kl}\gamma_{kl} \\ m_{kl}n_{k-1,\,l} &= \sigma_{kl}\gamma_{kl} \end{aligned}\right\} \tag{36.51}$$

$$1 + m_{kl}\alpha_{k-1,\,l} + \beta_{kl}n_{k,\,l-1} = (p_{kl} - \varkappa\sigma_{kl} - \varkappa\tau_{kl})\,\gamma_{kl}. \tag{36.52}$$

Inserting (36.50) into (36.51) and (36.52), we obtain

$$\left.\begin{aligned} \sigma_{kl} &= c_{kl}d_{k-1,\,l}\gamma_{k-1,\,l} \\ \tau_{kl} &= b_{kl}a_{k,\,l-1}\gamma_{k,\,l-1} \end{aligned}\right\} \tag{36.53}$$

and

$$\gamma_{kl} = \frac{1}{p_{kl} - c_{kl}\,(a_{k-1,\,l} + \varkappa d_{k-1,\,l})\,\gamma_{k-1,\,l} - b_{kl}\,(\varkappa a_{k,\,l-1} + d_{k,\,l-1})\,\gamma_{k,\,l-1}}. \tag{36.54}$$

Equations (36.47) and (36.48) can be rewritten, using (36.50), (36.53), and (36.46), in the form

$$\begin{aligned} z_{kl} = \gamma_{kl}\,[c_{kl}\,(z_{k-1,\,l} + d_{k-1,\,l}\gamma_{k-1,\,l}\delta_{kl}) + \\ + b_{kl}\,(z_{k,\,l-1} + a_{k,\,l-1}\gamma_{k,\,l-1}\mu_{kl}) + F_{kl}], \end{aligned} \tag{36.55}$$

$$\varphi_{kl} = \gamma_{kl}\,(a_{kl}\varphi_{k+1,\,l} + d_{kl}\varphi_{k,\,l+1}) + z_{kl}, \tag{36.56}$$

where

$$\left.\begin{aligned} \delta_{kl} &= \varphi_{k-1,\,l+1} - \varkappa\varphi_{kl}, \\ \mu_{kl} &= \varphi_{k+1,\,l-1} - \varkappa\varphi_{kl}. \end{aligned}\right\} \tag{36.57}$$

If we use the boundary conditions for φ in the difference equations (36.43), we have

$$c_{1l} = 0, \qquad b_{k1} = 0, \qquad a_{nl} = 0, \qquad d_{km} = 0. \tag{36.58}$$

If we make use of (36.58), Equations (36.54-36.56) will give values for γ_{kl}, z_{kl} and φ_{kl} at all interpolation points in the region. In particular, the values of γ_{kl}, z_{kl} and φ_{kl} on the diagonal passing through the point p_{11} are

$$\gamma_{11} = \frac{1}{p_{11}}, \qquad z_{11} = \gamma_{11}f_{11}, \qquad \varphi_{nm} = z_{nm}. \tag{36.59}$$

To find the first approximation for φ_{kl}, we give values for δ_{kl} and μ_{kl}.

The second approximation for φ is calculated according to

$$\gamma_{kl} = \frac{1}{p_{kl} - c_{kl}\,(a_{k-1,\,l} + \varkappa b_{k-1,\,l})\,\gamma'_{k-1,\,l} - d_{kl}\,(\varkappa a_{k,\,l+1} + b_{k,\,l+1})\,\gamma'_{k,\,l+1}}, \tag{36.60}$$

$$\left.\begin{aligned} z'_{kl} = \gamma'_{kl}\,[c_{kl}\,(z'_{k-1,\,l} + b_{k-1,\,l}\gamma'_{k-1,\,l}\delta'_{kl}) + \\ + d_{kl}\,(z'_{k,\,l+1} + a_{k,\,l+1}\gamma'_{k,\,l+1}\mu'_{kl}) + F_{kl}], \\ \varphi_{kl} = \gamma'_{kl}\,(a_{kl}\varphi_{k+1,\,l} + b_{kl}\varphi_{k,\,l-1}) + z'_{kl}, \end{aligned}\right\} \tag{36.61}$$

where

$$\left.\begin{aligned} \delta'_{kl} &= \varphi_{k-1,\,l-1} - \varkappa\varphi_{kl}, \\ \mu'_{kl} &= \varphi_{k+1,\,l+1} - \varkappa\varphi_{kl} \end{aligned}\right\}$$

are calculated from the first approximation for φ. In other words, in order to obtain the second approximation, we again write a set of equations such as (36.54-36.56) and we now solve it along the other diagonal of our region. In order to obtain the third approximation we again use (36.54-36.56), and so on.

It should be noted that successive approximations are obtained only for z_{kl} and φ_{kl} or for z'_{kl} and φ'_{kl}, whereas γ_{kl} and γ'_{kl} are calculated just once.

We recall that the set of difference equations (36.43) used in reactor calculations is obtained from a differential equation in Cartesian or cylindrical coordinates, that the diffusion coefficient is a piecewise continuous function, and that the coefficients of Equation (36.43) usually satisfy the conditions

$$c_{kl} \leqslant a_k, \qquad b_{kl} = d_{kl} \tag{36.62}$$

at all points not on or in the immediate vicinity of boundaries between different zones of the reactor.

With these assumptions concerning the coefficients it is easily shown that

$$(a_{kl} + d_{kl})\gamma_{kl} < 1, \qquad (a_{kl} + b_{kl})\gamma'_{kl} < 1, \tag{36.63}$$

and, further,

$$(c_{kl} + b_{kl})\gamma_{kl} < 1, \qquad (c_{kl} + d_{kl})\gamma'_{kl} < 1, \tag{36.64}$$

which implies that when calculating z_{kl} and φ_{kl} according to (36.55) and (36.56), or according to (36.61), the rounding-off errors made in $\underline{z}$ and φ will decay in propagating from point to point.

There has as yet been no theoretical study of the convergence of the iteration procedure of (36.55), (36.56) and (36.61) for values of κ chosen arbitrarily from the interval $0 \leq \kappa \leq 1$.

Calculations show, however, that with suitably chosen κ this successive approximation procedure converges more rapidly than the usual ones.*

The choice of κ depends on the type of the boundary conditions. If the function is given on the boundary of the region, it is better to choose κ close to unity, whereas if its derivative is given on the boundary, it is best to choose κ close to zero.**

If $\kappa = 1$ and $\delta = \mu = 0$, even the first approximation for the unknown function is close to the solution. In decreasing the separation between the iteration points, the convergence is made slightly worse.

S.M. Ermakov has used Equations (36.55) and (36.56) with $\kappa = 0$ for the diffusion equation with the function given on the boundary of the region and has proven the convergence of the iteration procedure on the assumption that the coefficients of the difference equation satisfy the condition

$$c_{kl}a_{k-1,l} + b_{kl}d_{k,l-1} \leqslant 2 \quad \text{for} \quad p_{kl} = 4.$$

In order to make statements as to the convergence of the iteration procedure given by (36.55), (36.56) and (36.61), one can, in general, use ordinary ϵ-methods (see [34]).

37. Finite-Difference Balance Equations

When solving finite-difference diffusion equations it is helpful if one has control relations which can be used to estimate the accuracy to which the finite-difference equations have been solved. We will call such con-

*For instance, in order to solve Poisson's equation to within an accuracy of 1% inside a square with 100 interpolation points and with a given function on the boundary, it is sufficient to go to the eighth approximation.

**Actual calculations have shown that if the boundary conditions of the problem contain derivatives of the unknown function, it is best to choose κ variable and depending on the coordinates of the interpolation point (κ_{kl}) for both iteration procedures. When using Equations (36.54-36.56) one must set $\kappa_{kl} = 0$ at points where $a_{kl} = 0$ or $d_{kl} = 0$, and $\kappa_{kl} = 1$ at all other points of the region, whereas when using Equations (36.60) and (36.61), one must set $\kappa_{kl} = 0$ where $a_{kl} = 0$ or $b_{kl} = 0$, and $\kappa_{kl} = 1$ at all other points. In obtaining the first approximation, however, one must set $\kappa_{kl} = 1$ everywhere.

trol relations finite-difference balance equations.

Let us first consider the one-dimensional case in which a discontinuity point lies on a $r = r_k$ interpolation point. We then use Equations (31.6) and (31.11), namely:

$$I_{k+1/2} - I_{k-1/2} - (\Sigma\Delta r)_k \varphi_k r_k^\alpha = -(f\Delta r)_k r_k^\alpha, \tag{37.1}$$

where

$$I_{k+1/2} = \mu_k (\varphi_{k+1} - \varphi_k). \tag{37.2}$$

Summing both parts of (37.1) over all k, we obtain

$$\eta = \sum_{k=1}^{m-1} [(\Sigma\Delta r)_k \varphi_k - (f\Delta r)_k] r_k^\alpha - I_{m-1/2} = 0, \tag{37.3}$$

where $k = m$ is the index of the iteration point on the extrapolated reactor boundary. In (37.3) we have used the fact that $I_{1/2} = 0$. Since $\varphi_m = 0$, we have

$$I_{m-1/2} = -\mu_{m-1}\varphi_{m-1}.$$

Therefore

$$\eta = \sum_{k=1}^{m-1} [(\Sigma\Delta r)_k \varphi_k - (f\Delta r)_k] r_k^\alpha + \mu_{m-1}\varphi_{m-1} = 0. \tag{37.4}$$

In view of the fact that in approximate calculations $\eta = \tilde{\eta} \neq 0$, it is convenient to introduce

$$\Delta = \frac{\tilde{\eta}}{\sum_{k=1}^{m-1} (f\Delta r)_k r_k^\alpha} \cdot 100\%. \tag{37.5}$$

This may serve as a relative criterion of the accuracy to which the finite-difference diffusion equation has been solved. In the one-dimensional case when a discontinuity point lies on an internal point of an interval $r = r_{k+1/2}$, we must use (31.17) and (31.21) to obtain the balance equations:

$$\left.\begin{aligned} I_{k+1/2} - I_{k-1/2} - (\Sigma_k\varphi_k - f_k) r_k^\alpha \Delta r_k = 0, \\ I_{k+1/2} = \mu_k(\varphi_{k+1} - \varphi_k). \end{aligned}\right\} \tag{37.6}$$

$$\eta = \sum_{k=1}^{m-1} (\Sigma_k \Delta r_k \varphi_k - f_k \Delta r_k) r_k^\alpha + \mu_{m-1}\varphi_{m-1} = 0. \tag{37.7}$$

$$\Delta = \frac{\tilde{\eta}}{\sum_{k=1}^{m-1} f_k r_k^\alpha \Delta r_k} \cdot 100\%. \tag{37.8}$$

Let us now consider the improved finite-difference diffusion equations. For the first case the balance equation is

$$\eta = \sum_{k=1}^{m-1} \left\{ \left[(\Sigma\Delta r)_k + \frac{\alpha}{8r_k} [\Sigma\Delta r^2]_k \right] \varphi_k - F_k \right\} r_k^\alpha + $$
$$+ \sum_{k=1}^{m-1} \frac{1}{8} \left[\frac{\Sigma}{D} \cdot \Delta r^2 \right]_k \cdot I_k + \mu_{m-1} \cdot \varphi_{m-1} = 0, \tag{37.9}$$

where

$$I_h = \frac{1}{2}\left(I_{h+1/2} + I_{h-1/2}\right).$$

the second sum in (37.9) is taken only over discontinuity points. The expression for F_k in (37.9) is given by (32.13). We define the relative error by

$$\Delta = \frac{\tilde{\eta}}{\sum\limits_{h=1}^{m-1} F_h r_h^\alpha} \cdot 100\%. \tag{37.10}$$

Let us now proceed to the finite-difference balance equations for a two-dimensional region. We then first consider (33.9), namely:

$$(J_{k+1/2,\,l} - J_{k-1/2,\,l})\,\Delta z_l + (Y_{k,\,l+1/2} - Y_{k,\,l-1/2})\,r_k^\alpha \Delta r_k - (\Sigma \Delta r\, \Delta z)_{kl}\varphi_{kl}\,r_k^\alpha = $$
$$= -(f\Delta r\, \Delta z)_{kl}\,r_k^\alpha. \tag{37.11}$$

Summing this over all $\underline{k}$ and l, we have

$$\eta = \sum_k \sum_l \left[(\Sigma \Delta r\, \Delta z)_{kl}\,\varphi_{kl} - (f\Delta r\, \Delta z)_{kl}\right] r_k^\alpha - \sum_l J_{m-1/2,\,l}\,\Delta z_l - $$
$$- \sum_k Y_{p-1/2,\,h}\,r_k^\alpha\,\Delta r_k = 0. \tag{37.12}$$

where

$$\left.\begin{array}{l} J_{m-1/2,\,l} = -\mu_{m-1/2}^l\, r_{m-1/2}^\alpha\,\varphi_{m-1,\,l}, \\[2mm] Y_{k,\,p-1/2} = -\mu_{p-1/2}^h\,\varphi_{k,\,p-1}, \end{array}\right\} \tag{37.13}$$

and $\underline{m}$ and $\underline{p}$ are the indices of the interpolation points at $r = R_e$ and $z = H_e$. Therefore,

$$\Delta = \frac{\tilde{\eta}}{\sum\limits_k' \sum\limits_l (f\Delta r \Delta z)_{kl} r_k^\alpha} \cdot 100\%. \tag{37.14}$$

In considering the balance equation for the second form, using Equation (33.16), namely

$$(J_{k+1/2,\,l} - J_{k-1/2,\,l})\,\Delta z_l + (Y_{k,\,l+1/2} - Y_{k,\,l-1/2})\,\Delta r_k r_k^\alpha - \Sigma_{kl}\varphi_{kl}\,\Delta r_k\,\Delta z_l \cdot r_k^\alpha = $$
$$= -f_{kl}\,\Delta r_k\,\Delta z_l\,r_k^\alpha, \tag{37.15}$$

we obtain

$$\eta = \sum_k \sum_l (\Sigma_{kl}\,\Delta r_k\,\Delta z_l\,\varphi_{kl} - f_{kl}\,\Delta r_k\,\Delta z_l)\,r_k^\alpha - \sum_l J_{m-1/2,\,l}\,\Delta z_l - $$
$$- \sum_k Y_{p-1/2,\,h}\,r_k^\alpha\,\Delta r_k = 0, \tag{37.16}$$

and the relative error of the calculation can be written

$$\Delta = \frac{\tilde{\eta}}{\sum\limits_k \sum\limits_l f_{kl}\,\Delta r_k \Delta z_l \cdot r_k^\alpha} \cdot 100\%. \tag{37.17}$$

A similar procedure will give the balance equation for the improved set of finite-difference diffusion equations.

38. The Exact Balance Equation

In solving finite-difference diffusion equations we have agreed to use the finite-difference balance equation as a control equation.

If the finite-difference balance equations are fulfilled, this means that the set of finite-difference diffusion equations are solved to a given accuracy. This does not, however, characterize the accuracy with which we have approximated the solution of the original differential equation. It is therefore useful to introduce certain integral control relations which we will also call balance equations. In view of the fact that these relations will be obtained directly from the differential diffusion equation, they are in a certain sense exact and can be used at the last stage to evaluate the error which has arisen due to the numerical methods. It should, however, be borne in mind that to satisfy these relations is a necessary, but hardly a sufficient, condition for accuracy.

Consider the diffusion equation in the form

$$\nabla D \nabla \varphi - \Sigma \varphi = -f(\vec{r}), \tag{38.1}$$

where $D(\vec{r})$, $\Sigma(\vec{r})$ and $f(\vec{r})$ are given functions.

Let us integrate (38.1) over the whole volume of G of the reactor bearing in mind that $\varphi(\vec{r})$ vanishes on the extrapolated boundary S_e. We then obtain

$$\int_G \Sigma \varphi \, d\vec{r} - \int_{S_e} (D \nabla \varphi)_{\vec{n}_0} \, dS = \int_G f \, d\vec{r}, \tag{38.2}$$

where $(\vec{\Phi})_{\vec{n}_0}$ is the projection of $\vec{\Phi}$ on the normal to the surface S_e. We introduce the notation

$$\varepsilon = \int_G f \, d\vec{r} - \int_G \Sigma \varphi \, d\vec{r} + \int_{S_e} (D \nabla \varphi)_{\vec{n}_0} \, dS. \tag{38.3}$$

We shall say that the neutron balance is maintained in the reactor for a given group if

$$\epsilon = 0. \tag{38.4}$$

If $\epsilon \neq 0$, we may characterize the accuracy of our calculation by

$$\delta = \frac{\varepsilon}{\int_G f \, d\vec{r}} \cdot 100\%. \tag{38.5}$$

The balance equation (38.4) can be written and applied to any isolated part of the reactor. Indeed, if we are interested in the neutron balance for a given group in the volume G_1, bounded by the surfaces S_1 and S_2, we have

$$\varepsilon = \int_{G_1} f_1 \, d\vec{r} - \int_{G_1} \Sigma \varphi \, d\vec{r} + \int_{S_1} (D \nabla \varphi)_{\vec{n}_1^0} \, dS + \int_{S_2} (D \nabla \varphi)_{\vec{n}_2^0} \, dS.$$

It should be borne in mind here that the normals to the surfaces S_1 and S_2 are directed outwardly from the volume G_1.

Chapter X

THE NET METHOD

In many cases it is convenient to use the net method in the classical form of a set of triangles [31, 43, 78] to solve the slowing-down equations. This method is particularly effective when the slowing-down length is small compared with the geometrical dimensions of the various reactor zones. This condition is usually satisfied in thermal reactors.

Essentially the method consists of writing the differential slowing-down equation as a difference equation which is then solved for the unknown function on some energy interval in terms of previously found values of this function on a preceding interval. The successive solution of the difference equations leads to the desired space-energy distribution of the neutron flux. The adjoint reactor equations are solved similarly.

It should be borne in mind that this method can be used to obtain the solution only under certain conditions which relate the energy interval to the space interval [28, 34, 46, 48, 69, 75, 77, 100, 103, 112]. This condition is the condition for the stability of the solution by the finite-difference procedure [43, 46, 75].

In the present chapter we shall consider finite-difference schemes of the triangle type both for the basic and adjoint reactor equations.

39. Definition of the Method. Basic Reactor Equations

In Chapters VI and VII, when we were considering the different multigroup methods, we defined procedure III, which is easily written in the form

$$\varphi^j = \frac{\xi\Sigma_s^{j-1}}{\xi\Sigma_s^j}\left[p^j\varphi^{j-1} + \mu^j\nabla D^j\nabla\varphi^{j-1} + S^j\right], \tag{39.1}$$

where p^j, μ^j and S^j are given in Table 16.

TABLE 16

Effective Functions and Constants in the Multigroup Set of Reactor Equations for Procedure III

	Weak absorption	Strong absorption
p^j	$1 - \frac{\Sigma_c^j}{\xi\Sigma_s^j}\Delta u_j$	$e^{-\int_{u_{j-1}}^{u_j}\frac{\Sigma_c}{\xi\Sigma_s}du}$
μ^j	$\frac{\Delta u_j}{\xi\Sigma_s^j}$	$\frac{1}{\xi\Sigma_s^j}\int_{u_{j-1}}^{u_j} e^{-\int_{u'}^{u_j}\frac{\Sigma_c}{\xi\Sigma_s}du}\,du'$
S^j	$\frac{Q(\vec{r})}{\xi\Sigma_s^{j-1}}\int_{u_{j-1}}^{u_j}\chi(u)du$	$\frac{Q(\vec{r})}{\xi\Sigma_s^{j-1}}\int_{u_{j-1}}^{u_j}\chi(u)\,e^{-\int_{u'}^{u_j}\frac{\Sigma_c}{\xi\Sigma_s}du}\,du'$

In order to use this procedure in practice, one must choose some kind of approximation method to calculate the expressions on the right side of (39.1). The most accurate of these methods is the continuous calculation method, which we have already used several times in obtaining the finite-difference diffusion equations.

Applied to Equations (39.1), this method is defined as follows.

We write (39.1) in the form

$$\nabla D^j \nabla \varphi^{j-1} + \Sigma^j \varphi^{j-1} = f^j, \tag{39.2}$$

where

$$\left.\begin{aligned} \Sigma^j = \frac{P^j}{\mu^j}, \quad f^j = \frac{\vartheta^j}{\mu^j}\varphi^j - \frac{S^j}{\mu^j}, \\ \vartheta^j = \frac{\xi\Sigma_s^j}{\xi\Sigma_s^{j-1}}. \end{aligned}\right\} \tag{39.3}$$

Let us now represent (39.2) in finite-difference form, using the simplest methods of Chapter VIII. Then for the first procedure, in which the discontinuities of D^j, Σ^j and f^j lie on a $r = r_k$ interpolation point, we have

$$a_k \varphi_{k+1}^{j-1} - b_k \varphi_k^{j-1} + c_k \varphi_{k-1}^{j-1} = F_k^j, \tag{39.4}$$

where

$$\left.\begin{aligned} a_k = \left(1 + \alpha\frac{\Delta r_{k+1/2}}{2r_k}\right)\frac{D_{k+1/2}^j}{\Delta r_{k+1/2}}, \quad b_k = a_k + c_k - (\Sigma^j \Delta r)_k, \\ c_k = \left(1 - \alpha\frac{\Delta r_{k-1/2}}{2r_k}\right)\frac{D_{k-1/2}^j}{\Delta r_{k-1/2}}, \quad F_k^j = (f^j \Delta r)_k, \end{aligned}\right\} \tag{39.5}$$

and

$$D^j = \frac{1}{3\Sigma_{tr}^j}.$$

Consider the expression

$$(f^j \Delta r)_k = \left(\frac{\vartheta^j \Delta r}{\mu^j}\right)_k \varphi^j - \left(\frac{S^j \Delta r}{\mu^j}\right)_k. \tag{39.6}$$

With this in mind, we write (39.4) in the form

$$\varphi_k^j = \frac{1}{\left(\frac{\vartheta^j \Delta r}{\mu^j}\right)_k}\left[a_k \varphi_{k+1}^{j-1} - b_k \varphi_k^{j-1} + c_k \varphi_{k-1}^{j-1} + \left(\frac{S^j \Delta r}{\mu^j}\right)_k\right]. \tag{39.7}$$

This is the triangle scheme generalized for the case of the continuous calculation.

Let us now consider the second procedure for representing the diffusion equation in finite-difference form, when the discontinuities lie on internal points $r = r_{k+1/2}$ of the interpolation intervals:

$$a_k \varphi_{k+1}^{j-1} - b_k \varphi_k^{j-1} + c_k \varphi_{k-1}^{j-1} = F_k^j, \tag{39.8}$$

where

$$a_k = \frac{1 + \alpha\frac{\Delta r_k}{2r_k}}{\left(\frac{\Delta r}{D^j}\right)_{k+1/2}}, \quad b_k = a_k + c_k - \Sigma_k^j \cdot \Delta r_k, \tag{39.9}$$

(more)

$$c_k = \frac{1 - \alpha \frac{\Delta r_k}{2r_k}}{\left(\frac{\Delta r}{D^j}\right)_{k-1/2}}, \quad F_k^j = f_k^j \Delta r_k. \tag{39.9}$$

(continued)

Recalling the definition of f in (39.3), we have

$$f_k^j \Delta r_k = \frac{\vartheta_k^j \cdot \Delta r_k}{\mu_k^j} \varphi_k^j - \frac{S_k^j \cdot \Delta r_k}{\mu_k^j}. \tag{39.10}$$

Using this equation, (39.8) is finally written in the form

$$\varphi_k^j = \frac{\mu_k^j}{\vartheta_k^j \cdot \Delta r_k} \cdot (a_k \varphi_{k+1}^{j-1} - b_k \varphi_k^{j-1} + c_k \varphi_{k-1}^{j-1}) + \frac{S_k^j}{\vartheta_k^j}. \tag{39.11}$$

We have thus arrived at two different triangle schemes which can be used with the continuous calculation method.

We note further that if, in spherical regions, we introduce the function $r_k \varphi_k^j$, which we shall again call φ_k^j, the form of the coefficients in (39.5) and (39.9) simplifies considerably.

Thus, in the first procedure,

$$\left.\begin{aligned} a_k &= \frac{D_{k+1/2}^j}{\Delta r_{k+1/2}}, \\ b_k &= a_k + c_k + \frac{\Delta r_{k+1/2}}{r_k}\left(a_k - \frac{\Delta r_{k-1/2}}{\Delta r_{k+1/2}} c_k\right) - (\Sigma^j \Delta r)_k, \\ c_k &= \frac{D_{k-1/2}^j}{\Delta r_{k-1/2}} \end{aligned}\right\} \tag{39.12}$$

and in the second,

$$\left.\begin{aligned} a_k &= \frac{1}{\left(\frac{\Delta r}{D^j}\right)_{k+1/2}}, \quad b_k = a_k + c_k + \frac{\Delta r_k}{r_k}(a_k - c_k) - \Sigma_k^j \Delta r_k, \\ c_k &= \frac{1}{\left(\frac{\Delta r}{D^j}\right)_{k-1/2}}. \end{aligned}\right\} \tag{39.13}$$

We must in addition consider the question of the boundary conditions for the finite-difference triangle scheme.

At the outer extrapolated boundary of the reactor we must set

$$\varphi_m^j = 0. \tag{39.14}$$

There are two possible boundary conditions for the center of the reactor, namely $\left.\frac{d\varphi^j}{dr}\right|_{r=0} = 0$ or $\varphi^j|_{r=0} = 0$. In the first case it is convenient to choose the interpolation points so that $r = 0$ lies at the center of the first Δr interval. Then the value of the function at the interpolation point whose index is $k = 0$, which lies at $r = -\Delta r/2$, is given by

$$\varphi_0^j = \varphi_1^j. \tag{39.15}$$

In the case of spherical symmetry it is sufficient to write

$$\varphi_0^j = 0. \tag{39.16}$$

In conclusion, let us consider some possible simplifications in the calculation of the p^j and μ^j. The criterion of stability, which we shall discuss below, will as a rule cause the energy interval Δu_j to be quite small. On this interval we may therefore consider the function $\Sigma_c/\xi\Sigma_s$ to be constant and equal to $\Sigma_c^j/\xi\Sigma_s^j$, where $\Sigma_c^j = \frac{1}{\Delta u_j}\int\limits_{u_{j-1}}^{u_j} \Sigma_c du$. Then for strong absorption we have

$$p^j = e^{-\frac{\Sigma_c^j}{\xi\Sigma_s^j}\Delta u_j}, \quad \mu^j = \frac{1-p^j}{\Sigma_c^j}, \quad S^j = Q\mu^j\frac{\chi^j}{\vartheta^j}. \tag{39.17}$$

These equations are convenient to use in actual calculations by the net method.

40. The Net Method Applied to the Solution of the Adjoint Equations

In the multigroup representation of the adjoint reactor equations for procedure III, we obtained an expression which we can write

$$\varphi^{*j-1} = p^j\varphi^{*j} + \mu^{*j}\nabla D^j\nabla\varphi^{*j} + S^{*j}, \tag{40.1}$$

where p^j, D^j and S^{*j} are given in Table 17.

TABLE 17

Effective Functions and Constants in the Multigroup Set of Adjoint Reactor Equations for Procedure III

	Weak absorption	Strong absorption
p^j	$1-\frac{\Sigma_c^j}{\xi\Sigma_s^j}\Delta u_j$	$e^{-\int\limits_{u_{j-1}}^{u_j}\frac{\Sigma_c}{\xi\Sigma_s}du}$
μ^{*j}	$\frac{\Delta u_j}{\xi\Sigma_s^j}$	$\frac{1}{\xi\Sigma_s^j}\int\limits_{u_{j-1}}^{u_j} e^{-\int\limits_{u_{j-1}}^{u'}\frac{\Sigma_c}{\xi\Sigma_s}du} du'$
S^{*j}	$Q^*(\vec{r})\int\limits_{u_{j-1}}^{u_j}\frac{\Sigma_f}{\xi\Sigma_s}du$	$Q^*(\vec{r})\int\limits_{u_{j-1}}^{u_j}\frac{\Sigma_f}{\xi\Sigma_s}e^{-\int\limits_{u_{j-1}}^{u'}\frac{\Sigma_c}{\xi\Sigma_s}du} du'$

Equation (40.1) can be written:

$$\nabla D^j\nabla\varphi^{*j} + \Sigma^j\varphi^{*j} = f^{*j}, \tag{40.2}$$

where

$$f^{*j} = \frac{1}{\mu^{*j}}(\varphi^{*j-1} - S^{*j}), \quad \Sigma^j = \frac{p^j}{\mu^{*j}}. \tag{40.3}$$

Now (40.2) is identical with (39.2) when the function φ^{*j} on the left side of (40.2) is replaced by φ^{*j-1}. Therefore, all the results of the previous section carry over automatically to the present case.

Thus, in considering the simplest finite difference schemes by the continuous calculation method, we will have

$$\varphi_k^{*j-1} = \frac{1}{\left(\frac{\Delta r}{\mu^{*j}}\right)_k} \left[a_k \varphi_{k+1}^{*j} - b_k \varphi_k^{*j} + c_k \varphi_{k-1}^{*j} + \left(\frac{S^{*j}\Delta r}{\mu^{*j}}\right)_k \right], \tag{40.4}$$

for the first scheme and

$$\varphi_k^{*j-1} = \frac{\mu_k^{*j}}{\Delta r_k}(a_k \varphi_{k+1}^{*j} - b_k \varphi_k^{*j} + c_k \varphi_{k-1}^{*j}) + \frac{S_k^{*j}}{\mu_k^{*j}} \tag{40.5}$$

for the second.

The quantities a_k, b_k and c_k in (40.4) and (40.5) are given by Equations (39.5) and (39.9), while for the special schemes used for spherical regions, they are given by (39.12) and (39.13).

Just as with the basic reactor equations, the small size of Δu_j makes it possible to make the following approximations for strong absorption of neutrons:

$$p^j = e^{-\frac{\Sigma_c^j}{\xi \Sigma_s^j}\Delta u_j}, \quad \mu^{*j} = \frac{1-p^j}{\Sigma_c^j}, \quad S^{*j} = Q^* \frac{\Sigma_f^j}{\Sigma_c^j}(1-p^j). \tag{40.6}$$

41. Stability of Finite-Difference Systems

It is well known that when solving partial differential equations numerically by finite difference methods, one must define the region of stability. This region is characterized by some relation between the parameters Δr and Δu [28, 34, 46, 48, 69, 75, 77, 100, 103, 112]. Several methods have been proposed to study this stability. The simplest of them is Neiman's method (see [100]), which we shall use in the future.

Consider the triangle difference scheme for a plane unbounded region whose physical properties are those of one of the zones of the reactor. Then the finite-difference equations will be of the form

$$\varphi_k^j = \frac{\xi \Sigma_s^{j-1}}{\xi \Sigma_s^j} \left[p^j \varphi_k^{j-1} + \frac{D^j \mu^j}{\Delta x^2} (\varphi_{k+1}^{j-1} - 2\varphi_k^{j-1} + \varphi_{k-1}^{j-1}) \right], \tag{41.1}$$

where we have assumed for simplicity that $\Delta x = \Delta r$ is independent of $\underline{r}$ and that there exist no fission sources. Equation (41.1) is conveniently written in the form

$$q_k^j = p^j q_k^{j-1} + \frac{\Delta\tau^*}{\Delta x^2}(q_{k+1}^{j-1} - 2q_k^{j-1} + q_{k-1}^{j-1}), \tag{41.2}$$

where

$$q_k^j = \xi \Sigma_s^j \varphi_k^j, \quad \Delta\tau^* = D^j \mu^j. \tag{41.3}$$

Here we have also assumed that $\Delta\tau^*$ is independent of the index $\underline{j}$.*

Let q_k^j be an exact solution of the finite-difference scheme, and let $\tilde{q}_k^j$ be an approximate sl lution. We write $\epsilon_k^j = q_k^j - \tilde{q}_k^j$ for the rounding-off error at the net point (x_k, u_{j-1}). Then the propagation of ϵ_k^j to the other points of the net is described by the equation

*For a fixed zone this can always be arranged by properly choosing the Δu_j intervals.

$$\varepsilon_k^j = p^j \varepsilon_k^{j-1} + \frac{\Delta\tau^*}{\Delta x^2}(\varepsilon_{k+1}^{j-1} - 2\varepsilon_k^{j-1} + \varepsilon_{k-1}^{j-1}). \tag{41.4}$$

Let us consider the error function on the initial line j = 0 to be written in terms of a Fourier integral. We shall consider only the harmonic

$$\varepsilon_k^0 = e^{i\alpha x_k}, \tag{41.5}$$

and will attempt to find the solution of (41.4) in the form

$$\varepsilon_k^j = \eta_j e^{i\alpha k\Delta x}, \tag{41.6}$$

where

$$x_k = k\Delta x.$$

We thus insert (41.6) into (41.4), arriving at

$$\eta_j = \left(p^j - 4\frac{\Delta\tau^*}{\Delta x^2}\sin^2\alpha\frac{\Delta x}{2}\right)\eta_{j-1} \tag{41.7}$$

with the condition

$$\eta_0 = 1. \tag{41.8}$$

In order that the calculation be stable, it is necessary that the expression in parentheses in (41.7) satisfy the inequality

$$\left|p^j - 4\frac{\Delta\tau^*}{\Delta x^2}\sin^2\alpha\frac{\Delta x}{2}\right| \leqslant 1 \tag{41.9}$$

or, equivalently,

$$-1 \leqslant p^j - 4\frac{\Delta\tau^*}{\Delta x^2}\sin^2\alpha\frac{\Delta x}{2} \leqslant 1. \tag{41.10}$$

Now (41.10) will be fulfilled if we set

$$\Delta\tau^* \leqslant \frac{1+p^j}{2}\frac{\Delta x^2}{2}. \tag{41.11}$$

This is the required condition for stability. When there is no neutron absorption in slowing down, (41.11) becomes

$$\Delta\tau \leqslant \frac{\Delta x^2}{2}, \tag{41.12}$$

where

$$\Delta\tau = \int_{u_{j-1}}^{u_j} \frac{D}{\xi\Sigma_s}\,du.$$

Thus, if (41.11) is fulfilled, the rounding-off error will not increase from step to step, and the finite difference equation (41.2) can then be used in practice.

In many-zone systems, the steps Δu and Δx must be chosen so that (41.11) is fulfilled in each zone. Thus, when performing the calculation using the triangle scheme, the conditions for the stability of the calculation must first be fulfilled. It can be shown that (41.11) is valid also for spherical and cylindrical geometry.

It is known that there exist stability conditions for all finite-difference equations. In particular, this is true with respect to the multigroup finite-difference diffusion equations studied in Chapter VIII. Let us consider, for instance, procedure I for handling the multigroup representation of the slowing-down equations. These equations are easily written in a form analogous to (41.2), namely:

$$q_k^j - \frac{\Delta\tau^*}{\Delta x^2}(q_{k+1}^j - 2q_k^j + q_{k-1}^j) = p^j q_k^{j-1}, \tag{41.13}$$

where for simplicity we have again eliminated fission sources.

As in the triangle scheme, we will write the equation for the propagation of the error in the form

$$\varepsilon_k^j - \frac{\Delta\tau^*}{\Delta x^2}(\varepsilon_{k+1}^j - 2\varepsilon_k^j + \varepsilon_{k-1}^j) = p^j \varepsilon_k^{j-1}. \tag{41.14}$$

We again attempt to solve this equation with the Condition (41.5) in the form

$$\varepsilon_k^j = \eta_j e^{i\alpha k\Delta x}, \tag{41.15}$$

and the η_j are then given by

$$\eta_j = \frac{p^j \eta_{j-1}}{\left(1 + 4\frac{\Delta\tau^*}{\Delta x^2}\sin^2\alpha\frac{\Delta x}{2}\right)}. \tag{41.16}$$

It is clear that for stability we must have

$$-\left(1 + 4\frac{\Delta\tau^*}{\Delta x^2}\sin^2\alpha\frac{\Delta x}{2}\right) < p^j < \left(1 + 4\frac{\Delta\tau^*}{\Delta x^2}\sin^2\alpha\frac{\Delta x}{2}\right). \tag{41.17}$$

Now (41.17) is fulfilled for all values of $\Delta\tau^*/\Delta x^2$. But this means that the calculation will then always be stable.

Similarly, it is easily shown that stability is again obtained for any value of the ratio $\Delta\tau^*/\Delta x^2$.

In conclusion, we remark that the results of this section are easily carried over to the adjoint reactor equations.

42. The Accuracy of Finite-Difference Systems

Having represented the basic and adjoint reactor equations in the form of a multigroup set of finite-difference equations, we must now evaluate the resulting error in the eigenvalue problem. An accurate evaluation of the error is possible only for the simplest cases. It turns out, however, that these evaluations can be reliably used as rough estimates also in more complicated problems of practical interest.

In the present section we will consider some simple problems in which the error can be accurately evaluated [31, 70, 95, 99].

Consider a nonreflected spherical reactor whose basic equations are (12.20), namely:

$$\left.\begin{aligned} q(\vec{r}, 0) &= Q(\vec{r}), \\ \nabla D\nabla\varphi - \Sigma_c\varphi &= \frac{\partial q}{\partial u}, \\ \nabla D_T\nabla\Phi - \Sigma_{cT}\Phi &= -q(\vec{r}, u_T). \end{aligned}\right\} \tag{42.1}$$

Here for simplicity we assume that the fission takes place at a neutron energy $E = E_0$ (or $u = 0$), so that $\chi(u) = \delta(u)$. Let us assume that the solution of (42.1) vanishes at $r = 0$ and $r = 1$, where $r = 1$ is the external extrapolated reactor boundary.

As in Chapter VI, we will show that the exact eigenvalue of Equations (42.1) is given by

$$K_{\text{eff}} = \nu_f \left(\int_0^{u_T} \frac{\Sigma_f}{\xi \Sigma_s} e^{-\int_0^u \frac{\Sigma_c}{\xi \Sigma_s} du'} du + \frac{1}{L_{fT}^2} \frac{e^{-\pi^2 \tau_T - \int_0^{u_T} \frac{\Sigma_c}{\xi \Sigma_s} du}}{\pi^2 + \frac{1}{L_{cT}^2}} \right), \tag{42.2}$$

where we have written $\kappa_1 = \pi$, $L_{cT}^2 = D_T/\Sigma_{cT}$, and $L_{fT}^2 = D_T/\Sigma_{fT}$.

Let us write (42.1) in the finite-difference form using the triangle scheme for the slowing-down equation. We then have

$$\left. \begin{aligned} q_k^0 &= \nu \left(\sum_{j=0}^m \alpha_j q_k^j + \Sigma_{fT} \Phi_k \right), \\ q_k^j &= p^j q_k^{j-1} + \frac{\Delta\tau_j^*}{\Delta x^2} \mathbf{M}(q_k^{j-1}), \\ \mathbf{M}(\Phi_k) - \frac{\Delta x^2}{L_{cT}^2} \Phi_k &= -\frac{1}{D_T} q_k^m, \end{aligned} \right\} \tag{42.3}$$

where

$$\mathbf{M}(y_k) = y_{k+1} - 2y_k + y_{k-1}, \quad \Delta\tau_j^* = D^j \mu^j,$$

the μ^j and p^j are given in Table 16, and the α_j are constants of integration.

The solution of (42.3) is of the form [95, 106]

$$\left. \begin{aligned} q_k^j &= b^j \sin k\pi\Delta x, \\ \Phi_k^j &= c \sin k\pi\Delta x. \end{aligned} \right\} \tag{42.4}$$

Let us insert (42.4) into Equations (42.3) and use the relations

$$\mathbf{M}(y_k) = -\xi y_k,$$

where y_k is a solution of (42.4) and $\xi = 4\sin^2 \frac{\pi \Delta x}{2}$. We then obtain

$$\left. \begin{aligned} 1 &= \nu \left(\sum_{j=0}^m \alpha_j b^j + \Sigma_{fT} \cdot c \right), \\ b^j &= b^{j-1} \left(p^j - \frac{\Delta\tau^*}{\Delta x^2} \xi \right), \quad c \left(\xi + \frac{\Delta x^2}{L_{cT}^2} \right) = b^m \frac{1}{D_T}. \end{aligned} \right\} \tag{42.5}$$

From this it follows that

$$\left. \begin{aligned} b^j &= \prod_{l=1}^j \left(p^l - \frac{\Delta\tau_l^*}{\Delta x^2} \xi \right), \\ c &= \prod_{l=1}^m \left(p^l - \frac{\Delta\tau_l^*}{\Delta x^2} \xi \right) \Big/ \left[D_T \left(\xi + \frac{\Delta x^2}{L_{cT}^2} \right) \right]. \end{aligned} \right\} \tag{42.6}$$

We now insert (42.6) into the first of (42.3). This gives

$$1=\nu\left[\sum_{j=0}^{m}\alpha_j\prod_{l=1}^{j}\left(p^l-\frac{\Delta\tau_l^*}{\Delta x^2}\xi\right)+\frac{1}{L_{fT}^2}\frac{\prod_{l=1}^{j}\left(p^l-\frac{\Delta\tau_l^*}{\Delta x^2}\xi\right)}{\xi+\frac{\Delta x^2}{L_{cT}^2}}\right]; \tag{42.7}$$

from which we obtain

$$\widetilde{K}_{\text{eff}}=\nu_f\left[\sum_{j=0}^{m}\alpha_j\prod_{l=1}^{j}\left(p^l-\frac{\Delta\tau_l^*}{\Delta x^2}\xi\right)+\frac{1}{L_{fT}^2}\frac{\prod_{l=1}^{j}\left(p^l-\frac{\Delta\tau_l^*}{\Delta x^2}\xi\right)}{\xi+\frac{\Delta x^2}{L_{cT}^2}}\right]. \tag{42.8}$$

This equation makes it possible to obtain the exact eigenvalue of the set of finite-difference equations. The relative errr is then given by

$$\epsilon=\frac{\widetilde{K}_{\text{eff}}-K_{\text{eff}}}{K_{\text{eff}}}\cdot 100\%. \tag{42.9}$$

There is no difficulty in using (42.8) and (42.2) to calculate $\widetilde{K}_{\text{eff}}$ and K_{eff}.

We can expand (42.9) into a Taylor's series, obtaining

$$\varepsilon=\frac{K_{\text{eff}}}{\nu_f}\frac{\pi^4\Delta x^2}{3}\left[\left(\frac{4}{\pi^2+\frac{1}{L_{cT}^2}}+\frac{1+6\mu}{4}\tau_T\right)\theta_1+\frac{1+6\mu}{4}\bar{\tau}\theta_2\right]. \tag{42.10}$$

Similarly, in procedure I, we have

$$\varepsilon=\frac{K_{\text{eff}}}{\nu_f}\frac{\pi^4\Delta x^2}{3}\left[\left(\frac{4}{\pi^2+\frac{1}{L_{cT}^2}}+\frac{\tau_T}{4}\right)\theta_1+\frac{\bar{\tau}}{4}\theta_2\right] \tag{42.11}$$

and in procedure II we have

$$\varepsilon=\frac{K_{\text{eff}}}{\nu_f}\frac{\pi^4\Delta x^2}{3}\left[\left(\frac{4}{\pi^2+\frac{1}{L_{cT}^2}}+\frac{1-6\mu}{4}\tau_T\right)\theta_1+\frac{1-6\mu}{4}\bar{\tau}\theta_2\right], \tag{42.12}$$

where $\mu=\Delta\tau/\Delta x^2$,

$$\theta_1=\frac{C}{C+D},\qquad \theta_2=\frac{D}{C+D},\qquad C=\frac{e^{-\pi^2\tau_T-\int_0^{\tau}\frac{d\tau}{L_c^2}}}{L_{fT}^2\left(\pi^2+\frac{1}{L_{cT}^2}\right)},$$

$$D=\int_0^{\tau_T}\frac{e^{-\pi^2\tau-\int_0^{\tau}\frac{d\tau}{L_c^2}}}{L_f^2}\,d\tau,$$

$$\bar{\tau}=\left(\int_0^{\tau_T}\tau\cdot\frac{e^{-\pi^2\tau-\int_0^{\tau}\frac{d\tau}{L_c^2}}}{L_f^2}\,d\tau\right)\Big/\left(\int_0^{\tau_T}\frac{e^{-\pi^2\tau-\int_0^{\tau}\frac{d\tau}{L_c^2}}}{L_f^2}\,d\tau\right).$$

Similar estimates can be obtained for the error in the eigenvalue also for the adjoint reactor equations.

Thus, for a nonreflected spherical reactor we have obtained an exact evaluation of the error in the eigenvalue at K_{eff}. We recall in this connection that every spherical many-zone reactor can be associated with some nonreflected reactor whose dimensions are chosen so that K_{eff} of the actual system is the same as K_{eff} of the "equivalent" nonreflected reactor.

The error in K_{eff} for the equivalent reactor can be obtained as described above. It is found that in the majority of cases this value can be used for the initial multigroup reactor.

There exist, however, some cases in which one needs a more accurate knowledge of the error made in solving the basic equations. This is particularly true of reactors in which the spatial distribution of the neutrons is very different from sinusoidal. This usually occurs in regions close to the boundary between the core and reflector. For such cases it is a good idea to attempt to find the error in K_{eff} by performing additional calculations with an increased number of steps both in Δr and Δu [34, 56, 57].

Chapter XI

PERTURBATION THEORY

In solving nuclear reactor problems it is often necessary to find very small deviations in the eigenvalue K_{eff} due to small variations in the physical parameters. Essentially the problem is the following. From given variations of the physical parameters of a critical reactor, one must find those changes which must be realized in K_{eff} in order that the reactor remain critical.

Attempts at the direct calculation of δK_{eff} as the difference between the eigenvalues of the "perturbed" and "unperturbed" reactors cannot always lead to satisfactory results, since the difference in K_{eff} is usually of the same order of magnitude as the uncertainty in the calculation.

Therefore, small variations in K_{eff} are more reasonably calculated by using some method which will lead directly to δK_{eff} from the variation in the physical parameters. Such a method for calculating intermediate and thermal reactors was first suggested by E. Wigner (see [141]) and was later developed by A.S. Romanovich, L.N. Usachev [82], R. Erlich and H. Hurwitz [101], S. Glasstone and M. Edlund [23], and others [127, 132].

For the one-velocity kinetic equations this method was developed by N.A. Dmitriev (see [82]) and K. Fuchs [105]. The most general formulation of perturbation theory has been given by L.N. Usachev [82].

The present chapter considers the fundamental propositions of perturbation theory and gives formulas for δK_{eff}.

43. Basic Considerations

Before discussing perturbation theory as applied to nuclear reactors, let us consider this theory in terms of linear operators. We thus turn our attention to a linear homogeneous equation of the form

$$\mathbf{L}f = 0, \tag{43.1}$$

where f, a solution of (43.1), belongs to the class of vector functions $\{f\}$ whose elements satisfy certain definite conditions, and $\mathbf{L}$ is an operator acting on the elements of this set.

If $\mathbf{L}$ is not a self-adjoint operator, we shall consider also the equation adjoint to (43.1), namely

$$\mathbf{L}^* f^* = 0, \tag{43.2}$$

where f^* is a solution of (43.2) belonging to the set $\{f^*\}$.

Let us recall in addition that any function in $\{f\}$ and any function in $\{f^*\}$, which may not, in general, be solutions of (43.1) and (43.2), satisfy the general functional equation

$$(f^*, \mathbf{L}f) = (f, \mathbf{L}^* f^*). \tag{43.3}$$

Let us now consider another equation

$$\mathbf{L}' f' = 0, \tag{43.4}$$

where f' again belongs to $\{f\}$, so that $\mathbf{L}'$ can be written in the form

$$\mathbf{L}' = \mathbf{L} + \delta\mathbf{L}. \tag{43.5}$$

Let us agree to call **L**' the perturbed operator, **L** the unperturbed one, and **δL** the perturbation operator. It is clear that any linear operator **L**' can always be written in the form given by (43.5). Let us now insert (43.5) into (43.4). We then have

$$\mathbf{L}f' + \delta\mathbf{L}f' = 0. \tag{43.6}$$

The expression $\mathbf{L}f'$ in this equation will not, in general, vanish.

Now let us take the scalar product of (43.6) with f^*, a function which is a solution of (43.2). We then have

$$(f^*, \mathbf{L}f') + (f^*, \delta\mathbf{L}f') = 0. \tag{43.7}$$

Since the vector functions f' and f^* belong to $\{f\}$ and $\{f^*\}$, respectively, we may write

$$(f^*, \mathbf{L}f') = (f', \mathbf{L}^*f^*) = 0. \tag{43.8}$$

Here we have also used the fact that f^* is a solution of (43.2).

Using (43.8), we can rewrite (43.7) in the form

$$(f^*, \delta\mathbf{L}f') = 0. \tag{43.9}$$

This functional equation is the fundamental one for finding the change in the eigenvalue of the problem. Let us consider it in more detail. To do this, we write **L** in the form of a sum

$$\mathbf{L} = \mathbf{M} + \lambda\mathbf{N}, \tag{43.10}$$

where **M** and **N** are new operators, and λ is the eigenvalue of the problem.

It is important to note that according to (43.10) we may write (43.1) in the form

$$\mathbf{L}f = \mathbf{M}f + \lambda\mathbf{N}f = 0 \tag{43.11}$$

which is solved in terms of the eigenvalue λ.

If we now write

$$\mathbf{M}' = \mathbf{M} + \delta\mathbf{M}, \quad \mathbf{N}' = \mathbf{N} + \delta\mathbf{N} \text{ and } \lambda' = \lambda + \delta\lambda,$$

Equation (43.4) becomes

$$\mathbf{L}'f' = \mathbf{M}'f' + \lambda'\mathbf{N}'f' = 0. \tag{43.12}$$

At the same time,

$$\delta\mathbf{L}f' = \delta\mathbf{M}f' + \lambda\delta\mathbf{N}f' + \delta\lambda\mathbf{N}f'. \tag{43.13}$$

Let us now insert (43.13) into (43.9). This gives

$$(f^*, \delta\mathbf{M}f') + \lambda(f^*, \delta\mathbf{N}f') + \delta\lambda(f^*, \mathbf{N}f') = 0. \tag{43.14}$$

Finally, solving this for δλ, we arrive at

$$\delta\lambda = -\frac{(f^*, \delta\mathbf{M}f') + \lambda(f^*, \delta\mathbf{N}f')}{(f^*, \mathbf{N}f')}. \tag{43.15}$$

Now this equation can be written in the form

$$\delta\lambda = -\frac{1}{(f^*, Nf')} (f^*, \delta L f'), \tag{43.16}$$

where

$$\delta L f' = \delta M f' + \lambda \delta N f'.$$

Equations (43.15) and (43.16) are necessary functional equations for the variation of the eigenvalue of the homogeneous problem. If we assume that the variations $\delta\mathbf{M}$ and $\delta\mathbf{N}$ are small and that the solutions of (43.1) and (43.4) differ negligibly, the function f' in (43.16) can everywhere be replaced by f, which gives the following formula for small perturbations:

$$\delta\lambda = -\frac{(f^*, \delta L f)}{(f^*, Nf)}. \tag{43.17}$$

In order to calculate $\delta\lambda$ from (43.17) we must have a solution either of (43.1) or (43.2).

This illustrates the simplicity of perturbation theory for evaluating many small effects which are difficult to attack by other methods.

44. A Very Simple Example of the Application of Perturbation Theory

As a particularly simple example of the use of perturbation theory, consider the one-dimensional diffusion equation of Chapter I. We write the diffusion equation in the form

$$Lf = \left| \begin{matrix} \frac{df_2}{dx} + (\lambda - \Sigma) f_1 \\ D\frac{df_1}{dx} - f_2 \end{matrix} \right| = 0. \tag{44.1}$$

The adjoint of (44.1), as has already been shown, is

$$L^* f^* = \left| \begin{matrix} -\frac{dDf_2^*}{dx} + (\lambda - \Sigma) f_1^* \\ -\frac{df_1^*}{dx} - f_2^* \end{matrix} \right| = 0. \tag{44.2}$$

The solutions of (44.1) and (44.2) belong to the set $\{f\} \equiv \{f^*\}$, whose elements are continuous in G and satisfy the boundary conditions

$$f_1(0) = f_1(1) = 0,$$
$$f_1^*(0) = f_1^*(1) = 0.$$

In addition to the unperturbed Equation (44.1), consider the perturbed equation

$$L'f' = \left| \begin{matrix} \frac{df_2'}{dx} + (\lambda' - \Sigma') f_1' \\ D'\frac{df_1'}{dx} - f_2' \end{matrix} \right| = 0. \tag{44.3}$$

Further, let us write (44.1) in the form

$$Lf = Mf + \lambda N f, \tag{44.4}$$

where

$$\mathbf{M}f = \begin{vmatrix} \frac{df_2}{dx} - \Sigma f_1 \\ D\frac{df_1}{dx} - f_2 \end{vmatrix}, \quad \mathbf{N}f = \begin{vmatrix} f_1 \\ 0 \end{vmatrix}. \tag{44.5}$$

Similarly, let us write (44.3) in the form

$$\mathbf{L}'f' = \mathbf{M}'f' + \lambda'\mathbf{N}'f'. \tag{44.6}$$

We now consider the variations

$$\delta\mathbf{M} = \mathbf{M}' - \mathbf{M} \text{ and } \delta\mathbf{N} = \mathbf{N}' - \mathbf{N},$$

which are of the form

$$\delta\mathbf{M}f' = \begin{vmatrix} -\delta\Sigma f_1' \\ \delta D\frac{df_1'}{dx} \end{vmatrix}, \quad \delta\mathbf{N}f' = \begin{vmatrix} 0 \\ 0 \end{vmatrix}. \tag{44.7}$$

We can then write

$$\mathbf{L}'f' = \mathbf{L}f' + \delta\mathbf{L}f', \tag{44.8}$$

where

$$\delta\mathbf{L}f' = \delta\mathbf{M}f' + \delta\lambda\mathbf{N}'f' + \lambda\delta\mathbf{N}f'. \tag{44.9}$$

Taking the scalar product of (44.8) with $\boldsymbol{f}^*$, we obtain

$$\delta\lambda = -\frac{(f^*, \delta\mathbf{M}f') + \lambda(f^*, \delta\mathbf{N}f')}{(f^*, \mathbf{N}'f')}. \tag{44.10}$$

Further, we rewrite (44.10) in the form

$$\delta\lambda = -\frac{\int_0^1 \left(-\delta\Sigma f_1^* f_1' + \delta D f_2^* \frac{df_1'}{dx}\right) dx}{\int_0^1 f_1^* f_1' \, dx}. \tag{44.11}$$

We have thus arrived at an expression for the change in the eigenvalue λ. If we now replace $\boldsymbol{f}'$ by $\boldsymbol{f}$ in (44.11) we arrive at the small-perturbation formula

$$\delta\lambda = -\frac{\int_0^1 \left(-\delta\Sigma f_1^* f_1 + \delta D f_2^* \frac{df_1}{dx}\right) dx}{\int_0^1 f_1^* f_1 \, dx}. \tag{44.12}$$

The formulas of perturbation theory can be obtained in a similar way for more complicated equations.

45. Derivation of the General Formulas for δK_{eff}

Let us now apply perturbation theory to reactor equations. For convenience, we shall choose $\lambda = 1/K_{eff}$ as the eigenvalue of the problem.

In view of the extreme importance of perturbation theory in calculating many different effects in nuclear reactors, particularly small ones, let us start our considerations from the most general considerations.

As was shown in Chapter I, the set of basic reactor equations can be written

$$\mathbf{L}f = \left| \begin{array}{l} \vec{\Omega}\nabla\varphi + \Sigma\varphi - \int\limits_{u-r}^{u} du' \int d\Omega' \Sigma_s \varphi f(\mu_0, u-u') - \frac{1}{4\pi K_{eff}} \chi(u) Q \\ \vec{\Omega}\nabla\Phi + \Sigma_T\Phi - \int d\Omega' \Sigma_{sT}\Phi g(\mu_0) - \int\limits_{u_T - r}^{u_T} du' \int d\Omega' \Sigma_s \varphi F(\mu_0 u') \end{array} \right| = 0, \qquad (45.1)$$

where

$$Q = \nu_f \int d\Omega \left(\int\limits_{-\infty}^{u_T} \Sigma_f \varphi \, du + \Sigma_{fT} \Phi \right).$$

The solution of (45.1) is a vector function whose components are given by the column

$$f = \left| \begin{array}{c} \varphi \\ \Phi \end{array} \right|,$$

and which belongs to the set $\{f\}$.

As for the adjoint equations, again according to Chapter I they can be written

$$\mathbf{L}^* f^* = \left| \begin{array}{l} -\vec{\Omega}\nabla\varphi^* + \Sigma\varphi^* - \Sigma_s \int\limits_{u}^{u+r} du' \int d\Omega' \varphi^* f(\mu_0, u'-u) - \\ \quad - \Sigma_s \int d\Omega \Phi^* F(\mu_0, u') - \frac{1}{4\pi K_{eff}} \Sigma_f \cdot Q^* \\ -\vec{\Omega}\nabla\Phi^* + \Sigma_T \Phi^* - \Sigma_{sT} \int d\Omega' \Phi^* g(\mu_0) - \\ \quad - \frac{1}{4\pi K_{eff}} \Sigma_{fT} Q^* \end{array} \right| = 0, \qquad (45.2)$$

where

$$Q^* = \nu_f \int\limits_{-\infty}^{u_T} du \int d\Omega \chi(u) \varphi^*.$$

The solution of this equation is a vector

$$f^* = \left| \begin{array}{c} \varphi^* \\ \Phi^* \end{array} \right|$$

belonging to the set $\{f^*\}$ defined in Chapter I. We shall call Equation (45.1) the unperturbed equation.

Let us now consider, in addition to (45.1), the equation

$$
\mathbf{L}'f' = \left| \begin{array}{l} \vec{\Omega}\nabla\varphi' + \Sigma'\varphi' - \int\limits_{u-r'}^{u} du' \int d\Omega' \Sigma_s' \varphi' f'(\mu_0, u-u') - \\ \qquad - \frac{1}{4\pi K_{\mathrm{eff}}'} \chi(u) Q' \\ \vec{\Omega}\nabla\Phi' + \Sigma_T' \Phi' - \int d\Omega' \Sigma_{sT}' \Phi' g'(\mu_0) - \\ \qquad - \int\limits_{u_T - r'}^{u_T} du' \int d\Omega' \Sigma_s' \varphi' F'(\mu_0, u') \end{array} \right| = 0, \tag{45.3}
$$

where

$$
\psi' = \psi + \delta\psi. \tag{45.4}
$$

In the space $\mathbf{G} \times \mathbf{\Omega} \times \mathbf{U}$ the vector functions $\boldsymbol{f}$ and $\boldsymbol{f}'$ are solutions of (45.1) and (45.3), and they both belong to the same set $\{\boldsymbol{f}\}$.

This means that the formulas of perturbation theory can be obtained by using the equation

$$
(f^*, \delta\mathbf{L}f') = 0, \tag{45.5}
$$

where $\boldsymbol{f}^*$ and $\boldsymbol{f}'$ are solutions of (45.2) and (45.3), and where

$$
\delta\mathbf{L}f' = \left| \begin{array}{l} \delta\Sigma\varphi' - \delta \int\limits_{u-r'}^{u} du' \int d\Omega' \Sigma_s' f'(\mu_0, u-u') - \\ \qquad - \frac{1}{4\pi K_{\mathrm{eff}}} \chi(u)\, \delta Q - \delta \frac{1}{K_{\mathrm{eff}}} \frac{\chi(u)}{4\pi} Q' \\ \delta\Sigma_T \Phi' - \delta \int d\Omega' \Sigma_{sT}' \Phi' g'(\mu_0) - \\ \qquad - \delta \int\limits_{u_T - r'}^{u_T} du' \int d\Omega' \Sigma_s' \varphi' F'(\mu_0, u') \end{array} \right| \tag{45.6}
$$

Recalling the definition of the scalar product for vectors $\boldsymbol{f}$ and $\boldsymbol{f}^*$ from the sets $\{\boldsymbol{f}\}$ and $\{\boldsymbol{f}^*\}$, as given in Chapter I, Equation (45.5) can be written in the form

$$
\begin{aligned}
(f^*, \delta\mathbf{L}f') = \int\limits_G d\vec{r} \int d\Omega \Bigg\{ \int\limits_{-\infty}^{u_T} du\, \varphi^* \Big[\delta\Sigma\varphi' - \\
- \delta \int\limits_{u-r'}^{u} du' \int d\Omega' \Sigma_s' \varphi' f'(\mu_0\, u-u') - \\
- \frac{1}{4\pi K_{\mathrm{eff}}} \chi(u)\, \delta Q - \delta \frac{1}{K_{\mathrm{eff}}} \frac{\chi(u) Q'}{4\pi} \Big] + \\
+ \Phi^* \Big[\delta\Sigma_T \Phi' - \delta \int d\Omega' \Sigma_{sT}' \Phi' g'(\mu_0) - \\
- \delta \int\limits_{u_T - r'}^{u_T} du' \int d\Omega' \Sigma_s' \varphi' F'(\mu_0, u') \Big] \Bigg\} = 0.
\end{aligned} \tag{45.7}
$$

We now solve this for the variation $\delta \frac{1}{K_{eff}}$,* arriving at

$$\delta \frac{1}{K_{eff}} = \frac{\nu_f}{\int\limits_G d\vec{r}\, Q^* Q'} \int\limits_G d\vec{r} \int d\Omega \left\{ \int\limits_{-\infty}^{u_T} du\, \varphi^* \left[\delta\Sigma\varphi' - \right.\right.$$

$$- \delta \int\limits_{u-r'}^{u} du' \int d\Omega' \Sigma_s' \varphi' f' (\mu_0, u-u') -$$

$$\left. - \frac{1}{4\pi K_{eff}} \nu(u)\, \delta Q \right] + \Phi^* \left[\delta\Sigma_T \Phi' - \delta \int d\Omega' \Sigma_{sT}' \Phi' g' (\mu_0) - \right.$$

$$\left.\left. - \delta \int\limits_{u_T - r'}^{u_T} du' \int d\Omega' \Sigma_s' \varphi' F' (\mu_0, u') \right] \right\}. \qquad (45.8)$$

If we now assume that the physical parameters of the reactor undergo only small perturbations, the solutions of (45.1) and (45.3) will also be small. We may therefore replace f' in (45.8) by f. We then arrive at the small-perturbation formula

$$\delta \frac{1}{K_{eff}} = \frac{\nu_f}{\int\limits_G d\vec{r} Q^* Q} \int\limits_G d\vec{r} \int d\Omega \left\{ \int\limits_{-\infty}^{u_T} du\, \varphi^* \left[\delta\Sigma\varphi - \right.\right.$$

$$- \delta \int\limits_{u-r'}^{u} du' \int d\Omega' \Sigma_s' \varphi f' (\mu_0, u-u') -$$

$$\left. - \frac{1}{4\pi K_{eff}} \chi(u)\, \delta Q \right] + \Phi^* \left[\delta\Sigma_T \Phi - \delta \int d\Omega' \Sigma_{sT}' \Phi g' (\mu_0) - \right.$$

$$\left.\left. - \delta \int\limits_{u_T - r'}^{u_T} du' \int d\Omega' \Sigma_s' \varphi F' (\mu_0, u') \right] \right\}. \qquad (45.9)$$

Let us now analyze this equation. To do this we write it in the form

$$\delta \frac{1}{K_{eff}} = \frac{\nu_f}{\int\limits_G d\vec{r} Q^* Q} (f^*, \delta \mathbf{L} f), \qquad (45.10)$$

where

$$\delta \mathbf{L} f = \left| \begin{matrix} \delta\Sigma\varphi - \delta \int\limits_{u-r'}^{u} du' \int d\Omega' \Sigma_s' \varphi f' (\mu_0, u-u') - \\ - \frac{1}{4\pi K_{eff}} \chi(u)\, \delta Q \\ \delta\Sigma_T \Phi - \delta \int d\Omega' \Sigma_{sT}' \Phi g' (\mu_0) - \\ - \delta \int\limits_{u_T - r'}^{u_T} du' \int d\Omega' \Sigma_s' \varphi F' (\mu_0, u') \end{matrix} \right| \qquad (45.11)$$

---

* The transition from $\delta \frac{1}{K_{eff}}$ to δK_{eff} presents no difficulty, and is given by

$$\delta K_{eff} = - K_{eff} \cdot K_{eff}' \cdot \delta \left(\frac{1}{K_{eff}} \right).$$

The vector $\delta \mathbf{L} f$ has two components, the first of which is

$$(\delta \mathbf{L} f)_1 = \delta\Sigma\varphi - \delta \int\limits_{u-r'}^{u} du' \int d\Omega' \Sigma_s' \varphi f' (\mu_0, u - u') - \frac{1}{4\pi K_{\text{eff}}} \chi(u)\, \delta Q \tag{45.12}$$

and represents the number of neutrons arising in or leaving an element of phase space $(\vec{r}, \vec{\Omega}, u)$ due to changes in the scattering, absorbing, and fission properties of the reactor, and the second of which is

$$(\delta \mathbf{L} f)_2 = \delta\Sigma_T\Phi - \delta \int d\Omega' \Sigma_{sT}' \Phi g' (\mu_0) - \delta \int\limits_{u_T - r'}^{u_T} du' \int d\Omega' \Sigma_s' \varphi F'' (\mu_0, u') \tag{45.13}$$

and represents the number of neutrons arising in and leaving the volume element of phase space due to variations in the scattering and absorbing properties at thermal energies.

If, for instance, the scattering and fission properties of the medium have not changed, $\delta \mathbf{L} f$ takes on the particularly simple form

$$\delta \mathbf{L} f = \begin{vmatrix} \delta\Sigma_c \varphi \\ \delta\Sigma_{cT}\Phi \end{vmatrix}, \tag{45.14}$$

where Σ_c is the cross section for radiative capture.

The components of this vector are experimentally measurable quantities, namely the numbers of neutrons absorbed per second.

Let us assume that at two different points $\vec{r}$ and $\vec{r}_1$ of the reactor we can record the number of neutrons of lethargy u and with direction $\vec{\Omega}$ absorbed additionally as a result of small perturbations. We will also assume that the number of additional absorptions is the same at both points. One may ask whether the neutron absorption at these points also leads to equal changes in the reactivity of the reactor. The answer to this question is given by (45.10), which states that the perturbation is proportional to f^*, the adjoint vector function whose components are given by the column

$$\begin{vmatrix} \varphi^* \\ \Phi^* \end{vmatrix}.$$

Thus, if f^* is the same at $\vec{r}$ and $\vec{r}_1$, the changes in the reactivity will also be the same.

Calculations show, however, that as a rule f^* changes greatly both from point to point and as a function of lethargy, so that these changes must be included when calculating the reactivity. Similar statements can be made with respect to the changes in the scattering and fission properties of the medium.

Thus, the function f^* in (45.10) is an additional weight function by which we must multiply the experimentally observed number of neutrons absorbed. In the literature f^* is usually called the neutron importance function, or simply the importance. A detailed analysis of the various terms in Equation (45.10) has been given by L.N. Usachev [82].

46. δK_{eff} in the Diffusion Approximation

Let us now obtain the perturbation theory formulas for the reactor equations in the diffusion approximation. With this end in view, we consider the basic and adjoint reactor equations in the form

$$
\mathbf{L}f = \left| \begin{array}{c}
\nabla\vec{\varphi}_1 + \Sigma\varphi_0 - \int\limits_{u-r}^{u} du'\Sigma_s\varphi_0 f_0(u-u') - \frac{1}{K_{\text{eff}}}\chi(u)Q \\
\nabla\varphi_0 + 3\Sigma\vec{\varphi}_1 - 3\int\limits_{u-r}^{u} du'\Sigma_s\vec{\varphi}_1 f_1(u-u') \\
\nabla\vec{\Phi}_1 + \Sigma_{cT}\Phi_0 - \int\limits_{u_T-r}^{u_T} du'\Sigma_s\varphi_0 F_0(u') \\
\nabla\Phi_0 + 3\Sigma_{trT}\vec{\Phi}_1 - 3\int\limits_{u_T-r}^{u_T} du'\Sigma_s\vec{\varphi}_1 F_1(u')
\end{array} \right| = 0, \qquad (46.1)
$$

where

$$
f = \left| \begin{array}{c} \varphi_0 \\ \vec{\varphi}_1 \\ \Phi_0 \\ \vec{\Phi}_1 \end{array} \right|
$$

and

$$
\mathbf{L}^*f^* = \left| \begin{array}{c}
-\nabla\vec{\varphi}_1^* + \Sigma\varphi_0^* - \Sigma_s \int\limits_{u}^{u+r} \varphi_0^* f_0(u'-u)\,du' - \Sigma_s\Phi_0^* F_0(u) - \\
-\frac{1}{K_{\text{eff}}}\Sigma_f Q^* \\
-\nabla\varphi_0^* + 3\Sigma\vec{\varphi}_1^* - 3\Sigma_s \int\limits_{u}^{u+r} \vec{\varphi}_1^* f_1(u'-u)\,du' - 3\Sigma_s\vec{\Phi}_1^* F_1(u) \\
-\nabla\vec{\Phi}_1^* + \Sigma_{cT}\Phi_0^* - \frac{1}{K_{\text{eff}}}\Sigma_{fT} Q^* \\
-\nabla\Phi_0^* + 3\Sigma_{trT}\vec{\Phi}_1^*
\end{array} \right| = 0, \qquad (46.2)
$$

where

$$
f^* = \left| \begin{array}{c} \varphi_0^* \\ \vec{\varphi}_1^* \\ \Phi_0^* \\ \vec{\Phi}_1^* \end{array} \right| .
$$

It is assumed further that the solutions of (46.1) and (46.2) belong to the sets $\{f\}$ and $\{f^*\}$, defined in Section 11.

Together with (46.1), we consider

$$
\mathbf{L}'f' = \left| \begin{array}{c}
\nabla\vec{\varphi}_1' + \Sigma'\varphi_0' - \int\limits_{u-r'}^{u} du'\Sigma_s'\varphi_0' f_0'(u-u') - \frac{1}{K_{\text{eff}}'}\chi(u)Q' \\
\nabla\varphi_0' + 3\Sigma'\vec{\varphi}_1' - 3\int\limits_{u-r'}^{u} du'\Sigma_s'\vec{\varphi}_1' f_1'(u-u') \\
\nabla\vec{\Phi}_1' + \Sigma_{cT}'\Phi_0' - \int\limits_{u_T-r'}^{u_T} du'\Sigma_s'\varphi_0' F_0'(u') \\
\nabla\Phi_0' + 3\Sigma_{trT}'\vec{\Phi}_1' - 3\int\limits_{u_T-r'}^{u_T} du'\Sigma_s'\vec{\varphi}_1' F_1'(u')
\end{array} \right| = 0. \qquad (46.3)
$$

It is clear that the solution of (46.3) will belong to $\{f\}$. Therefore the formula for $\delta \frac{1}{K_{eff}}$ can be obtained by using the functional equation

$$(\boldsymbol{f}^*, \delta \mathbf{L} \boldsymbol{f}') = 0, \tag{46.4}$$

where $\boldsymbol{f}^*$ and $\boldsymbol{f}'$ are solutions of (46.2) and (46.3), and

$$\delta \mathbf{L} f' = \begin{vmatrix} \delta\Sigma\varphi_0' - \delta \int\limits_{u-r'}^{u} du' \Sigma_s' \varphi_0' f_0' (u-u') - \\ - \frac{1}{K_{eff}} \chi(u)\, \delta Q - \delta \frac{1}{K_{eff}} \chi(u)\, Q' \\ 3\delta\Sigma\vec{\varphi}_1' - 3\delta \int\limits_{u-r'}^{u} du' \Sigma_s' \vec{\varphi}_1' f_1' (u-u') \\ \delta\Sigma_{cT}\, \Phi_0' - \delta \int\limits_{u_T-r'}^{u_T} du' \Sigma_s' \varphi_0' F_0'(u') \\ 3\delta\Sigma_{trT} \vec{\Phi}_1' - 3\delta \int\limits_{u_T-r'}^{u_T} du' \Sigma_s' \vec{\varphi}_1' F_1'(u') \end{vmatrix} = 0. \tag{46.5}$$

Inserting this last equation into (46.4), we obtain

$$\begin{aligned} (f^*, \delta \mathbf{L} f') = \int\limits_G d\vec{r} \Big\{ \int\limits_{-\infty}^{u_T} \Big[\varphi_0^* \Big(\delta\Sigma\varphi_0' - \\ - \delta \int\limits_{u-r'}^{u} du' \Sigma_s' \varphi_0' f_0' (u-u') - \frac{1}{K_{eff}} \chi(u)\, \delta Q - \\ - \delta \frac{1}{K_{eff}} \chi(u)\, Q' \Big) + \vec{\varphi}_1^* \Big(3\delta\Sigma\vec{\varphi}_1' - 3\delta \int\limits_{u-r'}^{u} du' \Sigma_s' \vec{\varphi}_1' f_1' (u-u') \Big) \Big] du + \\ + \Phi_0^* \Big(\delta\Sigma_{cT}\, \Phi_0' - \delta \int\limits_{u_T-r'}^{u_T} du' \Sigma_s' \varphi_0' F_0'(u') \Big) + \\ + \vec{\Phi}_1^* \Big(3\delta\Sigma_{trT} \vec{\Phi}_1' - 3\delta \int\limits_{u_T-r'}^{u_T} du' \Sigma_s' \vec{\varphi}_1' F_1'(u') \Big) \Big\} = 0. \end{aligned} \tag{46.6}$$

Now solving for $\delta \frac{1}{K_{eff}}$, we arrive at

$$\begin{aligned} \delta \frac{1}{K_{eff}} = \frac{\nu_f}{\int\limits_G d\vec{r}\, Q^* Q'} \cdot \int\limits_G d\vec{r} \Big\{ \int\limits_{-\infty}^{u_T} \Big[\varphi_0^* \Big(\delta\Sigma\varphi_0' - \delta \int\limits_{u-r'}^{u} du' \Sigma_s' \varphi_0' f_0' (u-u') - \\ - \frac{1}{K_{eff}} \chi(u)\, \delta Q \Big) + \vec{\varphi}_1^* \Big(3\delta\Sigma\vec{\varphi}_1' - 3\delta \int\limits_{u-r'}^{u} du' \Sigma_s' \vec{\varphi}_1' f_1' (u-u') \Big) \Big] du + \\ + \Phi_0^* \Big(\delta\Sigma_{cT}\, \Phi_0' - \delta \int\limits_{u_T-r'}^{u_T} du' \Sigma_s' \varphi_0' F_0'(u') \Big) + \\ + \vec{\Phi}_1^* \Big(3\delta\Sigma_{trT} \vec{\Phi}_1' - 3\delta \int\limits_{u_T-r'}^{u_T} du' \Sigma_s' \vec{\varphi}_1' F_1'(u') \Big) \Big\}. \end{aligned} \tag{46.7}$$

The formula for the case of small perturbations is obtained from this one by replacing $\boldsymbol{f}'$ by $\boldsymbol{f}$.

47. δK_{eff} in the Diffusion-Age Approximation

In this paragraph we will obtain one of the most generally used formulas for δK_{eff} in the diffusion-age approximation.

Consider the vector form of the basic reactor equations:

$$\mathbf{L}f = \begin{vmatrix} \nabla\vec{\varphi}_1 + \Sigma_c\varphi_0 + \frac{\partial}{\partial u}\xi\Sigma_s\varphi_0 - \frac{1}{K_{eff}}\chi(u)Q \\ \nabla\varphi_0 + 3\Sigma_{tr}\vec{\varphi}_1 \\ \nabla\vec{\Phi}_1 + \Sigma_{cT}\Phi_0 - \xi\Sigma_{sT}\varphi_{0T} \\ \nabla\Phi_0 + 3\Sigma_{trT}\vec{\Phi}_1 \end{vmatrix} = 0, \tag{47.1}$$

where

$$f = \begin{vmatrix} \varphi_0 \\ \vec{\varphi}_1 \\ \Phi_0 \\ \vec{\Phi}_1 \end{vmatrix}.$$

The solution of (47.1) belongs to the set $\{f\}$ defined in Section 11. The adjoint reactor equations are

$$\mathbf{L}^*f^* = \begin{vmatrix} -\nabla\vec{\varphi}_1^* + \Sigma_c\varphi_0^* - \xi\Sigma_s\frac{\partial\varphi_0^*}{\partial u} - \frac{1}{K_{eff}}\Sigma_f Q^* \\ -\nabla\varphi_0^* + 3\Sigma_{tr}\vec{\varphi}_1^* \\ -\nabla\vec{\Phi}_1^* + \Sigma_{cT}\Phi_0^* - \frac{1}{K_{eff}}\Sigma_{fT}Q^* \\ -\nabla\Phi_0^* + 3\Sigma_{trT}\vec{\Phi}_1^* \end{vmatrix} = 0, \tag{47.2}$$

where

$$f^* = \begin{vmatrix} \varphi_0^* \\ \vec{\varphi}_1^* \\ \Phi_0^* \\ \vec{\Phi}_1^* \end{vmatrix}$$

and we assume that the solution of (47.2) belongs to $\{f^*\}$.

We now consider the perturbed reactor equation

$$\mathbf{L}'f' = \begin{vmatrix} \nabla\vec{\varphi}_1' + \Sigma_c'\varphi_0' + \frac{\partial}{\partial u}\xi\Sigma_s'\varphi_0' - \frac{1}{K_{eff}'}\chi(u)Q' \\ \nabla\varphi_0' + 3\Sigma_{tr}'\vec{\varphi}_1' \\ \nabla\vec{\Phi}_1' + \Sigma_{cT}'\Phi_0' - \xi\Sigma_{sT}'\varphi_{0T}' \\ \nabla\Phi_0' + 3\Sigma_{trT}'\vec{\Phi}_1' \end{vmatrix} = 0. \tag{47.3}$$

As before, we obtain $\delta\frac{1}{K_{eff}}$ using the functional equation

$$(f^*,\ \delta \mathbf{L} f') = 0, \tag{47.4}$$

where

$$\delta \mathbf{L} f' = \begin{vmatrix} \delta\Sigma_c\varphi_0' + \dfrac{\partial}{\partial u}\delta\xi\Sigma_s\varphi_0' - \dfrac{1}{K_{\text{eff}}}\chi(u)\delta Q - \delta\dfrac{1}{K_{\text{eff}}}\chi(u)Q' \\ 3\delta\Sigma_{tr}\vec{\varphi}_1' \\ \delta\Sigma_{c\mathrm{T}}\Phi_0' - \delta\xi\Sigma_{s\mathrm{T}}\varphi_{0\mathrm{T}}' \\ 3\delta\Sigma_{tr\mathrm{T}}\vec{\Phi}_1' \end{vmatrix} = 0. \tag{47.5}$$

Combining these last two equations, we arrive at

$$(f^*, \delta \mathbf{L} f') = \int_G d\vec{r} \left\{ \int_{-\infty}^{u_{\mathrm{T}}} \left[\varphi_0^* \left(\delta\Sigma_c\varphi_0' + \frac{\partial}{\partial u}\delta\xi\Sigma_s\varphi_0' - \frac{1}{K_{\text{eff}}}\chi(u)\delta Q - \right.\right.\right.$$
$$\left. - \delta\frac{1}{K_{\text{eff}}}\chi(u)Q' \right) + 3\delta\Sigma_{tr}\vec{\varphi}_1^*\vec{\varphi}_1' \Big] du + \Phi_0^*(\delta\Sigma_{c\mathrm{T}}\Phi_0' - \delta\xi\Sigma_{s\mathrm{T}}\varphi_{0\mathrm{T}}') +$$
$$\left. + 3\delta\Sigma_{tr\mathrm{T}}\vec{\Phi}_1^*\vec{\Phi}_1' \right\} = 0. \tag{47.6}$$

This solution is

$$\delta\frac{1}{K_{\text{eff}}} = \frac{\nu_f}{\int_G d\vec{r}\, Q^*Q'} \cdot \int_G d\vec{r} \left\{ \int_{-\infty}^{u_{\mathrm{T}}} du \left[\delta\Sigma_c\varphi_0^*\varphi_0' - \delta\xi\Sigma_s\varphi_0' \frac{\partial\varphi_0^*}{\partial u} - \frac{1}{K_{\text{eff}}}\varphi_0^*\chi(u)\delta Q + \right.\right.$$
$$\left. + 3\delta\Sigma_{tr}\vec{\varphi}_1^*\vec{\varphi}_1' \Big] + \delta\Sigma_{c\mathrm{T}}\Phi_0^*\Phi_0' + 3\delta\Sigma_{tr\mathrm{T}}\vec{\Phi}_1^*\vec{\Phi}_1' \right\}. \tag{47.7}$$

To obtain this formula we performed the following integration by parts:

$$\int_{-\infty}^{u_{\mathrm{T}}} \varphi_0^* \frac{\partial}{\partial u}\delta\xi\Sigma_s\varphi_0' du = \varphi_0^*\delta\xi\Sigma_s\varphi_0' \Big|_{-\infty}^{u_{\mathrm{T}}} - \int_{-\infty}^{u_{\mathrm{T}}} \delta\xi\Sigma_s\varphi_0' \frac{\partial\varphi_0^*}{\partial u} du =$$
$$= \delta\xi\Sigma_{s\mathrm{T}}\Phi_0^*\varphi_{0\mathrm{T}}' - \int_{-\infty}^{u_{\mathrm{T}}} \delta\xi\Sigma_s\varphi_0' \frac{\partial\varphi_0^*}{\partial u} du. \tag{47.8}$$

Let us eliminate $\vec{\varphi}_1$, $\vec{\varphi}_1^*$, $\vec{\Phi}_1^*$ and $\vec{\Phi}_1$ from (47.7) using the equations

$$\left.\begin{aligned} \vec{\varphi}_1 &= -D\nabla\varphi_0, & \vec{\varphi}_1^* &= D\nabla\varphi_0^*, \\ \vec{\Phi}_1 &= -D_{\mathrm{T}}\nabla\Phi_0, & \vec{\Phi}_1^* &= D_{\mathrm{T}}\nabla\Phi_0^*. \end{aligned}\right\} \tag{47.9}$$

Then (47.7) becomes

$$\delta\frac{1}{K_{\text{eff}}} = \frac{\nu_f}{\int_G d\vec{r}\, Q^*Q'} \int_G d\vec{r} \left\{ \int_{-\infty}^{u_{\mathrm{T}}} du \left[\delta\Sigma_c\varphi_0^*\varphi_0' - \delta\xi\Sigma_s\varphi_0' \frac{\partial\varphi_0^*}{\partial u} + \right.\right.$$
$$\left. + \delta D(\nabla\varphi_0^*, \nabla\varphi_0') - \frac{1}{K_{\text{eff}}}\varphi_0^*\chi(u)\delta Q \right] +$$
$$\left. + \delta\Sigma_{c\mathrm{T}}\Phi_0^*\Phi_0' + \delta D_{\mathrm{T}}(\nabla\Phi_0^*, \nabla\Phi_0') \right\}, \tag{47.10}$$

where

$$\delta D = -\frac{\delta(3\Sigma_{tr})}{(3\Sigma_{tr})^2}.$$

If we replace $\boldsymbol{f}'$ by $\boldsymbol{f}$ in (47.10), we obtain the formula for small perturbations.

48. Perturbations of the Boundary Conditions on the Surface of a Control Rod*

In calculating the reactivity equivalent of a control rod in the center of the core of a cylindrical reactor, one usually uses effective boundary conditions on the surface of the reactor (see Sections 54-56). As has already been mentioned, the boundary conditions in diffusion theory are

$$\left.\begin{aligned} \frac{\partial \varphi_0}{\partial r} &= \frac{1}{\gamma}\varphi_0 \\ \frac{\partial \Phi_0}{\partial r} &= \frac{1}{\gamma_T}\Phi_0 \end{aligned}\right\} \quad \text{for} \quad r = r_0, \tag{48.1}$$

where $\gamma = \gamma(u)$ is a function which can be calculated from the asymptotic solution of the kinetic equations, and r_0 is the radius of the rod.

Now we do not get sufficiently accurate results if we calculate δK_{eff} as the difference in the K_{eff} values of a reactor with and without control rods. We therefore turn to perturbation theory. Consider the basic and adjoint reactor equations (47.1) and (47.2) in the diffusion-age approximation. We write the perturbed reactor equations in the form

$$\mathbf{L}'\boldsymbol{f}' = \begin{vmatrix} \nabla\vec{\varphi}_1' + \Sigma_c\varphi_0' + \frac{\partial}{\partial u}\xi\Sigma_s\varphi_0' - \frac{1}{K_{eff}'}\chi(u)\,Q' \\ \nabla\varphi_0' + 3\Sigma_{tr}\vec{\varphi}_1' \\ \nabla\vec{\Phi}_1' + \Sigma_{cT}\Phi_0' - \xi\Sigma_{sT}\varphi_{0T}' \\ \nabla\Phi_0' + 3\Sigma_{trT}\vec{\Phi}_1' \end{vmatrix} = 0. \tag{48.2}$$

It is important to note that the solution to this equation does not belong to the set $\{\boldsymbol{f}\}$, since the components of $\boldsymbol{f}'$ satisfy the new condition (48.1) on the boundary of the rod $r = r_0$.

In this case, therefore, we cannot use the equation

$$(\boldsymbol{f}^*, \delta\mathbf{L}'\boldsymbol{f}') = 0 \tag{48.3}$$

(which is valid only if $\boldsymbol{f}$ and $\boldsymbol{f}'$ belong to the same set $\{\boldsymbol{f}\}$), so that we shall proceed in a different way.

Consider the functional

$$\begin{aligned}(\boldsymbol{f}^*, \mathbf{L}'\boldsymbol{f}') = \int_{G_0} d\vec{r} \Big\{ \int_{-\infty}^{u_T} \Big[\varphi_0^* \Big(\nabla\vec{\varphi}_1' + \Sigma_c\varphi_0' + \frac{\partial}{\partial u}\xi\Sigma_s\varphi_0' - \frac{1}{K_{eff}'}\chi(u)\,Q' \Big) + \\ + \vec{\varphi}_1^*(\nabla\varphi_0' + 3\Sigma_{tr}\vec{\varphi}_1') \Big] du + \Phi_0^*(\nabla\vec{\Phi}_1' + \Sigma_{cT}\Phi_0' - \xi\Sigma_{sT}\varphi_{0T}') + \\ + \vec{\Phi}_1^*(\nabla\Phi_0' + 3\Sigma_{trT}\vec{\Phi}_1') \Big\},\end{aligned} \tag{48.4}$$

where G_0 is the whole volume G except for the cylinder $0 \leq r \leq r_0$. Using the fact that

$$\left.\begin{aligned} \int_{G_0} d\vec{r}\varphi_0^*\nabla\vec{\varphi}_1' &= -\int_{G_0} d\vec{r}\,\vec{\varphi}_1'\nabla\varphi_0^* + \int_{S_0} ds\varphi_0^*(\vec{\varphi}_1')_{\vec{n}0} + \int_{S_{R_e}} ds\varphi_0^*(\vec{\varphi}_1')_{\vec{n}0}, \\ \int_{G_0} d\vec{r}\,\vec{\varphi}_1^*\nabla\varphi_0' &= -\int_{G_0} d\vec{r}\varphi_0'\nabla\vec{\varphi}_1^* + \int_{S_0} ds\varphi_0'(\vec{\varphi}_1^*)_{\vec{n}0} + \int_{S_{R_e}} ds\varphi_0'(\vec{\varphi}_1^*)_{\vec{n}0} \end{aligned}\right\} \tag{48.5}$$

---

*The methods developed in this section can be used to calculate δK_{eff} for reactors with nonmoderating screens. For this case, as is known, the effect of the screen can be accounted for by using a boundary condition in the form of (48.1).

(where S_0 is the surface of the cylinder of radius $r = r_0$, and S_{Re} is the outer extrapolated reactor surface) and similar relations for Φ_0, $\vec{\Phi}_1$ and Φ_0^*, $\vec{\Phi}_1^*$, we can rewrite (48.4) in the form

$$(f^*, L'f') = (f', L^*f^*) - \frac{1}{\nu_f}\delta\frac{1}{K_{eff}}\int_{G_0} d\vec{r}Q^*Q' + \int_{S_0} ds\left\{\int_{-\infty}^{u_T}[\varphi_0^*(\vec{\varphi}_1')_{\vec{n}0} + \varphi_0'(\vec{\varphi}_1^*)_{\vec{n}0}]\,du + \Phi_0^*(\vec{\Phi}_1')_{\vec{n}0} + \Phi_0'(\vec{\Phi}_1^*)_{\vec{n}0}\right\}. \qquad (48.6)$$

If the functions f^* and f' are now chosen as solutions of (47.2) and (48.2), this last equation can be rewritten

$$\delta\frac{1}{K_{eff}} = \frac{\nu_f}{\int_{G_0} d\vec{r}Q^*Q'}\int_{S_0} ds\left\{\int_{-\infty}^{u_T}[\varphi_0^*(\vec{\varphi}_1')_{\vec{n}0} + \varphi_0'(\vec{\varphi}_1^*)_{\vec{n}0}]\,du + \Phi_0^*(\vec{\Phi}_1')_{\vec{n}0} + \Phi_0'(\vec{\Phi}_1^*)_{\vec{n}0}\right\}. \qquad (48.7)$$

We now transform the right side of (48.7) using Conditions (48.1). Setting

$$\vec{\varphi}_1^* = 0, \quad \vec{\Phi}_1^* = 0 \quad \text{at} \quad \vec{r} = 0,$$

and using the conditions

$$\left.\begin{aligned} (\vec{\varphi}_1')_{\vec{n}0} &= -D\frac{\partial\varphi_0'}{\partial r} = -\frac{D}{\gamma}\varphi_0' \\ (\vec{\Phi}_1')_{\vec{n}0} &= -D_T\frac{\partial\Phi_0'}{\partial r} = -\frac{D_T}{\gamma_T}\Phi_0', \end{aligned}\right\} \text{ on } S_0, \qquad (48.8)$$

we represent (48.7) in the form

$$\delta\frac{1}{K_{eff}} = -\frac{2\pi r_0\nu_f}{\int_{G_0} d\vec{r}Q^*Q}\left[\int_{-\infty}^{u_T}\frac{D}{\gamma}\varphi^*\varphi'\,du + \frac{D_T}{\gamma_T}\Phi^*\Phi'\right]_{r=r_0}. \qquad (48.9)$$

In this equation we have also replaced $Q'(\vec{r})$ by $Q(\vec{r})$. This is valid since

$$\frac{\int_G d\vec{r}Q^*Q'}{\int_G d\vec{r}Q^*Q} = 1 + \varepsilon,$$

and ϵ is never greater than 1 or 2%. Equation (48.9) gives the desired value of $\delta\frac{1}{K_{eff}}$ in terms of the solutions f^* and f'.

This equation was first obtained by L.N. Usachev [82].

Let us now find an expression for $\delta\frac{1}{K_{eff}}$ for a rod at any point of a reactor core $(\vec{r} = \vec{r}_j)$. To do this we again use (48.7), which we transform using (48.8) to the form

$$\delta\frac{1}{K_{eff}} = -\frac{\nu_f}{\int_G d\vec{r}Q^*Q}\int_{S_0} ds\left\{\int_{-\infty}^{u_T}\left[\frac{D}{\gamma}\varphi^*\varphi' - D\frac{\partial\varphi^*}{\partial\rho}\varphi'\right]du + \frac{D_T}{\gamma_T}\Phi^*\Phi' - D_T\frac{d\Phi^*}{d\rho}\Phi'\right\}_{r=r_j}, \qquad (48.10)$$

where ρ is the running coordinate of the radius from a center at r_j, and as usual the prime denotes quantities associated with the perturbed state of the reactor (that is, with a control rod at r_j). In this equation, as before, G_0 can be replaced by G without significant error.

If we now assume that the control rod is sufficiently thin, φ' and Φ' are independent of the azimuth coordinate on the surface of the block.

Further, we assume that

$$\left.\begin{aligned} \frac{D}{\gamma}\varphi^* &\gg \frac{1}{2\pi r_0}\int_0^{2\pi r_0} D\frac{\partial\varphi^*}{\partial\rho}\,dl, \\ \frac{D_T}{\gamma_T}\Phi^* &\gg \frac{1}{2\pi r_0}\int_0^{2\pi r_0} D_T\frac{d\Phi^*}{d\rho}\,dl. \end{aligned}\right\} \tag{48.11}$$

and neglect the small terms in (48.10). We then obtain an equation analogous to (48.9), namely:

$$\delta\frac{1}{K_{eff}} = -\frac{2\pi r_0 \nu_f}{\int_G d\vec{r}\,Q^*Q}\left[\int_{-\infty}^{u_T}\frac{D}{\gamma}\varphi^*\varphi' du + \frac{D_T}{\gamma_T}\Phi^*\Phi'\right]_{r=r_j} \tag{48.12}$$

49. Important Applications of Perturbation Theory

Perturbation theory in general and the theory of small perturbations in particular has wide application to physical reactor calculations.

A nuclear reactor is essentially so complicated a structure that attempts at really accurate mathematical descriptions of the events taking place in it must overcome very great computational difficulties. Therefore, one usually selects from the very many processes taking place in a reactor those which are most important and involve only the fundamental principles of a chain reaction. The mathematical description of such a problem is no longer difficult, and the corresponding calculations of the basic reactor characteristics such as, for instance, K_{eff} or the critical mass, can be successfully carried out. As for the small effects, they are calculated by means of perturbation theory.

There is a large group of problems which can be handled by the theory of small perturbations. This includes, in particular, changes in the reactivity due to inhomogeneities in the scattering properties of the medium or the absorption and fission properties, as well as effects due to inelastic scattering, etc.

At the same time, however, there exist many important questions which cannot be solved by the theory of small perturbations. Most of these problems are related to large local perturbations in the energy distribution of the neutron density, as may occur when neutron absorption is large. This happens, for instance, when a control rod or a system of control rods is inserted into the reactor, or when there are experiments taking place involving absorbers, etc. In such a case the local neutron flux may be perturbed to an extent that cannot be neglected.

It is found that the calculation of δK_{eff} can often be improved by using methods which account approximately for the change in the neutron flux close to an absorber. Let us consider just one such method. Assume, for instance, that an absorber is introduced into the reactor at $r = r_j$. If we denote by $\varphi(\vec{r}, u)$ the original neutron flux in the reactor, and by $\varphi'(\vec{r}, u)$ the flux after the absorber has been introduced, the perturbed flux can be written in the form

$$\varphi'(\vec{r}, u) = \varphi(\vec{r}, u)\left[1 - \frac{\varphi(\vec{r}, u) - \varphi'(\vec{r}, u)}{\varphi(\vec{r}, u)}\right] \tag{49.1}$$

Let us now assume that

$$\frac{\varphi(\vec{r}, u) - \varphi'(\vec{r}, u)}{\varphi(\vec{r}, u)} = \frac{\varphi_\infty(\vec{r}, u) - \varphi'_\infty(\vec{r}, u)}{\varphi_\infty(\vec{r}, u)}, \tag{49.2}$$

is a reasonable approximation, where the index ∞ denotes the solutions for the problem in an infinite homogeneous medium with and without the absorber.

Obviously, the expression on the right side of this equation is a function of both u and the distance between the absorber and the point of observation, so that we may write it in the form*

$$G(|\vec{r}-\vec{r}_j|, u) = \frac{\varphi_\infty(\vec{r}, u) - \varphi'_\infty(\vec{r}, u)}{\varphi_\infty(\vec{r}, u)}. \tag{49.3}$$

Using this equation, we approximate (49.1) in the form

$$\varphi'(\vec{r}, u) = \varphi(\vec{r}, u)\,[1 - G(|\vec{r}-\vec{r}_j|, u)]. \tag{49.4}$$

This expression for the perturbed flux can now be used in the formulas of perturbation theory when treating the experimental observations.

Let us now turn to a consideration of the reactivity equivalent of a system of control rods in a core.

As was pointed out above, for such a case δK_{eff} is calculated by (48.9). It should be remarked that for simplicity it is convenient for us to calculate δK_{eff} for a single rod at the center of a reactor without end reflectors (the influence of end reflectors can then be included using the methods of Section 34). Only then do we evaluate the effect of the whole system of control rods located at various points $\vec{r}_j$ of the core.

Neglecting the interactions of the rods, or, equivalently, assuming them to be separated by large distances, the total effect of the system of control rods can be obtained by simply summing the individual effects.

Let us therefore consider (48.9) for a rod at the center r = 0 of the reactor and for a rod at $r = r_j$. We then have

$$\left.\begin{aligned}
\left(\delta\frac{1}{K_{eff}}\right)_0 &= -\frac{2\pi r_0 \nu_f}{\int\limits_G d\vec{r}Q^*Q}\left[\int\limits_{-\infty}^{u_T}\frac{D}{\gamma}\varphi^*\varphi' du + \left(\frac{D_T}{\gamma_T}\right)_0 \Phi^*\Phi'\right]_{r_0},\\
\left(\delta\frac{1}{K_{eff}}\right)_j &= -\frac{2\pi r_0 \nu_f}{\int\limits_G d\vec{r}Q^*Q}\left[\int\limits_{-\infty}^{u_T}\frac{D}{\gamma}\varphi^*\varphi' du + \left(\frac{D_T}{\gamma_T}\right)_j \Phi^*\Phi'\right]_{r_j}.
\end{aligned}\right\} \tag{49.5}$$

According to the above, we may write

$$\begin{aligned}
\left(\delta\frac{1}{K_{eff}}\right)_j = -\frac{2\pi r_0 \nu_f}{\int\limits_G d\vec{r}Q^*Q}\Big\{\int\limits_{-\infty}^{u_T}\left(\frac{D}{\gamma}\right)_j [1 - G(r_0, u)]\times\\
\times\varphi^*(r_j, u)\,\varphi(r_j, u)\,du + \left(\frac{D_T}{\gamma_T}\right)_j [1 - G_T(r_0)]\,\Phi^*(r_j)\,\Phi(r_j)\Big\},
\end{aligned} \tag{49.6}$$

where we have set

$$r - r_j = r_0.$$

Thus, if we have solved the basic and adjoint reactor equations for the unperturbed problem without rods, and have solved the basic equations for the problem with a rod in an infinite medium or in the center of the reactor, Equation (49.6) can be used to evaluate the reactivity equivalent of a rod at any point $r = r_j$ of the reactor core.

*We should note that the G function can be obtained approximately by solving the problems of a finite unperturbed reactor and a reactor perturbed by an absorber at its center.

Now from (49.6) we can obtain more simple expressions which, however, are valid only within certain limits. Let us assume that we can separate the variables according to

$$\left.\begin{aligned} \varphi(\vec{r}, u) &= \varphi(u)\,Q(\vec{r}), \quad \Phi(\vec{r}) = \Phi Q(\vec{r}), \\ \varphi^*(\vec{r}, u) &= \varphi^*(u)\,Q^*(\vec{r}), \quad \Phi^*(\vec{r}) = \Phi^* Q^*(\vec{r}). \end{aligned}\right\} \tag{49.7}$$

Then, inserting this into (49.5), we obtain

$$\left(\delta \frac{1}{K_{\text{eff}}}\right)_j = \left(\delta \frac{1}{K_{\text{eff}}}\right)_0 \frac{Q^*(\vec{r}_j)\,Q(\vec{r}_j)}{Q^*(0)\,Q(0)}. \tag{49.8}$$

It should be borne in mind, however, that (49.7) is usually valid only at points relatively far from the boundary between the reactor core and the reflector. For such points (49.7) is sufficiently accurate, and Equations (49.8) are satisfactory. As one approaches the reflector, however, the neutron energy spectrum begins to change significantly, so that for such points (49.8) is no longer valid. It then becomes necessary to use (49.6) as it stands. This equation can also be used to calculate the reactivity equivalent of control rods in the reflector, although to do this one must first find the corresponding G functions.

The use of the above approximate theory to calculate δK_{eff} due to a control rod close to a boundary with a reflector ($h < \sqrt{D/\Sigma}$), however, can lead to serious errors.

Let us now consider the interaction of the control rods, and how this influences the energy yield and reactivity of the reactor. We assume that there are control rods at the points $r = r_j$ ($j = 1, 2, \ldots, n$) of a reactor. If these rods are sufficiently far from each other, the perturbed neutron spectrum can be written approximately in the form

$$\left.\begin{aligned} \varphi'(\vec{r}, u) &= \varphi(\vec{r}, u)\left[1 - \sum_{j=1}^{n} G(|\vec{r} - \vec{r}_j|, u)\right], \\ \Phi'(\vec{r}) &= \Phi(\vec{r})\left[1 - \sum_{j=1}^{n} G_T(|\vec{r} - \vec{r}_j|)\right], \end{aligned}\right\} \tag{49.9}$$

where $\varphi(\vec{r}, u)$ and $\Phi(\vec{r})$ are the unperturbed neutron spectra in the reactor, and $G(|\vec{r} - \vec{r}_j|, u)$ and $G_T(|\vec{r} - \vec{r}_j|)$ are the G functions for the rods at the points $\vec{r} = \vec{r}_j$.*

Let us now calculate $Q(\vec{r})$, which describes the spatial distribution of the energy yield. It is easily seen that

$$Q(\vec{r}) = \nu_f \left\{ \int_{-\infty}^{u_T} \Sigma_f \varphi \left[1 - \sum_{j=1}^{n} G(|\vec{r} - \vec{r}_j|, u)\right] du + \right.$$
$$\left. + \Sigma_{fT}\Phi \left[1 - \sum_{j=1}^{n} G_T(|\vec{r} - \vec{r}_j|)\right]\right\}. \tag{49.10}$$

This equation gives an approximate idea of the energy yield in a reactor with the system of control rods included.

It now remains to calculate the effect on the reactivity due to the whole system of rods, including their interaction. It is easily seen that this is given by

$$\delta \frac{1}{K_{\text{eff}}} = -\frac{2\pi r_0 \nu_f}{\int\limits_G d\vec{r} Q^* Q} \cdot \sum_{h=1}^{n} \left\{ \int_{-\infty}^{u_T} \left(\frac{D}{\gamma}\right)_h \varphi_h^* \varphi_h \left[1 - \mathfrak{G}(\vec{r}_h, u)\right] du + \right.$$
$$\left. + \left(\frac{D_T}{\gamma_T}\right)_h \Phi_h^* \Phi_h \left[1 - \mathfrak{G}_T(\vec{r}_h)\right]\right\}, \tag{49.11}$$

---

*We are here assuming that all the rods are the same. If they are not, the G functions will depend on the form of the rods.

where

$$\varphi_k = \varphi(\vec{r}_k, u), \quad \Phi_k = \Phi(\vec{r}_k)$$

$$\mathfrak{G}(\vec{r}_k, u) = \sum_{j=1}^{n} G(|\vec{r}_k - \vec{r}_j|, u), \quad \mathfrak{G}_T(\vec{r}_k) = \sum_{j=1}^{n} G_T(|\vec{r}_k - \vec{r}_j|).$$

In conclusion, we note that the equations of perturbation theory can be used to evaluate the errors made in using approximate methods such as the diffusion or the diffusion-age approximation to perform the reactor calculations. This is done by varying the operators of the equations themselves, rather than the physical constants.

Chapter XII

HETEROGENEOUS EFFECTS IN NUCLEAR REACTORS

Our theoretical analysis has so far been restricted to homogeneous nuclear reactors.

As is known, heterogeneous reactors can, in most cases, also be treated by the numerical methods described. This is done by associating the heterogeneous reactor with an equivalent homogeneous reactor, whose effective parameters account in detail for the processes taking place in the lattice of the heterogeneous reactor. The basic difficulty lies in finding the space and energy distribution of the neutrons in the individual cells of the lattice. But once this is done, the results can be used to find the necessary effective constants.

In performing the calculations for a cell it is important to include the effect of self-shielding of the neutrons by the surface layers of a uranium block. It is found that the neutron flux is not the same at all radii, but decreases as one moves away from the center of the block. This means that the surface layers absorb neutrons more effectively than do the internal ones. Hence, with equal amounts of fuel, neutron capture in a heterogeneous reactor will be weaker than in a homogeneous one, in which the uranium is distributed uniformly throughout the core.

This situation is usually used when designing thermal-neutron reactors with low uranium concentrations. For such a reactor the self-shielding increases the probability of avoiding resonance capture by U^{238} nuclei at epithermal energies, which improves the quality of the reactor.*

In intermediate reactors, which use highly enriched uranium, most of the neutrons are captured at epithermal energies by U^{235} nuclei. The energies involved lie primarily within the region of "low" U^{238} resonances. Because of the relatively low concentration of U^{238}, the cross section for capture on these nuclei is small and therefore the probability of avoiding resonance capture by these nuclei can usually be calculated as for a homogeneous reactor (see Section 18). The situation is quite different when one considers capture by U^{235} nuclei. In this case, the cross section for capture (and fission) is more convenient to average after having subtracted out the largest resonance level. Many of the neutrons are absorbed with a probability given by this averaged capture cross section. In this process the self-shielding of the neutrons by the surface layers of the uranium block is very important. In the final analysis, the inclusion of self-shielding reduces to using a lower effective capture cross section than in the equivalent homogeneous medium. As for self-shielding of the U^{235} levels, it differs in no way from the same effect in U^{238}.

It follows from the above that when performing calculations for heterogeneous thermal reactors one must first calculate the effective resonance integral on the U^{238} nuclei and the effective capture and fission cross sections for the thermal neutron group. In intermediate reactors one must calculate the effective constants Σ_c^{eff} and Σ_f^{eff} for the slowing-down neutrons and Σ_{cT}^{eff} and Σ_{fT}^{eff} for the thermal neutrons. The remaining constants can be found by formal homogenization of the heterogeneous reactor cells. We should mention that the above effective constants can be obtained also by direct solution of the kinetic equations.

A method for calculating the effective resonance integral for "thin" blocks has been developed by Gurevich and Pomeranchuk [26], while a method for "thick" blocks has been developed by Wigner [23, 54]. V.V. Orlov has used Wigner's method to generalize both theories and obtain a formula for the effective resonance integral

*In thermal reactors the shielding is important for the great majority of U^{238} resonances which lie at low energies ($E \leq 10^4$ ev).

The shielding of the high-energy U^{238} resonance levels is so weak that the probability of avoiding resonance capture hardly differs from its value in a homogeneous reactor. This is true also for all U^{235} levels when the concentration of this isotope in a uranium block is low.

which can be used for uranium blocks of any thickness. T.Kh. Sedel'nikov* has worked out the theory of self-shielding for intermediate reactors, basing his work on the ideas of Gurevich and Pomeranchuk.

The work of Soviet scientists (S.M. Feinberg [84], A.D. Galanin [16, 18], G.A. Bat' [4], and others) has brought forth a new theory on which to base heterogeneous thermal neutron reactor calculations. S.M. Feinberg [84] has shown that the only assumptions necessary within this area are the following three.

1. The thermal neutron field is axially symmetric in the neighborhood of a block, so that the sources and sinks of the neutrons can be assumed linear.

2. An elementary diffusion equation can be used to describe the diffusion of thermal neutrons between the blocks.

3. The absorbing power of a uranium block is characterized by the logarithmic derivative of the neutron flux on the surface of the block.

With these assumptions the critical mass problem for a heterogeneous reactor can be reduced, using Green's function methods, to a homogeneous set of linear algebraic equations whose solution gives both the eigenvalue and eigenfunction of the problem.** This theory of heterogeneous calculations was later successfully applied to the solution of problems involving complex lattices [84].

It is important to note that the heterogeneous method gave a more rigorous basis to the earlier methods of effective homogenization and could, in many cases, be used to evaluate the errors involved in the earlier methods.

In the present chapter we shall derive the formulas for the effective resonance integral in a heterogeneous system, shall discuss self-shielding theory for intermediate reactors, shall obtain the asymptotic equations for the boundary conditions on the surface of an absorbing block, and shall analyze the Wigner-Seitz cell theory.***

50. The Gurevich-Pomeranchuk Theory of Resonance Absorption

Consider a cell of a heterogeneous thermal reactor consisting of a uranium block surrounded on all sides by a moderator.**** The neutrons produced in the fission process are slowed down to the U^{238} resonance energy in the cell of the reactor. If the distance between blocks of the heterogeneous lattice is sufficiently large, there is high probability that in further elastic scattering by the moderator nuclei the neutron will lose part of its energy and enter the uranium block at an energy below the resonance, thus avoiding resonance capture by U^{238}. In this way, resonance capture in a heterogeneous reactor is less probable than in a homogeneous one, where the neutrons are slowed down to resonance energies in a medium in which the capturing substance is uniformly dispersed, so that there is a high probability that they will be absorbed.

The Gurevich-Pomeranchuk theory [26] does not take into account neutron scattering by uranium or the elastic slowing-down process in the block associated with this phenomenon. It is therefore valid only for small blocks.

It is clear that resonance capture by a single level is small if the energy change in a single collision is much greater than the width of the resonance capture "danger zone" [26].***** Further, the actual resonance

*This work was performed in 1954.

**It should be mentioned, however, that this theory is not sufficiently good for blocks in which the neutrons are slowed down to a significant extent (a uranium block, for instance, cooled by water).

***All the calculations of the special functions tabulated in this chapter were performed by A.I. Vaskin.

****The author has found V.V. Orlov's comments on the theory of resonance capture quite valuable.

*****We shall consider the width of the "danger zone" to be an energy interval ΔE about the resonance energy E_j in which

$$\lambda_c \leqslant \bar{l}$$

where $\bar{l}$ is the mean dimension of the block [see Equation (50.47)] and λ_c is the capture mean free path (absorption length) in the block. We shall assume also that

$$\Delta E \ll E_j(1-\varepsilon), \qquad \varepsilon = \left(\frac{M-1}{M+1}\right)^2.$$

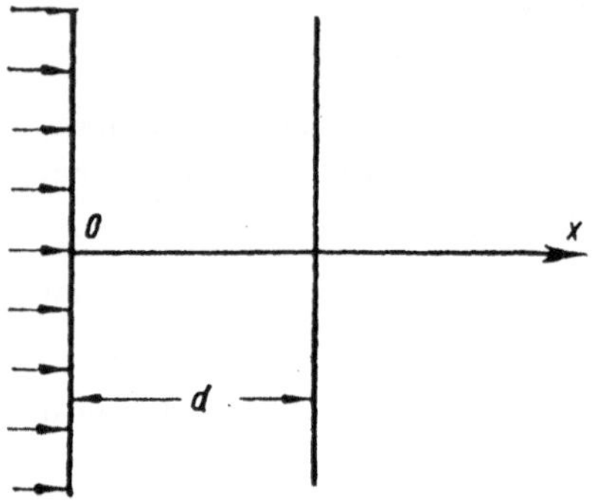

Fig. 7. A uranium layer of thickness d irradiated by a plane parallel neutron beam.

absorption will be less than the calculated one, since the Gurevich-Pomeranchuk theory does not take into account the perturbation of the resonance neutron field by the block.

Let us now calculate the probability of avoiding resonance capture for uranium blocks of various shapes. Let us first consider the simple case of a plane uranium layer of thickness d and surface area S immersed in the moderator. Consider a plane parallel beam of neutrons of density $\varphi(E)$ incident on this layer from the negative x direction (Fig. 7).

Let us calculate the number of neutrons which are incident on a unit area and are captured in the block per unit energy interval in passing through a thickness d. Let Σ_c be the macroscopic uranium capture cross section, and Σ_s the scattering cross section, given by

$$\Sigma_s = \Sigma_{sr} + \Sigma_{sn}, \tag{50.1}$$

where Σ_{sr} is the resonance cross section and Σ_{sn} accounts for nonresonance potential scattering.

We shall assume further that the uranium block is so thin that

$$\Sigma_{sn} d \ll 1.$$

With these assumptions it is clear that potential scattering can be ignored in the block so that we may set $\Sigma_{sn} = 0$ in (50.1).

The number of neutrons captured in a thickness dx per unit energy interval and per unit surface area will be

$$\Sigma_c \varphi \, dx. \tag{50.2}$$

On passing through a thickness x the flux of neutrons of energy E decays according to

$$\varphi = \varphi_0 (E) e^{-\Sigma x}, \tag{50.3}$$

where φ_0 is the unperturbed incident neutron flux, and $\Sigma = \Sigma_c + \Sigma_{sr}$ is the total cross section. We assume that the energy of the scattered neutron is not in the danger zone (ΔE).

With these assumptions, the number of neutrons absorbed in the uranium layer per unit energy interval is given by

$$p(E) = S\varphi_0 \int_0^d \Sigma_c e^{-\Sigma x} dx. \tag{50.4}$$

Integrating, we have

$$p(E) = S\varphi_0 \frac{\Sigma_c}{\Sigma} (1 - e^{-\Sigma d}). \tag{50.5}$$

This should now be integrated over all energies in the neighborhood of the j-th resonance. We thus consider

$$p_j = S \int_{E_j - \alpha}^{E_j + \alpha} \varphi_0 \frac{\Sigma_c}{\Sigma} (1 - e^{-\Sigma d}) \, dE, \tag{50.6}$$

where α is a constant of the order of magnitude of the resonance "danger zone." The functions φ_0 and Σ_c/Σ in the integrand of this expression vary slowly, so that we may take them outside the integral sign and assign them their values at the resonance energy E_j.* We then have

$$p_j = S\varphi_0(E_j)\frac{\Sigma_c^j}{\Sigma^j}\int\limits_{E_j-\alpha}^{E_j+\alpha}(1-e^{-\Sigma d})\,dE. \tag{50.7}$$

According to the Breit-Wigner formula (see, for instance, Akhiezer and Pomeranchuk [3])**

$$\Sigma_c = \frac{\Sigma_c^j}{1+\left(\frac{E-E_j}{\Gamma_j/2}\right)^2}, \quad \Sigma_{sr} = \frac{\Sigma_{sr}^j}{1+\left(\frac{E-E_j}{\Gamma_j/2}\right)^2}, \tag{50.8}$$

where Σ_c^j and Σ_{sr}^j are the capture and scattering cross sections at the maximum of the resonance, $\Sigma^j = \Sigma_c^j + \Sigma_{sr}^j$, and Γ_j is the width of the resonance, we rewrite (50.7) in the form

$$p_j = \Gamma_j S\varphi_0(E_j)\frac{\Sigma_c^j}{\Sigma^j}\int\limits_0^\infty\left(1-e^{-\frac{\Sigma^j d}{1+\xi^2}}\right)d\xi. \tag{50.9}$$

We have here introduced the new variable ξ, defined by

$$\xi = \frac{2}{\Gamma_j}(E-E_j) \tag{50.10}$$

and have extended the integral to infinity. We now rewrite (50.9) in the form

$$p_j = V\varphi_0(E_j)\frac{\pi\Sigma_c^j\Gamma_j}{2}f(\Sigma^j d), \tag{50.11}$$

where

$$f(\beta) = \frac{2}{\pi\beta}\int\limits_0^\infty\left(1-e^{-\frac{\beta}{1+\xi^2}}\right)d\xi,$$

and $V = Sd$ is the volume of the uranium layer. Noting that [97, 123]

$$\int\limits_0^\infty\left(1-e^{-\frac{\beta}{1+\xi^2}}\right)d\xi = \frac{\pi\beta}{2}e^{-\frac{\beta}{2}}\left[I_0\left(\frac{\beta}{2}\right)+I_1\left(\frac{\beta}{2}\right)\right], \tag{50.12}$$

where $I_0(x)$ and $I_1(x)$ are Bessel functions of an imaginary argument [68], we finally arrive at

$$f(\beta) = e^{-\frac{\beta}{2}}\left[I_0\left(\frac{\beta}{2}\right)+I_1\left(\frac{\beta}{2}\right)\right]. \tag{50.13}$$

Equation (50.11) can be used to find the neutron absorption in a uranium layer with one resonance level.

---

*We can take Σ_c/Σ outside the integral sign because $l\,\Sigma_c \gtrsim 1$ (by definition of ΔE), and $l\,\Sigma_{sn} \ll 1$, which means that within the limits of integration $\Sigma_{sn} \ll \Sigma_c$, so that $\frac{\Sigma_c}{\Sigma_c + \Sigma_{sr} + \Sigma_{sn}} \approx \frac{\Sigma_c}{\Sigma_c + \Sigma_{sr}} = \text{const.}$

**For simplicity, we assume in Equations (50.8) that we may set $\sqrt{E_j/E} = 1$ on the interval $E_j - \alpha \le E \le E_j + \alpha$.

Let us find the asymptotic behavior of $f(\beta)$ in the limits $\beta \ll 1$ and $\beta \gg 1$. When $\beta \ll 1$, we have

$$e^{-\frac{\beta}{2}}\left[I_0\left(\frac{\beta}{2}\right)+I_1\left(\frac{\beta}{2}\right)\right]=1+0(\beta). \tag{50.14}$$

In this case, therefore,

$$f(\beta)=1+0(\beta). \tag{50.15}$$

Let us now consider $\beta \gg 1$. Then

$$e^{-\frac{\beta}{2}}\left[I_0\left(\frac{\beta}{2}\right)+I_1\left(\frac{\beta}{2}\right)\right]=\frac{2}{\sqrt{\pi\beta}}\left[1+0\left(\frac{1}{\beta}\right)\right]. \tag{50.16}$$

This means that

$$f(\beta)=\frac{2}{\sqrt{\pi\beta}}\left[1+0\left(\frac{1}{\beta}\right)\right]. \tag{50.17}$$

All of the U^{238} resonance levels can now be separated into two groups – "low" resonances satisfying the condition

$$\Sigma^j d \gg 1 \qquad (j=1,\ 2,\ \ldots,\ j_0),$$

and "high" resonances satisfying the condition

$$\Sigma^j d \ll 1 \qquad (j=j_0+1,\ j_0+2,\ \ldots).$$

For these cases we can then use the asymptotic expressions (50.15) or (50.17). These we shall use in the further discussion of the Gurevich-Pomeranchuk theory.

Let us now go on to a calculation of the number of neutrons captured when the neutron flux is isotropic. We shall consider arbitrary geometry of the neutron block, assuming it to be bounded by some arbitrary convex surface S in the moderator (Fig. 8). Consider an elementary area dS whose center lies on a point M on the surface of the block.

Further, consider a ray whose direction is $\vec{\Omega}$ passing through M into the block. If we assume that the neutron flux is isotropic, that is if

$$\varphi(\vec{r},\vec{\Omega},E)=\frac{1}{4\pi}\varphi_0(E), \tag{50.18}$$

we can calculate the number of neutrons incident per unit time on the element of area dS in the direction $\vec{\Omega}$. This number is

$$\frac{|\Omega_n|\,dS}{4\pi}\varphi_0, \tag{50.19}$$

where $\Omega_n=(\vec{\Omega},\vec{n}^0)$ and $\vec{n}^0$ is the external normal to S.

The number of neutrons absorbed in a path length $d\xi$ per unit energy interval and per unit time is

$$\frac{|\Omega_n|\,dS}{4\pi}\Sigma_c\varphi\,d\xi. \tag{50.20}$$

Recalling that for neutrons with energy E the flux decays according to

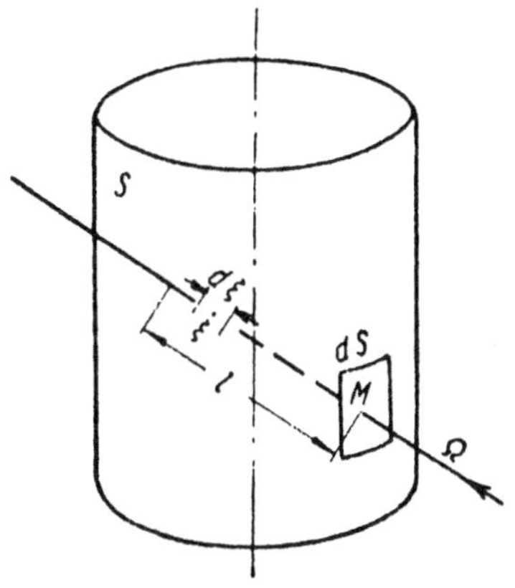

Fig. 8. The path of a neutron in an absorbing cylindrical block. l is the neutron path in the block; ξ is the coordinate along this line; dS is the element of area; $\vec{\Omega}$ is the unit vector in the neutron direction of motion.

$$\varphi=\varphi_0(E)\,e^{-\Sigma\xi}, \tag{50.21}$$

where $\varphi_0(E)$ is again the unperturbed neutron flux incident on the block, and that

$$\Sigma = \Sigma_c + \Sigma_{sr},$$

we find that the total number of neutrons entering the uranium block through the area dS in the $\vec{\Omega}$ direction and then absorbed by the block is

$$p(\vec{r},\vec{\Omega},E)=\frac{|\Omega_n|\,dS}{4\pi}\varphi_0(E)\int\limits_0^l \Sigma_c e^{-\Sigma\xi}\,d\xi. \tag{50.22}$$

Performing the indicated integration, we have

$$p(\vec{r},\vec{\Omega},E)=\frac{|\Omega_n|\,dS}{4\pi}\frac{\Sigma_c}{\Sigma}(1-e^{-\Sigma l})\varphi_0(E), \tag{50.23}$$

We now integrate over all energies in the neighborhood of the j-th resonance. This gives

$$p_j(\vec{r},\vec{\Omega})=\frac{|\Omega_n|\,dS}{4\pi}\int\limits_{E_j-\alpha}^{E_j+\alpha}\frac{\Sigma_c}{\Sigma}(1-e^{-\Sigma l})\,\varphi_0(E)\,dE, \tag{50.24}$$

where α is a quantity that has already been defined. As before, we will suppose that the functions $\varphi_0(E)$ and Σ_c/Σ vary slowly so that they can be taken outside the integral sign. This gives

$$p_j(\vec{r},\vec{\Omega})=\frac{|\Omega_n|\,dS}{4\pi}\frac{\Sigma_c^j}{\Sigma^j}\varphi_0(E_j)\int\limits_{E_j-\alpha}^{E_j+\alpha}(1-e^{-\Sigma l})\,dE. \tag{50.25}$$

We again use the Breit-Wigner formula to rewrite this equation in the form

$$p_j(\vec{r},\vec{\Omega})=\frac{\Gamma_j|\Omega_n|\,dS}{4\pi}\frac{\Sigma_c^j}{\Sigma^j}\varphi_0(E_j)\int\limits_0^\infty\left(1-e^{-\frac{\Sigma^j l}{1+\xi^2}}\right)d\xi. \tag{50.26}$$

According to (50.12) this can be written

$$p_j(\vec{r},\vec{\Omega})=\frac{\Gamma_j|\Omega_n|\,dS}{8}\frac{\Sigma_c^j}{\Sigma^j}\varphi_0(E_j)\,\Sigma^j l e^{-\frac{\Sigma^j l}{2}}\left[I_0\left(\frac{\Sigma_j l}{2}\right)+I_1\left(\frac{\Sigma_j l}{2}\right)\right]. \tag{50.27}$$

The equation we have derived refers to a unit solid angle about $\vec{\Omega}$. Let us now integrate over all possible directions on the unit hemisphere within the block. We then have

$$p_j(\vec{r})\,dS=\frac{\Gamma_j dS}{8}\frac{\Sigma_c^j}{\Sigma^j}\varphi_0(E_j)\int\limits_\Omega|\Omega_n|\Sigma^j l e^{-\frac{\Sigma^j l}{2}}\times$$
$$\times\left[I_0\left(\frac{\Sigma^j l}{2}\right)+I_1\left(\frac{\Sigma^j l}{2}\right)\right]d\Omega. \tag{50.28}$$

Integrating finally over S we obtain the total number of neutrons with the resonance energy E_j captured by the block, namely:

$$p_j = V\frac{\pi\Gamma_j}{2}\Sigma_c^j\varphi_0(E_j)\, f(\Sigma^j d), \tag{50.29}$$

where

$$f(\beta) = \frac{1}{4\pi V}\int_S dS \int_\Omega |\Omega_n|\, l\cdot e^{-\frac{\beta l'}{2}}\left[I_0\left(\frac{\beta l'}{2}\right)+I_1\left(\frac{\beta l'}{2}\right)\right] d\Omega, \tag{50.30}$$

$$V = \frac{1}{4\pi}\int_S dS \int_\Omega |\Omega_n|\, l\, d\Omega \tag{50.31}$$

(V is the volume of the block). We have assumed here that

$$l = dl',$$

where d is a characteristic length associated with the thickness of the block, and $l' = l'(\Omega)$ is a function only of direction (and, in general, of the location of M on S).

Equation (50.29) can be written

$$p_j = \frac{V}{\xi\Sigma_s^{\text{eff}}}\,\frac{\pi\Gamma_j\Sigma_c^j}{2E_j}\, f(\Sigma^j d), \tag{50.32}$$

where the neutron flux $\varphi_0(E)$ incident on the uranium block from the moderator is expressed in terms of unit slowing-down density (q = 1) by

$$\varphi_0(E) = \frac{1}{\xi\Sigma_s^{\text{eff}} E}, \tag{50.33}$$

where

$$\xi\Sigma_s^{\text{eff}} = \xi\Sigma_s^0 \frac{V^0}{V_{\text{cell}}}$$

(V^0 is the volume of moderator, and V_{cell} is the total volume of the cell). The superscript zero denotes quantities characterizing the moderator.

Let us now turn to the asymptotic behavior of $f(\beta)$ when $\beta \ll 1$ and $\beta \gg 1$.

First considering the case $\beta \ll 1$, Equation (50.14) gives

$$f(\beta) = 1 + 0(\beta).$$

For $\beta \gg 1$, on the other hand, we use (50.16). We thus obtain

$$f(\beta) = \frac{2}{\sqrt{\pi\beta}}\left\langle\frac{1}{\sqrt{l'}}\right\rangle\left[1+0\left(\frac{1}{\beta}\right)\right], \tag{50.34}$$

where

$$\left\langle\frac{1}{\sqrt{l'}}\right\rangle = \frac{1}{4\pi V}\int_S dS\int_\Omega d\Omega\, |\Omega_n|\, l\cdot\frac{1}{\sqrt{l'}}. \tag{50.35}$$

As before, we now separate all of the U^{238} resonances into two groups, for the first of which

$$\Sigma^j d \gg 1 \qquad (j = 1,\ 2,\ \ldots,\ j_0),$$

and for the second of which

$$\Sigma^j d \ll 1 \qquad (j = j_0 + 1,\ j_0 + 2,\ \ldots),$$

and this allows us to approximate the resonance absorption in the form

$$p = \frac{V}{\xi\Sigma_s^{\text{eff}}}\left(\sum_{j \geq j_0} \frac{\Sigma_c^j}{\Sigma^j}\frac{\pi\Sigma^j\Gamma_j}{2E_j} + \left\langle\frac{1}{\sqrt{l}}\right\rangle \sum_{j<j_0} \frac{\Sigma_c^j}{\Sigma^j}\frac{\sqrt{\pi\Sigma^j\Gamma_j^2}}{E_j}\right), \tag{50.36}$$

where, as before, $l = dl'$.

Transforming now from macroscopic to microscopic cross sections, we rewrite this equation in the form

$$p = \frac{N}{\xi\Sigma_s^{\text{eff}}} J^{\text{eff}}, \tag{50.37}$$

where $N = \rho_0^U V$ is the total number of uranium nuclei in the block, ρ_0^U is the number of metallic U^{238} nuclei per cm^3, and J_{eff} is the effective resonance integral; this latter is given by

$$J_{\text{eff}} = A + \frac{1}{\sqrt{d}} B, \tag{50.38}$$

where

$$A = \sum_{j \geq j_0} \frac{\sigma_c^j}{\sigma^j}\frac{\pi\sigma^j\Gamma_j}{2E_j}, \qquad B = \frac{1}{\sqrt{\rho_0^U}}\left\langle\frac{1}{\sqrt{l'}}\right\rangle \cdot \sum_{j<j_0} \frac{\sigma_c^j}{\sigma^j}\frac{\sqrt{\pi\sigma^j\Gamma_j^2}}{E_j}.$$

We should mention that the resonance parameters for U^{238} are still not very well known, so that when Equation (50.38) is used to calculate resonance capture, A and B are found from experiment. These constants were measured experimentally by M.B. Egiazarov, et al. [27] on cylindrical blocks of metallic uranium. They obtained

$$A = 5 \cdot 10^{-24}\ \text{cm}^2, \quad B = 9.5 \cdot 10^{-24}\ \text{cm}^{5/2}.$$

Thus, the final formula for J_{eff} is

$$J_{\text{eff}} = \left(5.0 + \frac{\alpha}{\sqrt{d}} \cdot 9.5\right) \cdot 10^{-24}, \tag{50.39}$$

where α is a geometric parameter defined by

$$\alpha = \frac{\left\langle\frac{1}{\sqrt{l'}}\right\rangle}{\left\langle\frac{1}{\sqrt{l'}}\right\rangle_{\text{cylinder}}}. \tag{50.40}$$

Equation (50.39) can be used only for blocks of natural metallic uranium. On the other hand, it can be used even for porous blocks. Then ρ_0^U must be replaced by ρ^U, the number of nuclei per cm^3 of porous uranium.

Writing

$$\rho^U = \varepsilon\rho_0^U,$$

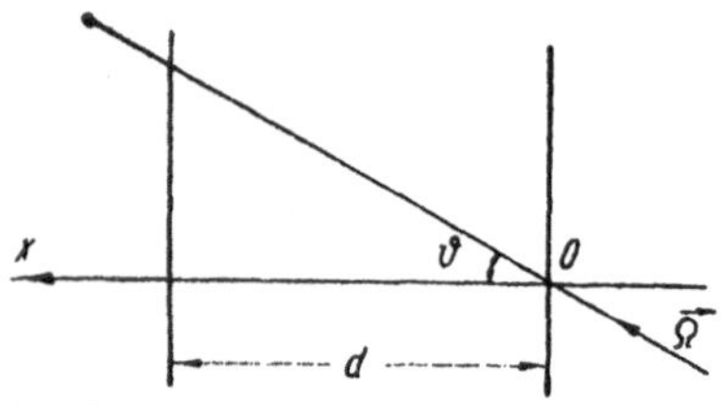

Fig. 9. Coordinate system for a plane uranium layer of thickness d. $\vec{\Omega}$ is the unit vector along the neutron velocity; ϑ is the angle between $\vec{\Omega}$ and the x axis.

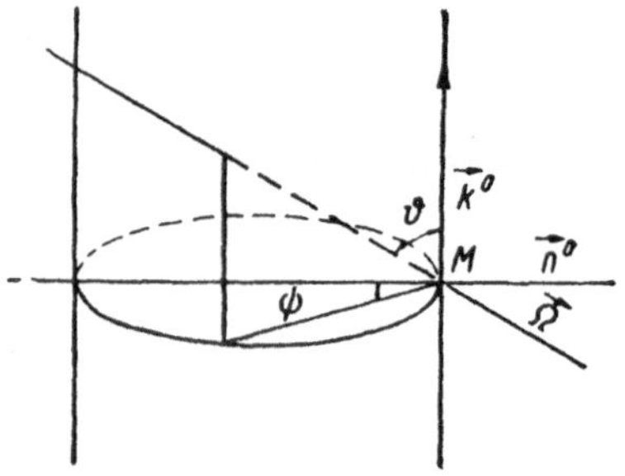

Fig. 10. Coordinate system for a cylindrical block. $\vec{\Omega}$ is the unit vector along the neutron velocity; $\vec{k}^0$ is the unit vector parallel to the generator of the cylinder; $\vec{n}^0$ is the unit vector outward normal; $\vartheta = (\vec{\Omega}, \vec{k}^0)$.

where $\epsilon = \rho^U/\rho_0^U$ is the porosity, we rewrite (50.37) in the form

$$p = \frac{N}{\xi \Sigma_s^{\text{eff}}} \cdot J_{\text{eff}},$$

where

$$J_{\text{eff}} = A + \frac{\alpha}{\sqrt{\epsilon d}} B, \quad N = \rho^U V.$$

Recalling the experimental values of A and B, we have finally

$$J_{\text{eff}} = \left(5.0 + \frac{\alpha}{\sqrt{\epsilon d}} 9.5\right) \cdot 10^{-24}. \tag{50.41}$$

We note at the same time that this equation can be used when the block is a mixture of uranium and structural elements which can be accounted for by changing the porosity of the uranium.

As has already been mentioned, the Gurevich-Pomeranchuk theory can be used for thin uranium blocks of any shape. Of particular interest, however, are plane layers and cylindrical blocks.

If the coordinates are chosen as shown in Figs. 9 and 10, we may use the relations indicated in Table 18.

We use the relations of Table 18 to obtain explicit expressions for $f(\beta)$. For the plane case we then have

$$f(\beta) = \int_0^1 e^{-\frac{\beta}{2\mu}} \left[I_0\left(\frac{\beta}{2\mu}\right) + I_1\left(\frac{\beta}{2\mu}\right) \right] d\mu, \tag{50.42}$$

and for the cylindrical case we have

$$f(\beta) = \frac{4}{\pi} \int_0^{\pi/2} \cos^2\psi \, d\psi \int_0^{\pi/2} e^{-\frac{\beta}{2}\frac{\cos\psi}{\sin\vartheta}} \times$$
$$\times \left[I_0\left(\frac{\beta}{2}\frac{\cos\psi}{\sin\vartheta}\right) + I_1\left(\frac{\beta}{2}\frac{\cos\psi}{\sin\vartheta}\right) \right] \sin\vartheta \, d\vartheta. \tag{50.43}$$

Figure 11 gives graphs of $f(\beta)$ (see also Table II of Appendix F). The numerical value of α for a plane is

$$\alpha = 0.685.$$

Table 19 gives the numerical values of A and B in the Gurevich-Pomeranchuk formula (50.38).

We now use the effective resonance integral to find the probability that a neutron is absorbed during the slowing-down process. This probability is

$$\langle\varphi\rangle = \prod_j \left(1 - \frac{p_j}{V_{\text{cell}}}\right), \tag{50.44}$$

where the product is extended over all resonances.

If we assume that $p_j/V_{\text{cell}} \ll 1$, we may use the approximation

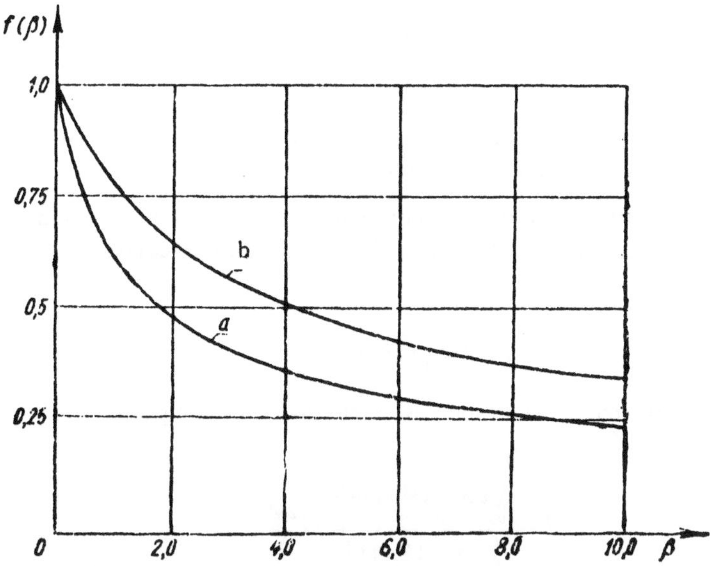

Fig. 11. Self-shielding $f(\beta)$ of a resonance level as a function of the total cross section Σ^j and block thickness d. $\beta = \Sigma^j d$. Curve a is for a plane layer; Curve b is for a cylinder.

TABLE 18

Expressions for $|\Omega_n|$, l' and $d\Omega$ in the Two Different Coordinate Systems

Shape of block	$\lvert\Omega_n\rvert$	l'	$d\Omega$
Plane	$\cos\vartheta$	$\frac{1}{\cos\vartheta}$	$\sin\vartheta\, d\vartheta$
Cylindrical	$\sin\vartheta\cdot\cos\psi$	$\frac{\cos\psi}{\sin\vartheta}$	$\sin\vartheta\, d\psi\, d\vartheta$

TABLE 19

The Values of A and B in the Formula for the Effective Resonance Integral

Shape of block	A	B
Plane	$5.0\cdot 10^{-24}$	$6.5\cdot 10^{-24}$
Cylindrical		$9.5\cdot 10^{-24}$

$$\langle\varphi\rangle = \prod_j e^{-\frac{p_j}{V_{cell}}} = e^{-\frac{1}{V_{cell}}\sum_j p_j}. \tag{50.45}$$

Writing

$$\sum_j p_j = p,$$

we finally arrive at

$$\langle\varphi\rangle = e^{-\frac{\rho_g^U}{\xi\Sigma_s} J_{\text{eff}}}, \tag{50.46}$$

where $\rho_g^U = N/V_{\text{cell}}$ is the number of U^{238} nuclei per cm^3 of homogenized mixture. We will make use of this formula in the future when dealing with heterogeneous reactors.

Finally, let us direct our attention to some shortcomings of the Gurevich-Pomeranchuk theory. First let us note the limited range of its applicability. This theory can be used to calculate resonance capture in small blocks for which

$$\Sigma_{s_n}\bar{l} \ll 1,$$

where $\bar{l}$ is the mean geometric thickness of the block as calculated according to

$$\bar{l} = \frac{\int\limits_s dS \int\limits_\Omega |\Omega_n| \, l d\Omega}{\int\limits_s dS \int\limits_\Omega |\Omega_n| \, d\Omega} = \frac{4V}{S}, \tag{50.47}$$

in which S is the total surface area of the block. It follows in particular that in a cylindrical block $\bar{l} = d$, while for a plane $\bar{l} = 2d$.

Further, the Gurevich-Pomeranchuk theory does not account for the decrease in the incident neutron flux due to multiple scattering in the vicinity of the block. Another deficiency is the assumption that the neutrons undergo only single scattering within the block within the energy of a single resonance. Actually, many of the resonances in U^{238} are such that

$$\Sigma_{sr} > \Sigma_c,$$

while the energy loss in collision with a nucleus is very small. As a result, a single collision may leave a neutron within the danger zone of the resonance.

We should mention that such an assumption lies at the basis not only of the Gurevich-Pomeranchuk theory, but of all theories of resonance capture. Any improvement in the theory in an effort to account for multiple scattering involves great computational difficulties related to the kinetic slowing-down equations for large and very variable cross sections.

In spite of these shortcomings, the Gurevich-Pomeranchuk theory has given the correct qualitative aspects of the formula for the effective resonance capture in small blocks. The experimental values of A and B lead to the semiempirical equation (50.39), which is in sufficiently good agreement with experiment. This is because all the above-mentioned restrictions were maintained in the experiments, and therefore they are to some extent effectively accounted for in the constants.

51. The Wigner Theory of Resonance Capture

As we have stated repeatedly, the Gurevich-Pomeranchuk theory can be used for small blocks in which we may neglect slowing down. In large blocks the elastic scattering is so important that it cannot be neglected. As a result of this scattering, an additional source appears in the block, and this must be accounted for when dealing with resonance capture.

The theory of resonance capture for uranium blocks whose dimensions are much larger than the scattering length in the block is due to E. Wigner (see, for instance, [23, 54]).

Consider a cylindrical block in a moderator and the cell it forms in a heterogeneous reactor. In the slowing-down process within the block, some of the neutrons whose energies lie in the danger zone of the resonance will be captured. Clearly the number of such neutrons is

$$p_j^{(1)} = V \int\limits_{E_j-\alpha}^{E_j+\alpha} \Sigma_c \varphi \, dE, \qquad (51.1)$$

where φ is the neutron flux at the resonance energy, and Σ_c is the capture cross section in the block.

If the resonance is sufficiently narrow, the energy dependence of the neutron flux is given by Wigner's formula (see Section 18)

$$\varphi = \frac{\Sigma_{sn}}{\Sigma} \varphi_0 = \frac{\Sigma_{sn}}{\Sigma} \frac{q}{E\xi\Sigma_s^{\mathrm{eff}}}, \qquad (51.2)$$

where $q = \xi\Sigma_s^{\mathrm{eff}} \varphi_0(E) \cdot E$, and q and $\xi\,\Sigma_s^{\mathrm{eff}}$ are determined by the properties of the moderator.

Inserting this into (51.1) and assuming that $q = 1$, we obtain

$$p_j^{(1)} = \frac{V}{\xi\Sigma_s^{\mathrm{eff}}} \frac{1}{E_j} \int\limits_{E_j-\alpha}^{E_j+\alpha} \Sigma_c \cdot \frac{\Sigma_{sn}}{\Sigma} dE. \qquad (51.3)$$

If we assume further that the potential scattering cross section Σ_{sn} is a slowly-varying function of the energy, while the capture cross section Σ_c and the U^{238} resonant scattering cross section Σ_{sr} are described by the Breit-Wigner formula, we may use (50.9) to rewrite this last expression in the form

$$p_j^{(1)} = \frac{V}{\xi\Sigma_s^{\mathrm{eff}}} \frac{\Gamma_j \Sigma_c^j}{2E_j} \int\limits_{-\infty}^{\infty} \frac{dx}{1 + h_j + x^2}, \qquad (51.4)$$

where

$$h_j = \frac{\Sigma_c^j + \Sigma_{sr}^j}{\Sigma_{sn}} = \frac{\Sigma_r^j}{\Sigma_{sn}} = \frac{\sigma_r^j}{\sigma_{sn}}.$$

$$\sigma_r^j = \sigma_c^j + \sigma_{sr}^j.$$

Recalling that

$$\int\limits_{-\infty}^{\infty} \frac{dx}{1 + h_j + x^2} = \frac{\pi}{\sqrt{1 + h_j}},$$

we finally arrive at

$$p_j^{(1)} = \frac{V}{\xi\Sigma_s^{\mathrm{eff}}} \frac{\pi \Gamma_j \Sigma_c^j}{2E_j} \frac{1}{\sqrt{1 + h_j}}. \qquad (51.5)$$

Noting that $p_j^{(1)}$ is proportional to the volume of the uranium block, we shall call it the volume absorption.

Let us now go on to a consideration of another factor in resonance absorption, namely surface absorption. Essentially what is involved is the following. As the neutrons are slowed down through the resonance energy in the uranium block, their flux is decreased, so that there appears a difference

$$\varphi' = \varphi_0 - \varphi, \qquad (51.6)$$

between the flux inside and outside the block, where φ is the flux of slowing-down neutrons in the block. Using (51.2), we may write φ' in the form

$$\varphi' = \frac{\Sigma_r}{\Sigma}\varphi_0 = \frac{\Sigma_r}{\Sigma}\cdot\frac{1}{E\xi\Sigma_s^{\text{eff}}}, \tag{51.7}$$

where we have set q = 1.

If φ' is isotropic, it is simple to calculate the flux $\delta\varphi$ of neutrons associated with the difference which is incident on the uranium block. We have

$$\delta\varphi = \varphi'\frac{1}{4\pi}\int_{-\pi/2}^{\pi/2}\cos\psi d\psi\int_0^{\pi}\sin^2\vartheta d\vartheta. \tag{51.8}$$

Making use of the fact that

$$\frac{1}{4\pi}\int_{-\pi/2}^{\pi/2}\cos\psi d\psi\int_0^{\pi}\sin^2\vartheta d\vartheta = \frac{1}{4},$$

we have

$$\delta\varphi = \frac{1}{4}\varphi'. \tag{51.9}$$

If we now assume that after the first collision with a nucleus in a block the neutron will either be absorbed or be scattered so as to avoid the resonant energy danger zone, the probabilities for capture and scattering are

$$w_c = \frac{\Sigma_c}{\Sigma}, \quad w_s = \frac{\Sigma_s}{\Sigma}, \tag{51.10}$$

so that

$$w_c + w_s = 1.$$

The number of neutrons absorbed as a result of incidence on S from the moderator is given for each resonance energy by

$$p_j^{(2)} = S\int_{E_j-\alpha}^{E_j+\alpha} w_c\delta\varphi dE, \tag{51.11}$$

where S is the surface area of the block. Using (51.9) and (51.7), this equation can be written

$$p_j^{(2)} = \frac{S}{4\xi\Sigma_s^{\text{eff}} E_j}\int_{E_j-\alpha}^{E_j+\alpha}\frac{\Sigma_c\Sigma_r}{\Sigma^2}dE. \tag{51.12}$$

Again using the Breit-Wigner formula as given in (50.8), this last equation can be written

$$p_j^{(2)} = \frac{1}{8}\frac{S}{\xi\Sigma_s^{\text{eff}}}\frac{\Gamma_j h_j}{E_j}\frac{\Sigma_c^j}{\Sigma_{sn}}\int_{-\infty}^{\infty}\left(\frac{1}{1+h_j+x^2}\right)^2 dx. \tag{51.13}$$

But since

$$\int_{-\infty}^{\infty}\left(\frac{1}{1+h_j+x^2}\right)^2 dx=\frac{\pi}{2\,(1+h_j)^{3/2}}\,, \tag{51.14}$$

we finally arrive at

$$p_j^{(2)}=\frac{1}{16}\,\frac{S}{\xi\Sigma_s^{\text{eff}}}\,\frac{\pi\Gamma_j h_j}{E_j\,(1+h_j)^{3/2}}\cdot\frac{\Sigma_c^j}{\Sigma_{sn}}\,. \tag{51.15}$$

This part of the resonance capture is proportional to S, and is therefore called the surface absorption. Thus, the total resonance capture by the j-th resonance level can be written

$$p_j=\frac{1}{\xi\Sigma_s^{\text{eff}}}\left(\frac{V\rho_0{}^{\text{U}}}{(1+h_j)^{1/2}}\,\frac{\pi\Gamma_j\sigma_c^j}{2E_j}+\frac{S}{16}\,\frac{\pi\Gamma_j h_j}{E_j\,(1+h_j)^{3/2}}\,\frac{\sigma_c^j}{\sigma_{sn}}\right). \tag{51.16}$$

Summing over all resonances, we have

$$p=\frac{N}{\xi\Sigma_s^{\text{eff}}}\,J_{\text{eff}} \tag{51.17}$$

where

$$J_{\text{eff}}=a+\frac{S}{M}\,b, \tag{51.18}$$

$$a=\sum_j\frac{1}{(1+h_j)^{1/2}}\,\frac{\pi\Gamma_j\sigma_c^j}{2E_j}\,,\qquad b=\frac{\varkappa^{-1}}{16}\sum_j\frac{\pi\Gamma_j h_j}{E_j\,(1+h_j)^{3/2}}\cdot\frac{\sigma_c^j}{\sigma_{sn}}$$

(N is the number of uranium nuclei in the block, M is the mass of uranium in a cell, and κ is the ratio between Avogadro's number and the atomic number of uranium).

The experimental values of a and b in Equation (51.18) for metallic U^{238} are [23]:

$$a = 9.25\cdot 10^{-24}\ \text{cm}^2,$$
$$b = 24.7\cdot 10^{-24}\ \text{g}.$$

The resulting Wigner formula is

$$J_{\text{eff}}=\left(9.25+\frac{S}{M}\cdot 24.7\right)\cdot 10^{-24}. \tag{51.19}$$

This equation can also be used for porous blocks.

Let us consider a block consisting of uranium and structural elements. In this case we can also obtain a formula for the resonance integral. To do this let us categorize all the resonance levels as either strong ($h_j \gg 1$) or weak ($h_j \ll 1$). We then have

$$a=\sum_{j<j_0}\sqrt{\frac{\sigma^j}{\sigma_{sn}\mu}}\cdot\frac{\pi\Gamma_j\sigma_c^j}{2E_j}+A,$$

$$b=\frac{\varkappa^{-1}}{16}\sum_{j<j_0}\sqrt{\frac{\sigma^j}{\sigma_{sn}\mu}}\cdot\frac{\sigma_c^j}{\sigma_{sn}\mu}\,\frac{\pi\Gamma_j}{E_j}\,,$$

where

$$A = \sum_{j>j_0} \frac{\pi \Gamma_j \sigma_c^j}{2E_j},$$

$$\mu = \frac{\Sigma_{sn}^U}{\Sigma_{sn}}.$$

Here Σ_{sn} is the macroscopic cross section for the mixture, and Σ_{sn}^U is the cross section for the uranium in the mixture.

Using the fact that A = 5.0 and the equations for a and b, we obtain

$$a = \left(5.0 + \frac{4.25}{\sqrt{\mu}}\right) \cdot 10^{-24}, \quad b = \frac{24.7}{\mu\sqrt{\mu}} \cdot 10^{-24}.$$

In this case, therefore, the effective resonance integral will be

$$J_{eff} = \left(5.0 + \frac{4.25}{\sqrt{\mu}} + \frac{S}{M} \cdot \frac{24.7}{\mu\sqrt{\mu}}\right) \cdot 10^{-24}. \quad (51.20)$$

The probability that the neutron will avoid resonance capture is then obtained from the relation

$$\langle \varphi \rangle = e^{-\frac{\rho_g^U}{\xi \Sigma_s^{eff}} \cdot J_{eff}}, \quad (51.21)$$

where ρ_g^U is the number of U^{238} nuclei per cm^3 of homogenized mixture. Wigner's theory also has several shortcomings. The fundamental difficulty is its narrow range of applicability. As follows from the construction itself, it can be used to calculate resonance capture only in large blocks in which

$$\Sigma_{sn}\bar{l} \gg 1.$$

For blocks so small that neutrons which undergo no collisions play an important role, Wigner's theory gives large errors.

In calculating the volume part of resonance capture, we used Wigner's formula (51.2) which is valid only for resonances whose half-width is much less than the mean lethargy loss.

Although high-energy resonances satisfy this condition, low resonances and very strong ones do not. For such resonances Wigner's formula must be replaced by another, for instance that of Greuling and Goertzel (see, for instance, Galanin [17]). As for Equation (51.19), these factors are to a large extent included in the empirical constants of this equation.

52. Improved Theory of Resonance Capture

V.V. Orlov has used the Gurevich-Pomeranchuk and Wigner theories to develop a general theory of resonance capture good for blocks of any size. In the limits of very small and very large blocks, Orlov's equations become those of the appropriate one of the two earlier theories. We shall now go on to a consideration of this improved theory [63]. Consider a uranium block bounded by an arbitrary convex surface S in a moderator of infinite extent.

The space energy neutron distribution in such a system is described by the Boltzmann integro-differential equation

$$\frac{\partial \varphi}{\partial s} + \Sigma \varphi = \int_{u-r}^{u} du' \int d\Omega' \, \Sigma_s \varphi f(\mu_0, u - u'), \quad (52.1)$$

where s is the distance along a fixed ray $\vec{\Omega}$.

Let us assume the neutron flux $\varphi(\vec{r}, \vec{\Omega}, u)$ to be unperturbed for energies above the resonance level. We shall consider this flux to be of the form $\frac{1}{4\pi}\varphi_0 = \text{const}$ and assume the danger zone of the resonance to be much less than the maximum lethargy loss. We can then move the quantity $\Sigma_s\varphi_0$ outside the integral sign in (52.1).

Bearing in mind the normalization of the scattering function $f(\mu_0, u - u')$, we have

$$\frac{\partial\varphi}{\partial s} + \Sigma\varphi = \frac{1}{4\pi}\Sigma_{sn}\varphi_0. \tag{52.2}$$

Let us now solve this equation at some internal point s of the block, considering $s = 0$ to be a point on its surface (Fig. 12). Integrating, we have

$$\varphi(\vec{r}, \vec{\Omega}, u) = \varphi\Big|_{s=0} e^{-\Sigma\cdot s} + \frac{1}{4\pi}\Sigma_{sn}\varphi_0 \int_0^s e^{-\Sigma(s-\xi)}\, d\xi, \tag{52.3}$$

where $\varphi|_{s=0}$ is the incident neutron flux on S. We shall assume that the incident flux is isotropic, so that $\varphi|_{s=0} = \frac{1}{4\pi}\varphi_0$. Then (52.3) becomes

$$\varphi = \frac{1}{4\pi}\varphi_0 \left[e^{-\Sigma s} + \frac{\Sigma_{sn}}{\Sigma}(1 - e^{-\Sigma s}) \right] \tag{52.4}$$

Let us now consider a cylindrical element of volume in the block, with base dS and generators parallel to $\vec{\Omega}$. We shall calculate the number of neutrons in this cylinder moving in the direction given by $\vec{\Omega}$ and crossing a plane parallel to dS per unit time.

It is clear that the number of such neutrons will be

$$\varphi|\Omega_n|\,dS.$$

Then the number of neutrons absorbed in the elementary cylinder is given by

$$p(\vec{r}, \vec{\Omega}, E)\,dS = |\Omega_n|\,dS \int_0^l \Sigma_c\varphi\, ds, \tag{52.5}$$

or, using (52.4),

$$p(\vec{r}, \vec{\Omega}, E)\,dS = \varphi_0 \frac{|\Omega_n|\,l\,dS}{4\pi}\left(\frac{\Sigma_c\Sigma_{sn}}{\Sigma} + \frac{\Sigma_c\Sigma_r}{\Sigma^2}\frac{1 - e^{-\Sigma l}}{l}\right). \tag{52.6}$$

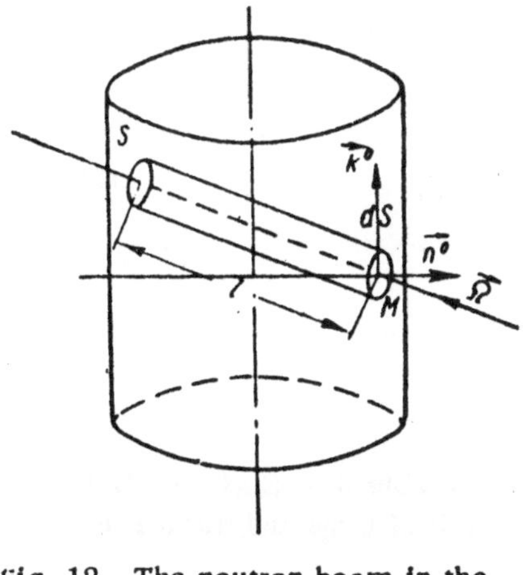

Fig. 12. The neutron beam in the cylindrical block. dS is the base of the elementary cylindrical volume whose generator is parallel to the unit vector $\vec{\Omega}$ in the direction of motion of the neutron; l is the neutron path length in the block; n^0 is the external normal; k^0 is the unit vector parallel to the generator of the cylindrical block.

We now integrate this over all energies in the neighborhood of the j-th resonance.

We then have

$$p_j(\vec{r}, \vec{\Omega})\,dS = \varphi_0 \frac{|\Omega_n|\,l\,dS}{4\pi}\left[\int_{E_j-\alpha}^{E_j+\alpha} \frac{\Sigma_c\Sigma_{sn}}{\Sigma}\,dE + \right.$$

$$\left. + \frac{1}{l}\int_{E_j-\alpha}^{E_j+\alpha} \frac{\Sigma_c\Sigma_r}{\Sigma^2}\Phi(\Sigma_{sn}l)\,dE\right], \tag{52.7}$$

where

$$\Phi(\Sigma_{sn}l)=\frac{\int\limits_{E_j-a}^{E_j+a}\frac{\Sigma_c\Sigma_r}{\Sigma^2}(1-e^{-\Sigma l})\,dE}{\int\limits_{E_j-a}^{E_j+a}\frac{\Sigma_c\Sigma_r}{\Sigma^2}\,dE}. \tag{52.8}$$

We will now show that Φ is indeed a function of only one variable. To do this, we transform to the independent variable $\underline{x}$ according to (50.10) and make use of the Breit-Wigner formula (50.8). This gives

$$\Phi(\Sigma_{sn}l)=\frac{\int\limits_{-\infty}^{\infty}\frac{1-e^{-\left(\frac{\Sigma_r^j l}{1+x^2}+\Sigma_{sn}l\right)}}{\left[1+\frac{\Sigma_{sn}}{\Sigma_r^j}(1+x^2)\right]^2}dx}{\int\limits_{-\infty}^{\infty}\frac{dx}{\left[1+\frac{\Sigma_{sn}}{\Sigma_r^j}(1+x^2)\right]^2}}. \tag{52.9}$$

If we now assume the resonances to be strong, that is,

$$\Sigma^j l \gg 1 \quad\text{and}\quad \frac{\Sigma_{sn}}{\Sigma^j}\ll 1,$$

we may approximate $1 + x^2$ in (52.9) by x^2. Transforming further to the new independent variable $\underline{y}$ according to

$$\frac{\Sigma_r^j l}{x^2}=\frac{1}{y^2}, \tag{52.10}$$

we arrive at

$$\Phi(\beta)=[1-\operatorname{erf}(\sqrt{\beta})](1+2\beta)+\frac{2\sqrt{\beta}}{\sqrt{\pi}}e^{-\beta}-2\beta, \tag{52.11}$$

where

$$\operatorname{erf}(x)=\frac{2}{\sqrt{\pi}}\int\limits_x^{\infty}e^{-y^2}dy.$$

We have thus shown that as an approximation Φ is a function of only the single variable $\beta = \Sigma_{sn}l$, and is therefore independent of the parameters of the resonance. It is easily shown in the limit of large and small β this function $\Phi(\beta)$ can be represented asymptotically by

$$\Phi(\beta)\to 1 \qquad (\beta\to\infty).$$
$$\Phi(\beta)\to 0 \qquad (\beta\to 0).$$

Figure 13 shows a graph of $\Phi(\beta)$ (see also Table III of Appendix F).

Let us now turn to (52.7) and integrate over all solid angles of the unit hemisphere within the block. This integration gives

$$p_j(\vec{r})\,dS = \frac{\varphi_0\,dS}{4\pi}\left[\int\limits_{E_j-a}^{E_j+a} \frac{\Sigma_c\,\Sigma_{sn}}{\Sigma}\,dE \int\limits_{\Omega} |\Omega_n|\,l\,d\Omega + \int\limits_{E_j-a}^{E_j+a} \frac{\Sigma_c\,\Sigma_r}{\Sigma^2}\,dE \cdot \int\limits_{\Omega} |\Omega_n|\,\Phi(\Sigma_{sn}\,l)\,d\Omega\right]. \qquad (52.12)$$

We now integrate over the whole surface S. Some simple mathematical operations lead to

$$p_j = \varphi_0\left[V\int\limits_{E_j-a}^{E_j+a} \frac{\Sigma_c\,\Sigma_{sn}}{\Sigma}\,dE + \frac{S}{4}\int\limits_{E_j-a}^{E_j+a} \frac{\Sigma_c\,\Sigma_r}{\Sigma^2}\,F(\Sigma_{sn}\,d)\,dE\right], \qquad (52.13)$$

where

$$\left.\begin{aligned} V &= \frac{1}{4\pi}\int\limits_S dS \int\limits_\Omega |\Omega_n|\,l\,d\Omega, \\ S &= \frac{1}{\pi}\int\limits_S dS \int\limits_\Omega |\Omega_n|\,d\Omega, \end{aligned}\right\} \qquad (52.14)$$

$$F(\Sigma_{sn}\,d) = \frac{\int\limits_S dS \int\limits_\Omega |\Omega_n|\,\Phi(\Sigma_{sn} l)\,d\Omega}{\int\limits_S dS \int\limits_\Omega |\Omega_n|\,d\Omega}. \qquad (52.15)$$

In the last of these, as previously, we write

$$l = d \cdot l'.$$

From the properties of Φ which we have discussed, we arrive at the following asymptotic representations of $F(\beta)$:

$$F(\beta) \longrightarrow 1 \quad \text{as } \beta \longrightarrow \infty,$$
$$F(\beta) \longrightarrow 0 \quad \text{as } \beta \longrightarrow 0.$$

Let us transform in (52.13) from the flux φ_0 to the slowing-down density. We then have

$$p_j = p_j^{(1)} + p_j^{(2)}\,F(\Sigma_{sn}\,d), \qquad (52.16)$$

where $p_j^{(1)}$ and $p_j^{(2)}$ are the volume and surface parts of the resonance capture, given by (51.3) and (51.12), respectively.

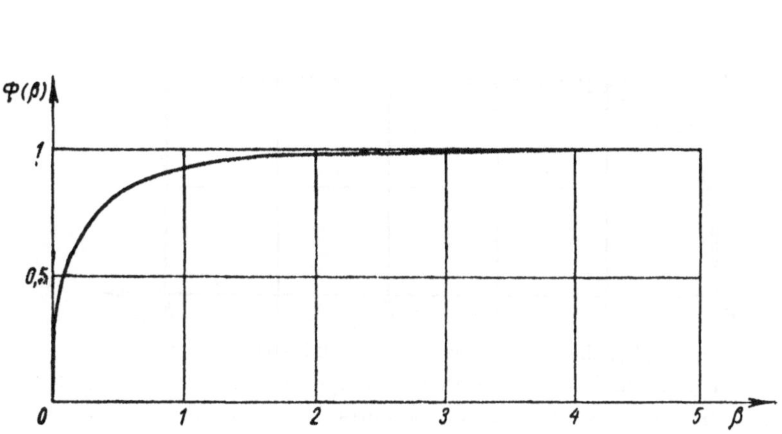

Fig. 13. The function $\Phi(\beta)$ which gives the probability that a neutron moving in the direction $\vec{\Omega}$ will pass through the block without collision. $\beta = \Sigma_{sn} l$ is the cross section for potential scattering in the block; l is the neutron path length in the block.

Using (51.5) and (51.15), we can transform (52.16) to a form similar to (51.16), namely:

$$p_j = \frac{1}{\xi\Sigma_s^{\text{eff}}}\left[\frac{\rho_0^{\text{U}} V}{(1+h_j)^{1/2}}\frac{\pi\Gamma_j\sigma_c^j}{2E_j} + \frac{S}{16}\frac{\pi\Gamma_j h_j}{E_j(1+h_j)^{3/2}}\frac{\sigma_c^j}{\sigma_{sn}}F(\Sigma_{sn}d)\right]. \tag{52.17}$$

Summing over all resonances we have*

$$p = \frac{N}{\xi\Sigma_s^{\text{eff}}} J_{\text{eff}}, \tag{52.18}$$

where

$$J_{\text{eff}} = a + \frac{S}{M} bF(\Sigma_{sn}d), \tag{52.19}$$

where a and b are of the same form as in (51.18). Let us analyze this expression. Recalling the asymptotic behavior of $F(\beta)$, we see that when $\Sigma_{sn}d \gg 1$ (52.19) goes over into Wigner's Formula (51.18). This means that the coefficients a and b in (52.19) and (51.18) are the same. We therefore have

$$J_{\text{eff}} = \left[9.25 + 24.7\frac{S}{M}F(\Sigma_{sn}d)\right]\cdot 10^{-24}. \tag{52.20}$$

If the block is a mixture of uranium and structural elements, this last equation can be rewritten, as in Wigner's theory,

$$J_{\text{eff}} = \left[5.0 + \frac{4.27}{\sqrt{\mu}} + \frac{24{,}7}{\mu\sqrt{\mu}}\frac{S}{M}F(\Sigma_{sn}d)\right]\cdot 10^{-24}, \tag{52.21}$$

where, as before,

$$\mu = \frac{\Sigma_{sn}^{\text{U}}}{\Sigma_{sn}}.$$

Again using the asymptotic behavior of $F(\beta)$, it can be shown that when $\Sigma_{sn}d \ll 1$, Equation (52.20) goes over into the Gurevich-Pomeranchuk Formula (50.39).

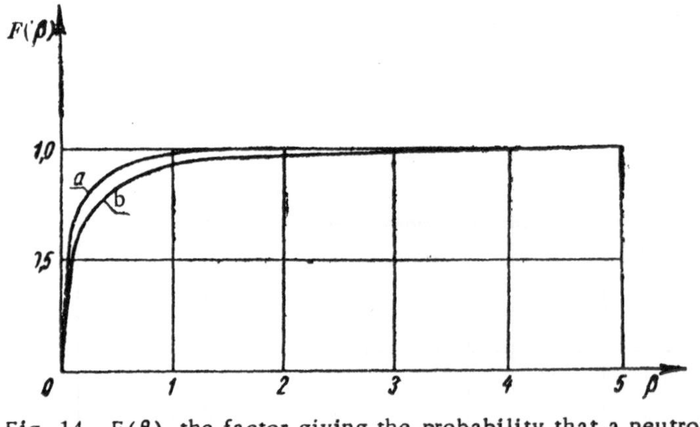

Fig. 14. $F(\beta)$, the factor giving the probability that a neutron will pass through a block without undergoing collision. $\beta = \Sigma_{sn}d$; Σ_{sn} is the corss section for potential scattering in the block; d is the thickness of the block. Curve (a) is for a plane layer, and Curve (b) is for a cylindrical block.

*A.P. Rudik arrived independently at the same result (see Galanin [20]).

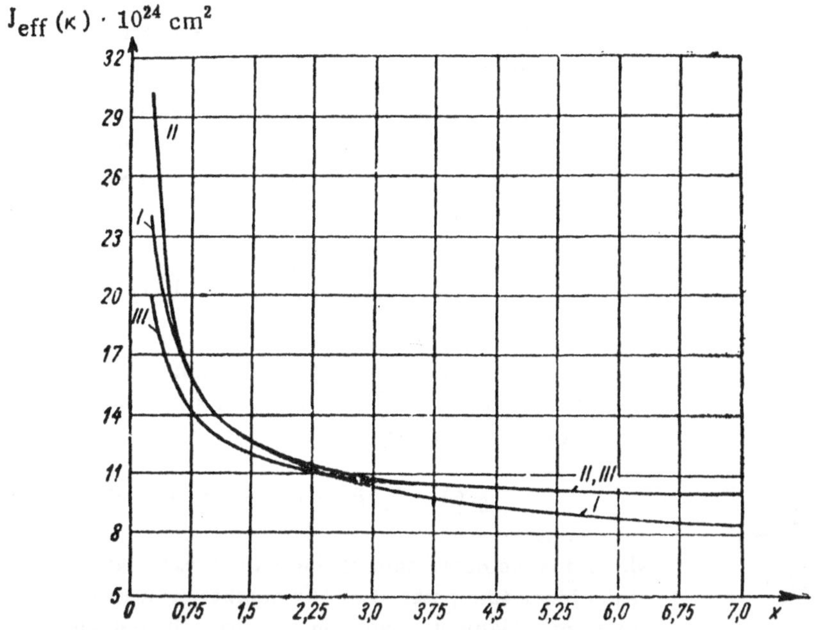

Fig. 15. The effective resonance integral J_{eff} for a cylindrical block of natural uranium: I) according to the Gurevich-Pomeranchuk theory; II) according to the Wigner theory; III) according to the Orlov theory.

Let us now consider the function $F(\Sigma_{sn}d)$ given by (52.15). For a plane uranium layer this function is

$$F(\Sigma_{sn} d) = 2 \int_0^1 \Phi\left(\frac{\Sigma_{sn} d}{\mu}\right) \mu \, d\mu, \tag{52.22}$$

while for a cylindrical block it is

$$F(\Sigma_{sn} d) = \frac{4}{\pi} \int_0^{\frac{\pi}{2}} \cos\psi \, d\psi \int_0^{\frac{\pi}{2}} \Phi\left(\Sigma_{sn} d \frac{\cos\psi}{\sin\vartheta}\right) \sin^2\vartheta \, d\vartheta, \tag{52.23}$$

where $\underline{d}$ is the thickness of the layer or the diameter of the cylinder, respectively. Figure 14 (see also Table IV in Appendix F) shows graphs of $F(\beta)$. Figure 15 shows graphs of J_{eff} as calculated according to the Gurevich-Pomeranchuk, the Wigner, and the Orlov theories (see also Table V in Appendix F).

53. Mutual Shielding of Rods

All of the above considerations have been based on the assumption that the lattice spacing of the rods is so large that the probability is vanishingly small that a neutron having passed through a rod will pass through another at the same energy. This is not the case for closely spaced rods in which the probability is high that a neutron will pass through several without slowing down. Then the effective resonance integral will be different for this system. The question of mutual shielding of uranium rods has been treated by Gurevich and Pomeranchuk [26], Sedel'nikov and Orlov [61], Petrov [65], and others. In the present section we shall consider the method developed by the author together with V.V. Orlov for calculating the resonance capture of neutrons in a plane lattice of uranium blocks.

Consider a lattice in a heterogeneous reactor to consist of plane uranium layers in the moderator. We consider a single cell of such a reactor. Of course, the neutron flux across the external boundary of the cell must vanish.

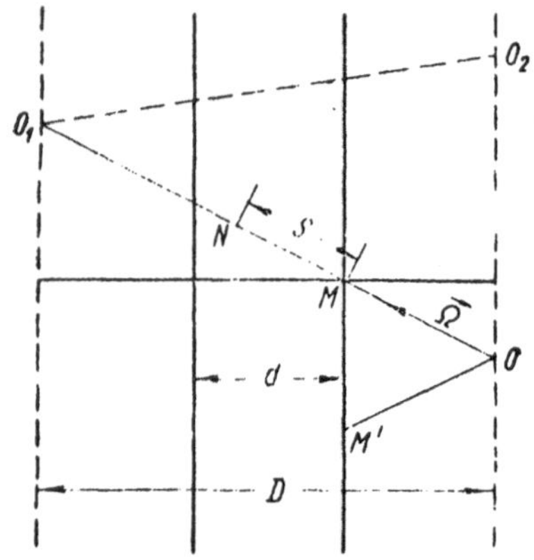

Fig. 16. Diagram of a plane cell of a heterogeneous reactor. d is the thickness of the block; D is the size of the Wigner-Seitz cells; $\vec{\Omega}$ is the unit vector along the neutron direction of motion.

Consider a neutron incident on the block from the moderator, travelling in the direction given by the unit vector $\vec{\Omega}$. Assuming the flux to be isotropic for the neutrons whose energy is greater than the resonance energy, the flux incident on the block is found by integrating the expression $\Sigma_s^0 e^{-\Sigma_s^0 \xi}$ along the broken line M'OM (Fig. 16). We assume in addition that when the neutrons hit the boundary of the cell they are "reflected" in accordance with the laws of geometric optics. Then integration gives

$$\varphi(M, \vec{\Omega}, E) = \frac{1}{4\pi} \varphi_0(E) (1 - e^{-\Sigma_s^0 L}), \tag{53.1}$$

where L is twice the distance from the uranium block to the boundary along the ray given by $\vec{\Omega}$. As the neutrons enter the block the flux in the $\vec{\Omega}$ direction will decay according to

$$\varphi(N, \vec{\Omega}, E) = \varphi(M, \vec{\Omega}, E) e^{-\Sigma s}, \tag{53.2}$$

where s is the coordinate of the point N calculated from the boundary of the block (the point M). The number of neutrons absorbed in a cylindrical element of volume with base dS and with generator parallel to $\vec{\Omega}$ in the block is

$$p(\vec{r}, \vec{\Omega}, E)\, dS = \varphi(M, \vec{\Omega}, E) |\Omega_n| dS \frac{\Sigma_c}{\Sigma} (1 - e^{-\Sigma l}) \tag{53.3}$$

Let us now calculate how many of the neutrons slowed down in the block are captured. To do this we first find the flux of neutrons in the $\vec{\Omega}$ direction which are slowed down to resonance energy in the block. The decay of the flux along the $\vec{\Omega}$ direction is clearly obtained by integrating $\Sigma_{sn} e^{-\Sigma \xi}$:

$$\varphi(N, \vec{\Omega}, E) = \frac{1}{4\pi} \varphi_0(E) \frac{\Sigma_{sn}}{\Sigma} (1 - e^{-\Sigma s}). \tag{53.4}$$

Multiplying by Σ_c and integrating over the interval $0 \leq s \leq l$, we find that, of the neutrons slowed down in the block, the fraction captured is

$$p(\vec{r}, \vec{\Omega}, E)\, dS = \varphi_0(E) \frac{|\Omega_n|\, dS}{4\pi} \frac{\Sigma_{sn} \Sigma_c}{\Sigma} \left(l - \frac{1 - e^{-\Sigma l}}{\Sigma} \right). \tag{53.5}$$

Adding this to (53.3), we obtain the total number of neutrons captured in a block on the "first shot":

$$p_0(\vec{\Omega}, E)\, dS = \varphi_0(E) \frac{|\Omega_n|\, l dS}{4\pi} \left(\frac{\Sigma_{sn} \Sigma_c}{\Sigma} + \frac{\Sigma_c \Sigma_r}{\Sigma^2} \frac{1 - e^{-\Sigma l}}{l} - \frac{\Sigma_c}{\Sigma} e^{-\Sigma_s^0 L} \frac{1 - e^{-\Sigma l}}{l} \right). \tag{53.6}$$

In obtaining this expression, we have made use of (53.1).

We now find the neutron flux leaving the block. To do this we use (53.2) and (53.4), setting $s = l$. This gives

$$\varphi(\vec{\Omega}, E) = \frac{1}{4\pi} \varphi_0(E) \left[(1 - e^{-\Sigma_s^0 L}) e^{-\Sigma l} + \frac{\Sigma_{sn}}{\Sigma} (1 - e^{-\Sigma l}) \right]. \tag{53.7}$$

The neutrons leaving the block are scattered in the moderator. Part of these reach the boundaries of the cell and are reflected along O_1O_2, again enter the block at M (in view of the fact that O_1O_2 and O_1O are equivalent directions, we are assuming for convenience that the neutrons are reflected along O_1O). The neutron flux incident on the block this time is

$$\varphi(\vec{\Omega}, E)\, e^{-\Sigma_s^0 L}.$$

On passing through the block this flux drops to

$$\varphi(\vec{\Omega}, E)\, e^{-\Sigma_s^0 L - \Sigma l}.$$

Let us continue this process in which the neutrons are reflected from the boundaries of the cells. Then the k-th time the neutrons are incident on the block, the flux is given by

$$\varphi(\vec{\Omega}, E) \cdot e^{-k\Sigma_s^0 L - (k-1)\Sigma l}.$$

If we recall that each time the neutrons go through the block, the fraction

$$\frac{\Sigma_c}{\Sigma}(1 - e^{-\Sigma l}),$$

is captured by the uranium nuclei, we find that the total number of neutrons captured in the k-th reflection from the boundary is given by

$$p_k(\vec{\Omega}, E)\, dS = \varphi(\vec{\Omega}, E)\, |\Omega_n|\, dS \frac{\Sigma_c}{\Sigma}(1 - e^{-\Sigma l})\, e^{-k\Sigma_s^0 L - (k-1)\Sigma l}. \tag{53.8}$$

Summing (53.8) over k [using (53.6)] and performing some operations, we arrive at

$$p(\vec{\Omega}, E)\, dS = \varphi_0(E) \frac{|\Omega_n|\, dS}{4\pi} \left[\frac{\Sigma_{sn}\Sigma_c}{\Sigma} l + \frac{\Sigma_c \Sigma_r}{\Sigma^2}(1 - e^{-\Sigma l}) \cdot \frac{1 - e^{-\Sigma_s^0 L}}{1 - e^{-\Sigma_s^0 L - \Sigma l}} \right]. \tag{53.9}$$

We now integrate over energies in the neighborhood of the j-th resonance, obtaining

$$p_j(\vec{r}, \vec{\Omega})\, dS = \varphi(E_j) \frac{|\Omega_n|\, l\, dS}{4\pi} \left[\int_{E_j - a}^{E_j + a} \frac{\Sigma_c \Sigma_{sn}}{\Sigma} dE + \right.$$
$$\left. + \int_{E_j - a}^{E_j + a} \frac{\Sigma_c \Sigma_r}{\Sigma^2} \Phi(\Sigma_{sn} l;\, \Sigma_s^0 L)\, dE \right], \tag{53.10}$$

where

$$\Phi(\Sigma_{sn} l;\, \Sigma_s^0 L) = \frac{\displaystyle\int_{E_j - a}^{E_j + a} \frac{\Sigma_c \Sigma_r}{\Sigma^2}(1 - e^{-\Sigma l}) \frac{1 - e^{-\Sigma_s^0 L}}{1 - e^{-\Sigma_s^0 L - \Sigma l}} dE}{\displaystyle\int_{E_j - a}^{E_j + a} \frac{\Sigma_c \Sigma_r}{\Sigma^2} dE}. \tag{53.11}$$

We shall show that for strong resonances Φ is independent of the parameters of the resonance, being a function only of the two variables

$$\alpha = \Sigma_s^0 L, \quad \beta = \Sigma_{sn} l.$$

To do this we transform to the variable of integration x by a formula similar to (50.10). This gives

$$\Phi=\frac{\displaystyle\int_{-\infty}^{\infty}\frac{1-e^{-\left(\frac{\Sigma^j l}{1+x^2}+\Sigma_{sn}l\right)}}{\left[1+\frac{\Sigma_{sn}}{\Sigma^j}(1+x^2)\right]^2}\,\frac{1-e^{-\Sigma_s^0 L}}{1-e^{-\Sigma_s^0 L-\left(\frac{\Sigma^j l}{1+x^2}+\Sigma_{sn}l\right)}}\,dx}{\displaystyle\int_{-\infty}^{\infty}\frac{dx}{\left[1+\frac{\Sigma_{sn}}{\Sigma^j}(1+x^2)\right]^2}}. \tag{53.12}$$

Here we have also used (50.8).

If, as was mentioned, we assume that the resonance is strong, so that

$$\Sigma^j l \gg 1, \quad \frac{\Sigma_{sn}}{\Sigma^j} \ll 1,$$

we can replace $1 + x^2$ by x^2. Introducing the new variable of integration y by

$$\frac{\Sigma^j l}{x^2}=\frac{1}{y^2},$$

we rewrite (53.12) in the form

$$\Phi(\alpha, \beta)=\frac{2\sqrt{\beta}}{\pi}\int_{-\infty}^{\infty}\frac{1-e^{-\left(\frac{1}{y^2}+\beta\right)}}{(1+\beta y^2)^2}\cdot\frac{1-e^{-\alpha}}{1-e^{-\alpha-\left(\frac{1}{y^2}+\beta\right)}}\,dy, \tag{53.13}$$

where

$$\alpha=\Sigma_s^0 L, \quad \beta=\Sigma_{sn} l.$$

Analysis of this equation shows that in the limit as $\alpha \to \infty$, the function $\Phi(\alpha, \beta)$ approaches $\Phi(\beta)$, which is defined in (52.11).

Let us now integrate (53.10) over all solid angles within a hemisphere inside the block and then over S. As before, we arrive easily at the expression

$$p_j=\varphi_0\left(V\int_{E_j-a}^{E_j+a}\frac{\Sigma_c\Sigma_{sn}}{\Sigma}dE+\frac{SF}{4}\int_{E_j-a}^{E_j+a}\frac{\Sigma_c\Sigma_r}{\Sigma^2}dE\right), \tag{53.14}$$

where

$$F=\frac{\int_S dS\int_\Omega |\Omega_n|\,\Phi\,d\Omega}{\int_S dS\int_\Omega |\Omega_n|\,d\Omega}. \tag{53.15}$$

As $\alpha = \Sigma_s^0 L \to \infty$, the function F approaches $F(\beta)$, the function defined in (52.15). Transforming from φ_0 to the slowing-down density in (53.14), we obtain

$$p_j=p_j^{(1)}+p_j^{(2)}F, \tag{53.16}$$

where $p_j^{(1)}$ and $p_j^{(2)}$ are, as before, the volume and surface parts of the absorption defined by (51.3) and (51.12), respectively. Using (51.5) and (51.15), we rewrite this in the form

$$p_j = \frac{1}{\xi\Sigma_s^{\text{eff}}} \left[\frac{\rho^{U} V}{(1+h)^{1/2}} \cdot \frac{\pi\Gamma_j\sigma_c^j}{2E_j} + \frac{S}{16} \frac{\pi\Gamma_j h_j}{E_j (1+h_j)^{3/2}} \cdot \frac{\sigma_c^j}{\sigma_{sn}} F \right]. \tag{53.17}$$

Summing over all resonances we obtain

$$p = \frac{N}{\xi\Sigma_s^{\text{eff}}} J_{\text{eff}}, \tag{53.18}$$

where

$$J_{\text{eff}} = a + \frac{S}{M} bF, \tag{53.19}$$

and a and b are again of the same form as in (51.18).

Recalling the behavior of F, we find asymptotically that

$$F \longrightarrow 1 \quad \text{as} \quad \Sigma_s^0 \bar{L} \longrightarrow \infty, \; \Sigma_{sn} \bar{l} \longrightarrow \infty,$$

where $\bar{L}$ and $\bar{l}$ are the average values of the appropriate quantities. In this limit, therefore, (53.19) is the same as Wigner's Formula (51.18). This gives us the values of a and b. As a result we have

$$J_{\text{eff}} = \left(9.25 + 24.7 \frac{S}{M} F \right) 10^{-24}. \tag{53.20}$$

This equation is basic to the calculation of the effective resonance integral when one takes into account the mutual shielding of the uranium blocks in the lattice of a heterogeneous reactor. When the blocks are composed both of uranium and structural elements, (53.20) is replaced, in analogy with (52.21), by

$$J_{\text{eff}} = \left(5.0 + \frac{4.25}{\sqrt{\mu}} + \frac{24.7}{\mu\sqrt{\mu}} \frac{S}{M} F \right) \cdot 10^{-24}. \tag{53.21}$$

Let us now study F. This function is of the form

$$F(\alpha, \beta) = 2 \int_0^1 \Phi \left(\frac{\alpha}{\mu}; \frac{\beta}{\mu} \right) \mu \, d\mu \tag{53.22}$$

or

$$F(\alpha, \beta) = \frac{2\sqrt{\beta}}{\pi} \int_0^1 \sqrt{\mu} \, d\mu \int_{-\infty}^{\infty} \frac{1 - e^{-\left(\frac{1}{y^2} + \frac{\beta}{\mu}\right)}}{\left(1 + \frac{\beta y^2}{\mu}\right)} \cdot \frac{1 - e^{-\frac{\alpha}{\mu}}}{1 - e^{-\frac{\alpha}{\mu} - \left(\frac{1}{y^2} + \frac{\beta}{\mu}\right)}} dy,$$

where $\alpha = \Sigma_s^0 (D - d)$, $\beta = \Sigma_{sn}^0 \cdot d$. Here D is the distance between the blocks and d the thickness of the uranium layer.

Figure 17 is a plot of $F(\alpha, \beta)$ (see also Table VI of Appendix F).

It should be noted that to develop a theory for resonance capture for a lattice of cylindrical blocks is quite difficult, mostly because for strong resonances one cannot define an equivalent cylindrical Wigner-Seitz cell in the lattice without seriously reducing the neutron flux incident on the block.* For some lattices, as has been pointed out by Gurevich and Pomeranchuk [26], it is nevertheless possible to develop an approximate theory of resonance capture with mutual shielding. These same authors obtained general formulas, and these were later used by Iu.V. Petrov [65] in the first concrete calculations for the functions which give the effect of mutual

*This was first pointed out by T.Kh. Sedel'nikov.

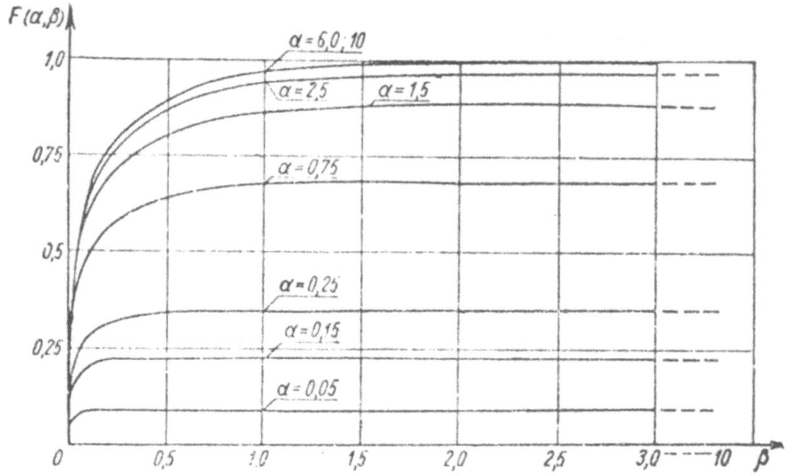

Fig. 17. $F(\alpha, \beta)$, the factor which accounts for the probability that a neutron can pass through the block without collision, and the mutual shielding of the blocks. $\alpha = \Sigma_s^0(D - d)$; $\beta = \Sigma_{sn}d$; Σ_s^0 is the scattering cross section in the moderator; Σ_{sn} is the potential scattering cross section in the block; D is the distance between blocks; d is the thickness of a block.

screening. Gurevich and Pomeranchuk base their work [26] on the possibility of describing resonance capture statistically in a close-packed lattice of a heterogeneous reactor.*

Consider a close-packed lattice of a heterogeneous reactor consisting of small blocks which are absorbing centers. The number of neutrons absorbed per unit time in a cell of this reactor is

$$p(E) = V_{cell}\Sigma_c^{eff}\varphi, \tag{53.23}$$

where φ is the mean neutron flux in the lattice, V_{cell} is the volume of the cell, and Σ_c^{eff} is the effective neutron capture cross section in the lattice. If φ and Σ_c^{eff} are known, this equation can be used to calculate resonance capture in the cell. To calculate φ we proceed as follows. It is known that when absorption takes place in the moderator, the neutron flux can be written in terms of the slowing-down density according to

$$\varphi = \frac{1}{\xi\Sigma_s^0 E}.$$

If at the same time neutrons are absorbed, the neutron flux in the moderator is given by Wigner's formula

$$\varphi = \frac{1}{\xi E}\frac{1}{\Sigma_s^0 + \Sigma_c^{eff}} \tag{53.24}$$

Using this expression, we rewrite (53.23) in the form

$$p(E) = \frac{V_{cell}}{\xi E}\cdot\frac{\Sigma_c^{eff}}{\Sigma_s^0 + \Sigma_c^{eff}}. \tag{53.25}$$

We have now calculated the effective capture cross section Σ_c^{eff} of the "equivalent" homogenized medium. It is clear that this function can be written as a product in the form $\Sigma_c^{eff} = f\zeta(E)$, where f is a geometric factor proportional to the probability that a neutron will collide with a block, and where ζ is the probability that a neu-

*We define a close-packed lattice by the condition $L \gg \lambda_s^0$, where L is the mean free path of a neutron for collision with a block, and $\lambda_s^0 = 1/\Sigma_s^0$ is the scattering length in the moderator.

tron incident on a block will be absorbed by it. The considerations can be simplified somewhat if the block is assumed to be a perfect absorber. Then $\zeta = 1$, and we may write

$$\Sigma_c^{\text{eff}} = f,$$

where $f = 1/L$, and L is the mean free path of a neutron for collision with the block, given by

$$\bar{l}/L = V/V^0, \tag{53.26}$$

where V is the volume of the uranium, V^0 is the volume of the moderator, and $\bar{l} = 4V/S$ is the mean free path of the neutron in the block.

We shall write

$$L = \omega\bar{l}.$$

Here $\omega = V^0/V$ is the ratio of the moderator volume to the uranium volume in a block. Proceeding to the general case of resonance capture, we have

$$\Sigma_c^{\text{eff}} = f\zeta(E). \tag{53.27}$$

According to its definition, $\zeta(E)$ is given by

$$\zeta(E) = \frac{1}{\pi S}\int_S dS \int_\Omega |\Omega_n|\, d\Omega \int_0^l \Sigma_c e^{-\Sigma\xi}\, d\xi. \tag{53.28}$$

We write this in the form

$$\zeta(E) = 1 - \frac{1}{\pi S}\int_S dS \int_\Omega |\Omega_n|\, e^{-\Sigma_c l}\, d\Omega. \tag{53.29}$$

Inserting this into (53.27), we have

$$\Sigma_c^{\text{eff}} = f\left(1 - \frac{1}{\pi S}\int_S dS \int_\Omega |\Omega_n|\, e^{-\Sigma_c l}\, d\Omega\right). \tag{53.30}$$

We now insert this, in turn, into (53.25), obtaining

$$p(E) = \frac{V_{\text{cell}}}{\xi E}\, \frac{1 - \frac{1}{\pi S}\int_S dS \int_\Omega |\Omega_n|\, e^{-\Sigma_c l}\, d\Omega}{\frac{1}{\eta} - \frac{1}{\pi S}\int_S dS \int_\Omega |\Omega_n|\, e^{-\Sigma_c l}\, d\Omega}, \tag{53.31}$$

where

$$\eta = \frac{1}{1 + \Sigma_s^0 L}.$$

Integrating (53.31) over all energies about the resonant energy E_j, we arrive at

$$p_j = \frac{V_{\text{cell}}}{\xi}\int_{E_j-\alpha}^{E_j+\alpha} \frac{1 - \frac{1}{\pi S}\int_S dS \int_\Omega |\Omega_n|\, e^{-\Sigma_c l}\, d\Omega}{\frac{1}{\eta} - \frac{1}{\pi S}\int_S dS \int_\Omega |\Omega_n|\, e^{-\Sigma_c l}\, d\Omega}\, \frac{dE}{E}. \tag{53.32}$$

We now separate the resonances into weak ($j > j_0$) and strong ($j < j_0$). For weak resonances we may use the expansion

$$e^{-\Sigma_c l} = 1 - \Sigma_c l + \ldots \tag{53.33}$$

Keeping only the first two terms, we perform the integration in (53.32), obtaining

$$p_j = \frac{V_{\text{cell}}}{\xi} \int_{E_j-\alpha}^{E_j+\alpha} \frac{1}{1+\frac{L}{l}\frac{\Sigma_s^0}{\Sigma_c}} \frac{dE}{E}. \tag{53.34}$$

If we now write

$$\omega \frac{\Sigma_s^0}{\Sigma_c} \gg 1,$$

we arrive at the simplification

$$p_j = \frac{N}{\xi\Sigma_s^{\text{eff}}} \int_{E_j-\alpha}^{E_j+\alpha} \sigma_c \frac{dE}{E}, \tag{53.35}$$

where N is the number of uranium nuclei in the block, and

$$\xi\Sigma_s^{\text{eff}} = \xi\Sigma_s^0 \frac{V_0}{V_{\text{cell}}}.$$

Using (50.8) and performing the integration in (53.35) as before, we obtain

$$p_j = \frac{N}{\xi\Sigma_s^{\text{eff}}} \frac{\pi\sigma_c^j\Gamma_j}{2E_j}. \tag{53.36}$$

Finally, we sum over all the weak resonances, and obtain the volume part of the resonance absorption, namely

$$p = \frac{N}{\xi\Sigma_s^{\text{eff}}} A, \tag{53.37}$$

where

$$A = \sum_{j \geqslant j_0} \frac{\pi\sigma_c^j\Gamma_j}{2E_j}.$$

We now go on to capture on the strong resonance levels. We first write (53.32) in the form

$$p_j = \frac{\eta V_{\text{cell}}}{\xi} \int_{E_j-\alpha}^{E_j+\alpha} \frac{1 - \frac{1}{\pi S}\int_S dS \int_\Omega |\Omega_n| e^{-\Sigma_c l}\, d\Omega}{1 - \frac{\eta}{\pi S}\int_S dS \int_\Omega |\Omega_n| e^{-\Sigma_c l}\, d\Omega} \frac{dE}{E},$$

where

$$\eta = \frac{1}{1 + L\Sigma_s^0}.$$

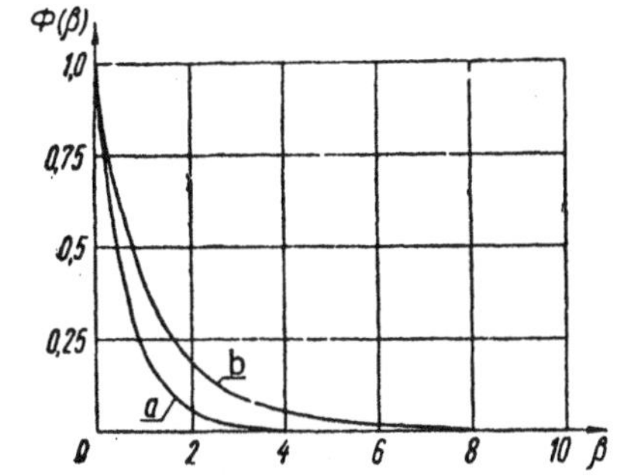

Fig. 18. The dependence of $\Phi(\beta)$ on the block thickness and ξ ($\beta = d/\xi^2$): a) plane layer; b) cylindrical block.

We again assume that the capture cross section in this expression is given by the Breit-Wigner formula. Then we have

$$p_j = \frac{V_{cell}\,\eta}{\xi}\,\frac{\sqrt{\Sigma^j \Gamma_j^2}}{E_j} \cdot \int\limits_0^\infty \frac{1 - \frac{1}{\pi S}\int\limits_S dS \int\limits_\Omega |\Omega_n|\, e^{-\frac{l}{\xi^2}}\, d\Omega}{1 - \frac{\eta}{\pi S}\int\limits_S dS \int\limits_\Omega |\Omega_n|\, e^{-\frac{l}{\xi^2}}\, d\Omega}\, d\xi. \quad (53.38)$$

This can be written

$$p_j = \frac{V}{\xi \Sigma_s^{eff}}\,\frac{\sqrt{\pi \Sigma^j \Gamma_j^2}}{E_j} \left\langle \frac{1}{\sqrt{l}} \right\rangle R(\eta), \quad (53.39)$$

where

$$R(\eta) = \frac{V_{cell}}{V}\,\frac{\Sigma_s^{eff}}{\left\langle \frac{1}{\sqrt{l}} \right\rangle}\,\eta \int\limits_0^\infty \frac{1 - \Phi\left(\frac{d}{\xi^2}\right)}{1 - \eta\Phi\left(\frac{d}{\xi^2}\right)}\, d\xi, \quad (53.40)$$

$$\Phi(\beta) = \frac{1}{\pi S}\int\limits_S dS \int\limits_\Omega |\Omega_n|\, e^{-\beta l'}\, d\Omega, \quad (53.41)$$

$l' = l/d$, and $\langle 1/\sqrt{l} \rangle$ is given by (50.35). The function R which enters into this expression is a function of only one variable. Indeed, noting that

$$\Sigma_s^{eff}\,\frac{V_{cell}}{V} = \frac{1-\eta}{\eta \bar{l}},$$

and changing the variable of integration in (53.40) according to

$$d/\xi^2 = 1/y^2,$$

we have

$$R(\eta) = \frac{1}{\sqrt{\pi}}\,\frac{\eta}{\frac{\bar{l}'}{\left\langle \frac{1}{\sqrt{l'}} \right\rangle}} \int\limits_0^\infty \frac{1 - \Phi\left(\frac{1}{y^2}\right)}{1 - \eta\Phi\left(\frac{1}{y^2}\right)}\, dy. \quad (53.42)$$

Let us now consider some specific geometries. For the plane geometry $\Phi(\beta)$ is

$$\Phi(\beta) = 2E_3(\beta), \quad (53.43)$$

while for the cylindrical,

$$\Phi(\beta) = \frac{4}{\pi}\int\limits_0^{\pi/2} \sin^2\vartheta\, d\vartheta \int\limits_0^1 e^{-\frac{\beta}{\sin\vartheta}\mu}\, \frac{\mu}{\sqrt{1-\mu^2}}\, d\mu. \quad (53.44)$$

Figure 18 shows graphs of $\Phi(\beta)$ (see also Table VII of Appendix F), and Fig. 19 shows graphs of $R(\eta)$ (see also Table VIII of Appendix F).

The function $R(\eta)$ was first calculated for cylindrical blocks by Petrov [65] by means of an infinite-series expansion of the integrand in (53.40).

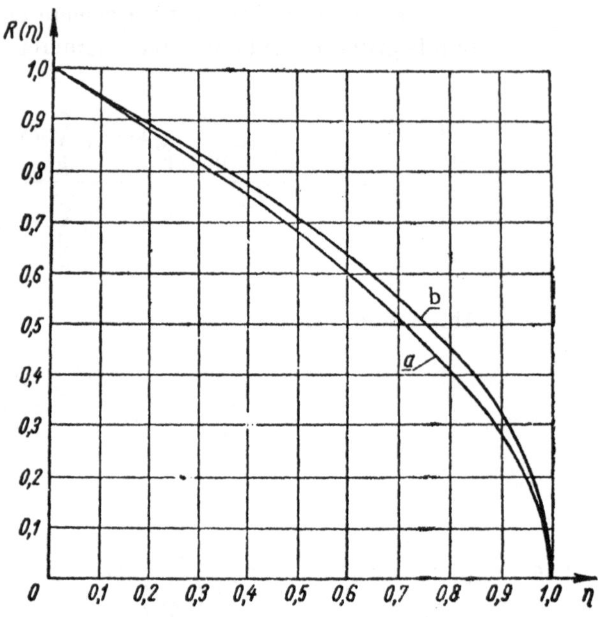

Fig. 19. Mutual shielding factor $R(\eta)$ for block resonances. $\eta = 1/(1 + L\Sigma_s^0)$; Σ_s^0 is the scattering cross section in the moderator; L is the mean free path of a neutron in the moderator for collision with a block; a) plane layer; b) cylindrical block.

We now sum (53.39) over all strong resonances. Then, using the fact that R is independent of the resonance parameters, we obtain

$$p = \frac{\rho^U V}{\xi \Sigma_s^{\text{eff}}} \frac{BR(\eta)}{d}, \tag{53.45}$$

where B is given by

$$B = \frac{1}{\sqrt{\rho^U}} \left\langle \frac{1}{\sqrt{l'}} \right\rangle \sum_{j<j_0} \frac{\sqrt{\pi \sigma^j \Gamma_j^2}}{E_j}.$$

Finally, adding (53.37) and (53.45) we arrive at the resonance capture formula

$$p = \frac{N}{\xi \Sigma_s^{\text{eff}}} J_{\text{eff}} \tag{53.46}$$

where

$$J_{\text{eff}} = A + \frac{1}{\sqrt{d}} BR(\eta). \tag{53.47}$$

Using the numerical values of A and B, we obtain a formula similar to (50.39), namely:

$$J_{\text{eff}} = \left[5.0 + \frac{\alpha}{\sqrt{d}} 9.5R(\eta) \right] \cdot 10^{-24}. \tag{53.48}$$

For porous blocks we have

$$J_{\text{eff}} = \left[5.0 + \frac{\alpha}{\sqrt{\varepsilon d}} 9.5R(\eta) \right] \cdot 10^{-24}. \tag{53.49}$$

In conclusion, we remark that $R(\eta)$ has the following asymptotic representations:

$$R(0) = 1, \quad R(1) = 0. \tag{53.50}$$

It follows from this that when the lattice spacing is large ($\Sigma_s^0 L \gg 1$), Equation (53.48) is the same as (50.39), the corresponding formula for an isolated block. If, however, the lattice spacing is small ($\Sigma_s^0 L \ll 1$), only the first term in (53.48) remains. Comparison of (53.20) and (53.48) shows that in this limit (53.48) is incorrect, since rather than $J_{eff} = 9.25$, it gives $J_{eff} = 5.0$. This is because this theory does not take into account neutrons which are slowed down in the uranium blocks themselves. In spite of this and other shortcomings, however, the theory is of methodological importance, since it gives a conceptually new approach to the analysis of mutual screening by absorbing elements in the nuclear reactors.

54. Heterogeneous Intermediate Reactors

In dealing with heterogeneous intermediate reactors one must usually consider uranium blocks highly-enriched with fissile isotope. In the final analysis this leads to a significant amount of self-shielding by the surface layers of the uranium blocks. It should be mentioned that although in thermal reactors self-shielding is important only for the strongest isolated U^{238} resonances, in intermediate reactors this factor is important even for U^{235}, and not only on isolated resonances, but over the whole energy interval where slowing-down neutrons are strongly absorbed. Self-shielding is important to an equal extent also in the thermal diffusion region. The effect of self-shielding is essentially a decrease in the neutron capture cross section as compared with homogeneous reactors.

Thus, a heterogeneous intermediate reactor can be handled by homogenization, the only difference being that the capture and fission cross sections obtained in the homogeneous calculation must be decreased to take account of self-shielding in the cell of the heterogeneous reactor; we shall call the factor by which these cross sections are decreased the shielding factor.

T.Kh. Sedel'nikov has extended the Gurevich-Pomeranchuk theory to take account of self-shielding in intermediate-neutron reactors. His method is the following.

Assume that the lattice spacing of the reactor is so large that one can neglect the mutual shielding of the uranium blocks. We can then calculate the total number of neutrons absorbed by a uranium block per unit energy interval. Let us start with the simplest case, a uranium layer in a moderator. Consider a plane parallel neutron flux φ_0 incident from the negative x direction on the surface S of a uranium layer. On passing through the block the flux of neutrons with energy E decays according to

$$\varphi(x, E) = \varphi_0(E)\, e^{-\Sigma x}. \tag{54.1}$$

The number of neutrons of energy E absorbed by the uranium block per unit time is:

$$p(E) = S \int_0^d \Sigma_c \varphi\, dx, \tag{54.2}$$

or, using (54.1),

$$p(E) = S\varphi_0(E) \frac{\Sigma_c}{\Sigma} (1 - e^{-\Sigma d}). \tag{54.3}$$

We write this in the form

$$p(E) = V\Sigma_c^{eff} \varphi_0. \tag{54.4}$$

where

$$\Sigma_c^{eff} = \Sigma_c b(\Sigma d), \tag{54.5}$$

$$b(\beta) = \frac{1 - e^{-\beta}}{\beta}, \tag{54.6}$$

and $V = Sd$ is the volume of the uranium layer. Recalling now that in a homogeneous reactor we have

$$p(E) = V\Sigma_c \varphi_0, \tag{54.7}$$

we see that in a heterogeneous reactor the neutron absorption will be lower, and the self-shielding factor is $b(\Sigma d)$. Thus the heterogeneity of the reactor reduces to calculating the effective capture cross section according to (54.5). After Σ_c^{eff} is found, the calculation can proceed as for a homogeneous reactor.

Let us now consider self-shielding by blocks of any convex geometry. We assume that the incident flux $\varphi(\vec{\Omega}, E)$ is isotropic. Then the number of neutrons of energy E incident per unit time on a surface element dS in the $\vec{\Omega}$ direction will be

$$\varphi_0 \cdot \frac{|\Omega_n|\, dS}{4\pi}.$$

The number of neutrons absorbed by the block along the $\vec{\Omega}$ ray is given by the integral

$$p(\vec{r}, \vec{\Omega})\, dS = \varphi_0 \cdot \frac{|\Omega_n|\, dS}{4\pi} \int_0^l \Sigma_c e^{-\Sigma \cdot \xi}\, d\xi, \tag{54.8}$$

where l is the path length, within the block, of a neutron moving in the $\vec{\Omega}$ direction. We write this in the form

$$p(\vec{r}, \vec{\Omega})\, dS = \frac{|\Omega_n|\, l\, dS}{4\pi} \Sigma_c b(\Sigma l)\, \varphi_0, \tag{54.9}$$

where, as before,

$$b(\beta) = \frac{1 - e^{-\beta}}{\beta}.$$

Let us integrate over all solid angles within the block, and then over S. This gives

$$p = V\Sigma_c^{eff} \varphi_0, \tag{54.10}$$

where

$$\Sigma_c^{eff} = \Sigma_c B(\Sigma d), \tag{54.11}$$

$$B(\beta) = \frac{\int_S dS \int_\Omega |\Omega_n|\, l b(\beta l')\, d\Omega}{\int_S dS \int_\Omega |\Omega_n|\, l\, d\Omega}, \tag{54.12}$$

$$V = \frac{1}{4\pi} \int_S dS \int_\Omega |\Omega_n|\, l\, d\Omega. \tag{54.13}$$

We have here written

$$l = dl'.$$

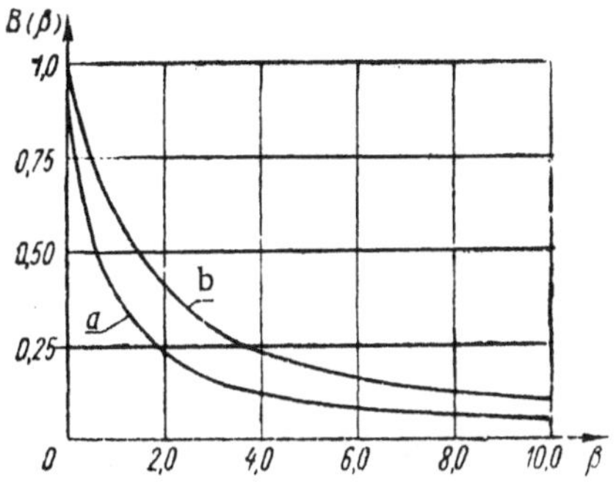

Fig. 20. The self-shielding factor B(β). $\beta = \Sigma d$; Σ is the total cross section in the block; d is the thickness of the block; a) plane layer; b) cylindrical block.

The self-shielding factor B(β) depends on the geometry of the block and is a function of the single variable

$$\beta = \Sigma d.$$

For a plane layer this function is

$$B(\beta) = \frac{\frac{1}{2} - E_3(\beta)}{\beta} \tag{54.14}$$

while for a cylindrical block it is

$$B(\beta) = \frac{4}{\pi} \int_0^{\pi/2} \cos\psi \, d\psi \int_0^{\pi/2} \sin^2\vartheta \, \frac{1 - e^{-\beta \frac{\cos\psi}{\sin\vartheta}}}{\beta} \, d\vartheta. \tag{54.15}$$

Figure 20 shows graphs of B(β) (see also Table IX, Appendix F).

It should be borne in mind that the fundamental shortcoming of this theory is that it fails to take into account the decrease in the neutron flux in the vicinity of the block. Whereas in the theory of resonance capture this fact is to a large extent accounted for by the experimentally determined constants in the formula for the effective resonance integral, in this theory it must be handled separately.

55. Effective Boundary Conditions for "Black" Bodies

In the preceding sections of this chapter we have assumed that the neutron sources are uniformly distributed about the absorbing blocks. This assumption is usually good for thin blocks which barely perturb the neutron spectrum, but it cannot be used when the neutrons are strongly absorbed by the blocks. This is the situation, for instance, for thermal neutrons scattered in the cell of a heterogeneous reactor, or for neutrons scattered in the neighborhood of a control rod.

It should be noted that in view of the high gradient in the neutron flux, the neutron field in the neighborhood of an absorbing block cannot be calculated by a formal application of diffusion theory. This means that in order to find the neutron distribution in the neighborhood of thin blocks one must proceed directly from the Boltzmann kinetic equation. Even in the simplest cases, however, this equation is very difficult to solve. As a result it is necessary to refine diffusion theory for specific cases. Such refinement consists of "correcting" the ordinary diffusion boundary conditions on the surface of the block by taking into account the exact solutions of the kinetic equations. In the present section we will consider this problem for the case of absolutely absorbing ("black") blocks in the plane and cylindrical geometries in an infinite scattering medium [20, 29, 54, 98, 121].

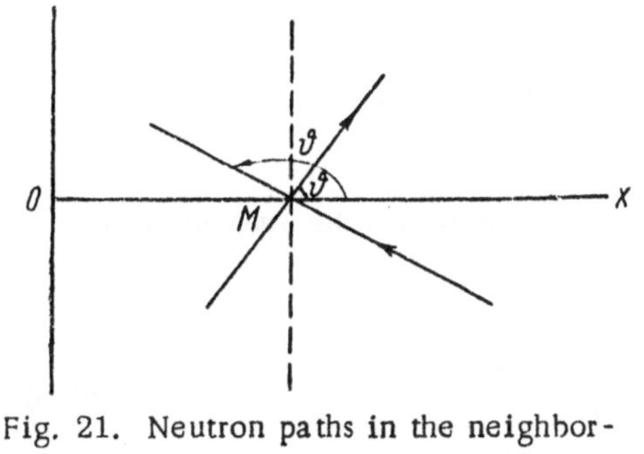

Fig. 21. Neutron paths in the neighborhood of a point M in the semi-infinite space $x \geq 0$. ϑ is the angle between the neutron velocity vector $\vec{\Omega}$ and the x axis.

Consider the one-velocity Boltzmann kinetic equation for a semi-infinite space bounded by an absolute absorber (or the vacuum) along the plane $x = 0$. In accordance with the cylindrical symmetry of the problem, the integro-differential equation is

$$\mu \frac{\partial \varphi}{\partial x} + \varphi = \frac{1}{2} \int_{-1}^{+1} \varphi \, d\mu \qquad (0 \leqslant x < \infty), \tag{55.1}$$

$$\varphi(0, \mu) = 0 \qquad (0 \leqslant \mu \leqslant 1), \tag{55.2}$$

where $\varphi = \varphi(x, \mu)$, $\mu = \cos\vartheta$, and the distance x is measured in units of the mean free path for scattering.

We write (55.1) in the form

$$\mu \frac{\partial \varphi}{\partial x} + \varphi = \frac{1}{2} \varphi_0 (x), \tag{55.3}$$

where

$$\varphi_0 (x) = \int_{-1}^{+1} \varphi (x, \mu) \, d\mu.$$

Let us consider further a point M whose coordinate is x, and let us calculate the number of neutrons per unit time per unit solid angle in the interval of solid angle $d\Omega$ about the direction $\vec{\Omega}$ which cross an area perpendicular to this direction (Fig. 21). We shall separate the range of ϑ into the two regions

$$0 \leqslant \vartheta \leqslant \frac{\pi}{2} \quad \text{and} \quad \frac{\pi}{2} \leqslant \vartheta \leqslant \pi.$$

For $\pi/2 \leq \vartheta \leq \pi$, the solution of (55.3) is written in the form

$$\varphi (x, \mu) = -\frac{1}{2\mu} \int_{x}^{\infty} \varphi_0 (\xi) \, e^{\frac{\xi - x}{\mu}} \, d\xi, \tag{55.4}$$

and for $0 \leq \vartheta \leq \pi/2$, it is written in the form

$$\varphi (x, \mu) = \frac{1}{2\mu} \int_{0}^{x} \varphi_0 (\xi) \, e^{-\frac{x - \xi}{\mu}} \, d\xi. \tag{55.5}$$

Then the neutron density at M

$$\varphi_0 (x) = \int_{-1}^{+1} \varphi (x, \mu) \, d\mu \tag{55.6}$$

can be written

$$\varphi_0 (x) = -\frac{1}{2} \int_{-1}^{0} \frac{d\mu}{\mu} \int_{x}^{\infty} \varphi_0 (\xi) \, e^{\frac{\xi - x}{\mu}} \, d\xi + \frac{1}{2} \int_{0}^{1} \frac{d\mu}{\mu} \int_{0}^{x} \varphi_0 (\xi) \, e^{-\frac{x - \xi}{\mu}} \, d\xi. \tag{55.7}$$

In the first of these integrals we transform to the new variable $\mu' = -\mu$, which transforms the equation to the form

$$\varphi_0 (x) = \frac{1}{2} \int_{0}^{1} \frac{d\mu}{\mu} \left\{ \int_{0}^{x} \varphi_0 (\xi) \, e^{-\frac{x - \xi}{\mu}} \, d\xi + \int_{x}^{\infty} \varphi_0 (\xi) e^{-\frac{\xi - x}{\mu}} \, d\xi \right\}. \tag{55.8}$$

Within the limits of integration, the exponents $(x - \xi)/\mu$ and $(\xi - x)/\mu$ are positive, so that we may write

$$\varphi_0 (x) = \frac{1}{2} \int_{0}^{\infty} \varphi_0 (\xi) \, d\xi \int_{0}^{1} \frac{e^{-\frac{|x - \xi|}{\mu}}}{\mu} \, d\mu. \tag{55.9}$$

Using the definition of $E_1(x)$ [35, 40, 68], we arrive at Peierls' integral equation

$$\varphi_0(x) = \frac{1}{2}\int_0^{\infty} \varphi_0(\xi) E_1(|x - \xi|)\, d\xi. \tag{55.10}$$

Wiener and Hopf were the first to show that this homogeneous integral equation has, in addition to the trivial solution, a solution of the form

$$\varphi_0(x) = \frac{1}{q^0(x)}[1 + q^0(x)\, x], \tag{55.11}$$

where $q^0(x)$ is a bounded function which, in the limit as $x \to \infty$, approaches a constant value

$$q^0(\infty) = q^0_\infty. \tag{55.12}$$

It is known that the asymptotic value q^0_∞ of $q^0(x)$ is approached with sufficient accuracy at just a few mean free paths. Thus, far from the boundary $x = 0$, the asymptotic solution of (55.10) can be written, up to a constant factor,

$$\varphi^0_0(x) = 1 + q^0_\infty x. \tag{55.13}$$

Let us now recall that in the absence of absorption the diffusion equation is of the form

$$d^2\varphi_0/dx^2 = 0, \tag{55.14}$$

whose solution is

$$\varphi_0(x) = 1 + cx, \tag{55.15}$$

where *c* is an arbitrary constant.

Comparing (55.13) with this equation, we see that if *c* is chosen equal to q^0_∞, the solution of the diffusion equation (55.14) will be the same as the asymptotic solution of the kinetic equation. Comparison of (55.13) and (55.15) for the purpose of finding *c* can, however, be replaced by giving an effective boundary condition at $x = 0$ which follows from (55.13):

$$\frac{1}{\varphi_0}\frac{d\varphi_0}{dx} = \frac{1}{\gamma^0} \quad (x = 0), \tag{55.16}$$

where

$$1/\gamma^0 = q^0_\infty.$$

It is useful to note that if *x* is measured in nondimensionless units, $1/\gamma^0$ is given by

$$\frac{1}{\gamma^0} = q^0_\infty \cdot \Sigma^0_s,$$

where Σ^0_s is the scattering cross section for $x \geq 0$.

Thus, the neutron scattering problem far from the plane of absorption can be solved by using the diffusion equation with the effective boundary condition given by (55.16). The problem therefore is to find q^0_∞. The asymptotic quantity q^0_∞ is found by direct solution of the integral equation (55.10) [144]. The solution gives

$$q^0_\infty = (0.7104)^{-1}. \tag{55.17}$$

Let us now consider a black cylinder in an infinite scattering medium. Then the Boltzmann integro-differential equation is

$$\sin\psi\left[\mu\frac{\partial\varphi}{\partial r}+\frac{1-\mu^2}{r}\frac{\partial\varphi}{\partial\mu}\right]+\varphi=\frac{1}{4\pi}\varphi_0(r) \tag{55.18}$$

with the condition

$$\varphi(r,\mu,\psi)=0,\quad r=a,\quad 0<\mu\leqslant 1, \tag{55.19}$$

where

$$\varphi_0(r)=\int_0^{2\pi}d\psi\int_{-1}^{+1}d\mu\varphi(r,\mu,\psi), \tag{55.20}$$

$\mu = \cos\vartheta$, and $\underline{a}$ is the radius of the cylindrical block. This can be written in the form of Peierls' integral equation [98, 128]

$$\varphi_0(r)=\int_a^\infty \varphi_0(r')K(r,r')r'dr', \tag{55.21}$$

whose kernel $K(r, r')$ is

$$K(r,r')=\frac{2}{\pi}\int_{|r-r'|}^{\sqrt{r^2-a^2}+\sqrt{r'^2-a^2}}\frac{K_i(t)\,dt}{\sqrt{4r^2r'^2-(r^2+r'^2-t^2)^2}}, \tag{55.22}$$

where

$$K_i(t)=\int_t^\infty K_0(x)\,dx,$$

and $K_0(x)$ is MacDonald's function [74, 78].

Far from the boundary the solution of (55.21) should also be of the same form as the solution of the appropriate diffusion equation, namely:

$$\varphi_0^0=1+q_\infty^0\cdot\ln\frac{r}{a}. \tag{55.23}$$

In this case, therefore, the effective boundary condition on the surface of the black cylinder can be written

$$\frac{1}{\varphi_0}\frac{d\varphi_0}{dr}=\frac{1}{\gamma^0}, \tag{55.24}$$

where

$$\frac{1}{\gamma^0}=\frac{q_\infty^0}{a}. \tag{55.25}$$

Here q_∞^0 is calculated from Peierls' integral equation [128]. In nondimensionless units (55.25) becomes

$$\frac{1}{\gamma^0(r_0)}=\frac{\Sigma_s^0}{\gamma_0(a)}, \tag{55.25'}$$

where r_0 is the geometric radius of the block. The ordinary diffusion equation whose boundary condition is given by (55.24) can now be used to calculate φ_0. All of the above considerations were closely related to the solution of the kinetic equations.

There are, however, also other approximation methods for finding q^0_∞, which are not related to the direct solution of the Peierls' equation. We mention only two of these, namely the variational method and the method of balance [5]. We shall deal in detail only with the simpler of these, the method of balance, which is due to A.L. Brudno. This method is essentially the following. The total number of neutrons absorbed by the block per unit time is

$$p = \int\limits_S dS \int\limits_\Omega |\Omega_n| \varphi^0(\vec{r}_s, \vec{\Omega})\, d\Omega, \tag{55.26}$$

where $|\Omega_n| \varphi^0(\vec{r}_s, \vec{\Omega})$ is the neutron flux per unit area across the surface S in the $\vec{\Omega}$ direction. If $\varphi^0_0(\vec{r})$ is the total neutron flux through the unit sphere whose center is at $\vec{r}$, the differential flux is found by integrating:

$$\varphi^0(\vec{r}_s, \vec{\Omega}) = \frac{1}{4\pi} \int\limits_0^\infty \varphi^0_0(\vec{r})\, e^{-\xi}\, d\xi, \tag{55.27}$$

where the integral is taken along the $\vec{\Omega}$ ray with origin at $\vec{r}_s$, and $\vec{r}$ and ξ are measured in units of the mean free path for scattering.

As $\varphi^0_0(\vec{r})$, we choose the solution

$$\varphi^0_0(\vec{r}) = 1 + q^0_\infty I(\vec{r}), \tag{55.28}$$

of the kinetic equation, where

$$I(\vec{r}) = \begin{cases} x \text{ for the plane geometry } (0 \leq x < \infty), \\ \ln\dfrac{r}{a} \text{ for the cylindrical geometry } (a \leq r < \infty). \end{cases} \tag{55.29}$$

We insert (55.28) into (55.27), obtaining

$$\varphi^0(\vec{r}_s, \vec{\Omega}) = \frac{1}{4\pi}\left[1 + q^0_\infty \int\limits_0^\infty I(\vec{r})\, e^{-\xi}\, d\xi\right]. \tag{55.30}$$

Then we can write (55.26) in the form

$$p = \frac{S}{4} + \frac{q^0_\infty}{4\pi} \int\limits_S dS \int\limits_\Omega |\Omega_n|\, d\Omega \int\limits_0^\infty I(\vec{r})\, e^{-\xi}\, d\xi. \tag{55.31}$$

Let us now find the relation between the kinetic flux p and the diffusion flux. If there is no absorption outside the block, the total diffusion flux through any closed surface surrounding the block must be equal to the number of neutrons absorbed by the block (the kinetic flux), so that:

$$p = \frac{1}{3} \int\limits_{S_R} \nabla_n \varphi^0_0\, dS, \tag{55.32}$$

where S_R is a surface sufficiently far from the block. If we choose for $\varphi^0_0(\vec{r})$ its asymptotic expression given by (55.28), we may choose S_R to be any surface, even the surface S of the block.

Using (55.28) and (55.31) we may write (55.32) in the form

$$\frac{S}{4} + \frac{q_\infty^0}{4\pi} \int_S dS \int_\Omega |\Omega_n| \, d\Omega \int_0^\infty I(\vec{r}) \, e^{-\xi} \, d\xi = \frac{q_\infty^0}{3} \int_S \nabla_n I \, dS. \tag{55.33}$$

Solving for q_∞^0, we obtain

$$q_\infty^0 = \left(\frac{4}{3S} \int_S \nabla_n I \, dS - \Phi \right)^{-1}, \tag{55.34}$$

where

$$\Phi = \frac{1}{\pi S} \int_S dS \int_\Omega |\Omega_n| \, d\Omega \int_0^\infty I(\vec{r}) \, e^{-\xi} \, d\xi$$

and, in particular, for symmetric blocks,

$$q_\infty^0 = \left(\frac{4}{3} \nabla_n I - \Phi \right)^{-1}. \tag{55.35}$$

For the case of the semi-infinite space we have

$$I = x = \mu\xi, \quad \mu = \cos\vartheta, \quad \nabla_n I = 1, \quad \int_0^\infty I e^{-\xi} \, d\xi = \mu.$$

Therefore

$$q_\infty^0 = \frac{1}{4} \left(\frac{1}{3} - \frac{1}{4\pi} \int_0^{2\pi} d\psi \int_0^1 \mu^2 \, d\mu \right)^{-1} = \frac{3}{2}.$$

Hence,

$$\gamma^0 = 2/3.$$

This value for γ^0 differs from the exact value $\gamma^0 = 0.7104$ by no more than 6%. Such an error is not too great when performing reactor calculations.

Let us now go over to a consideration of cylindrical blocks. In this case

$$I = \ln \frac{r}{a}, \quad \nabla_n I = \frac{1}{a}.$$

From simple geometrical considerations it follows that

$$\frac{r}{a} = \sqrt{1 + \frac{2\xi \sin\vartheta \cos\psi}{a} + \frac{\xi^2 \sin^2\vartheta}{a^2}}, \tag{55.36}$$

where ξ is the distance along the $\vec{\Omega}(\vartheta, \psi)$ ray from a point M on the surface S of the block. We then arrive at

$$\int_0^\infty I(\vec{r}) \, e^{-\xi} \, d\xi = T\left(\frac{2 \sin\vartheta \cos\psi}{a}; \; \frac{\sin^2\vartheta}{a^2} \right), \tag{55.37}$$

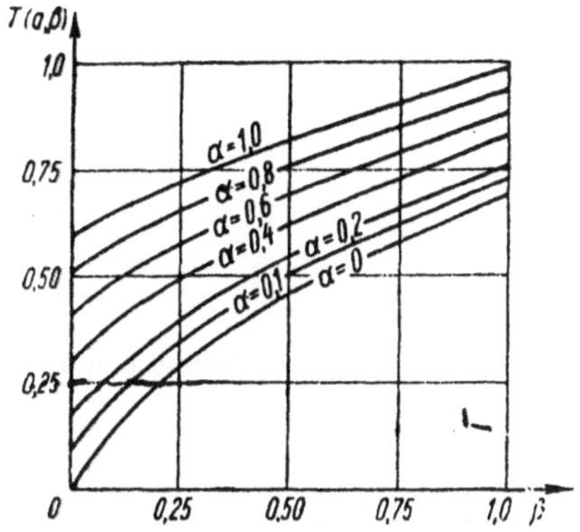

Fig. 22. T(α, β).

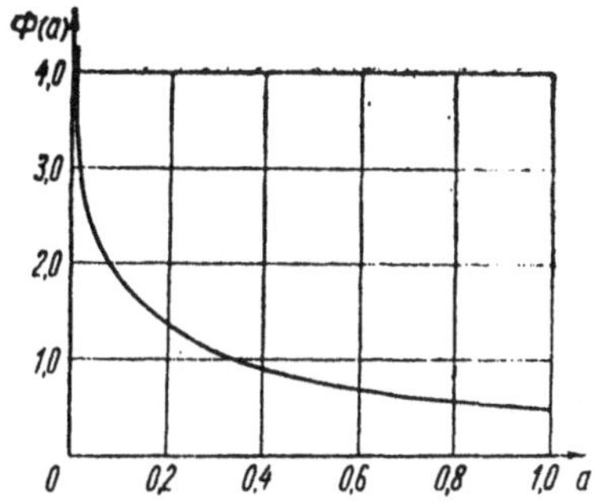

Fig. 23. Φ(a), where $\underline{a}$ is the dimensionless radius of the block.

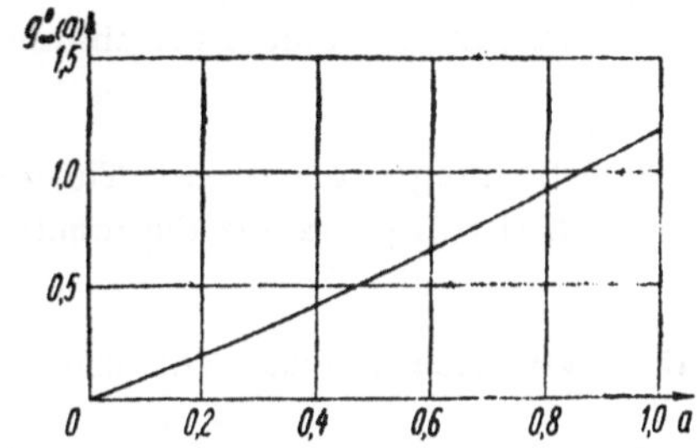

Fig. 24. q^0_∞ as a function of the dimensionless radius $\underline{a}$ of the block.

where

$$T(\alpha, \beta) = \int_0^\infty e^{-\xi} \ln \sqrt{1 + \alpha\xi + \beta\xi^2}\, d\xi. \qquad (55.38)$$

Figure 22 gives a graph of T(α, β) (see also Table X of Appendix F). Using (55.37), (55.35) becomes

$$q^0_\infty(a) = \left(\frac{4}{3a} - \Phi(a)\right)^{-1}, \qquad (55.39)$$

where

$$\Phi(a) = \frac{4}{\pi} \int_0^{\pi/2} \cos\psi\, d\psi \int_0^{\pi/2} \sin^2\vartheta\, d\vartheta\, T\left(\frac{2\sin\vartheta\cos\psi}{a}\,;\ \frac{\sin^2\vartheta}{a^2}\right). \qquad (55.40)$$

Figure 23 gives a graph of Φ(a), and Fig. 24 a graph of q^0_∞ (a) (see also Tables XI and XII of Appendix F).

Using (55.25), we obtain

$$\gamma^0 = 4/3 - a\Phi(a). \qquad (55.41)$$

Let us consider the two limiting cases

$$a = 0 \text{ and } a = \infty.$$

In the first case, for a $\ll$ 1 (infinitesimally thin cylinder), we have

$$\gamma^0 = 4/3.$$

This follows from the fact that T(0, 0) = 0. In the second case,

$$\gamma^0 = 2/3.$$

This equation is obtained by using the properties of T(α, β). Indeed, for a $\gg$ 1, we have

$$\ln \sqrt{1 + \frac{2\xi\sin\vartheta\cos\psi}{a} + \frac{\xi^2\sin^2\vartheta}{a^2}} = \frac{\xi\sin\vartheta\cos\psi}{a} + \dots,$$

so that

$$\gamma\left(\frac{2\sin\vartheta\cos\psi}{a};\ \frac{\sin^2\vartheta}{a^2}\right)=\frac{\sin\vartheta\cos\psi}{a}+\dots$$

This means that

$$\Phi(a)=2/3a+\dots$$

and then (55.40) gives the required result.

Recalling now that for $a = 0$ the exact solution is $\gamma^0 = 4/3$ [5], we find that the method of balance leads to a completely satisfactory result for blocks of any thickness.

56. Effective Boundary Conditions on the Surfaces of "Grey" Bodies

We now go on to the most important problem in the theory of heterogeneous reactors, namely the establishment of the effective boundary conditions on the surface of an imperfect absorber ("grey" body). Whereas in black blocks the externally incident neutrons are entirely absorbed, in grey blocks they lose only part of their initial intensity and undergo further scattering on nuclei of the moderator. Thus, in grey blocks the absorption is weaker than in black ones.

To properly account for this effect one must solve the appropriate Peierls' integral equations. This direct path is, however, very difficult. It is thus useful to develop simpler methods that will guarantee the required accuracy.*

It is found that such approximate methods can be established for thin cylindrical blocks. This question has been studied by D.F. Zaretskii and D.D. Odintsev [29].

Let us proceed, as do these authors, considering a cylindrical absorber in an infinite scattering medium.** Assuming that the block is "small," or that its diameter is much smaller than the mean free path in the scatterer, we may consider the neutron flux to be almost isotropic.

Let j be the diffusion flux through a cylindrical surface (S_R) of sufficiently large radius R surrounding the block, and let p be the total number of neutrons absorbed by the block per unit time. Then assuming that there is no absorption outside the block, we have

$$j/j^0 = p/p^0, \tag{56.1}$$

where the index zero denotes quantities for a perfectly absorbing block.

We rewrite (56.1) in the form

$$j=\frac{p}{p^0}j^0. \tag{56.2}$$

We will write the asymptotic fluxes φ_0 and φ_0^0 in the form

$$\varphi_0 = 1+Q_\infty \ln r \quad\text{and}\quad \varphi_0^0 = 1+Q_\infty^0 \ln r, \tag{56.3}$$

where r is measured in units of the mean free path in the scatterer. We calculate j from the formula

$$j=\frac{S_R}{3}\frac{d\varphi_0}{dr}.$$

For the two fluxes we have

*Such methods are reviewed by Bat' [5] and Galanin [20].

**As in the Gurevich-Pomeranchuk and Wigner theories, we will treat scattering in the block within the framework of the "first shot" method.

$$j = \frac{S_R}{3R} Q_\infty \text{ and } j^0 = \frac{S_R}{3R} Q^0_\infty. \tag{56.4}$$

Inserting this into (56.2), we have

$$Q_\infty = \frac{p}{p^0} Q^0_\infty. \tag{56.5}$$

If we now assume that for small radii of the cylindrical block the neutron flux incident on S, the surface of the block, is almost isotropic, we may write p and p^0 in the approximate form

$$p = \frac{S}{\pi} \int_0^{\pi/2} \cos\psi \, d\psi \int_0^{\pi/2} \sin^2\vartheta \, \frac{\Sigma_c}{\Sigma} \left(1 - e^{-\Sigma d \frac{\cos\psi}{\sin\vartheta}}\right) d\vartheta,$$

$$p^0 = \frac{S}{\pi} \int_0^{\pi/2} \cos\psi \, d\psi \int_0^{\pi/2} \sin^2\vartheta \, d\vartheta$$

so that the ratio of these quantities is

$$\frac{p}{p^0} = \frac{\Sigma_c}{\Sigma} A(\Sigma d), \tag{56.6}$$

where

$$A(\beta) = \frac{4}{\pi} \int_0^{\pi/2} \cos\psi \, d\psi \int_0^{\pi/2} \sin^2\vartheta \left(1 - e^{-\beta \frac{\cos\psi}{\sin\vartheta}}\right) d\vartheta. \tag{56.7}$$

With this expression for the ratio, (56.5) becomes

$$Q_\infty = \frac{\Sigma_c}{\Sigma} A(\Sigma d) \, Q^0_\infty. \tag{56.8}$$

The logarithmic derivatives on the surface of the rod are

$$\frac{1}{\varphi_0} \frac{d\varphi_0}{dr} = \frac{Q_\infty}{a(1 + Q_\infty \ln a)} \tag{56.9}$$

and

$$\frac{1}{\varphi^0_0} \frac{d\varphi^0_0}{dr} = \frac{Q^0_\infty}{a(1 + Q^0_\infty \ln a)},$$

where $r = a$ is the dimensionless radius of the rod. As before, we denote the desired logarithmic derivatives on the surface of the cylindrical block by $1/\gamma$ and $1/\gamma^0$. Then (56.9) and (56.8) lead to the fundamental Zaretskii-Odintsev relation

$$\gamma = \frac{1 - \frac{1}{\gamma^0} a \ln a \, (1 - hA)}{hA} \gamma^0, \tag{56.10}$$

where

$$h = \Sigma_c / \Sigma.$$

In conclusion, we make the following useful remark. If we go over to the notation we have been using for the asymptotic neutron density,

$$\left.\begin{aligned}\varphi_0 &= 1 + q_\infty \ln \frac{r}{a},\\ \varphi_0^0 &= 1 + q_\infty^0 \ln \frac{r}{a},\end{aligned}\right\} \tag{56.11}$$

It is easy to show that Q_∞ and Q_∞^0 can be given in terms of q_∞ and q_∞^0 by

$$Q_\infty = \frac{q_\infty}{1 - q_\infty \ln a}, \quad Q_\infty^0 = \frac{q_\infty^0}{1 - q_\infty^0 \ln a}. \tag{56.12}$$

Then (56.8) can be written

$$\frac{q_\infty}{1 - q_\infty \ln a} = hA \frac{q_\infty^0}{1 - q_\infty^0 \ln a}. \tag{56.13}$$

As for γ and γ^0, they are

$$\gamma = \frac{a}{q_\infty} \text{ and } \gamma^0 = \frac{a}{q_\infty^0}. \tag{56.14}$$

Eliminating q_∞ and q_∞^0 from the last two equations, we again arrive at (56.10). Finally, solving (56.13) for q_∞, we have

$$q_\infty = \frac{hA}{1 - q_\infty^0 (1 - hA) \ln a} q_\infty^0. \tag{56.15}$$

This equation can be used to calculate q_∞ for a grey block. We will also use it later for establishing the limits of applicability of the Zaretskii-Odintsov theory. We mention here only that for a given $q_\infty^0(a)$, (56.15) gives the correct expressions in the limits. Thus, as $\Sigma d \to 0$ we find that $q_\infty \to 0$, while as $\Sigma d \to \infty$ and $h \to 1$, we find that $q_\infty \to q_\infty^0$.

The Zaretskii-Odintsev theory is applicable only within certain limits and can be used only for small blocks. The author, together with T.Kh. Sedel'nikov, has succeeded in generalizing this theory.

Consider a uranium block in an infinite scattering medium. The total number of neutrons absorbed per unit time by the block is

$$p = \int_S d_S \int_\Omega |\Omega_n| \varphi(\vec{r}_0, \vec{\Omega}) \frac{\Sigma_c}{\Sigma} (1 - e^{-\Sigma l})\, d\Omega, \tag{56.16}$$

where $\varphi(\vec{r}_0, \vec{\Omega})$ is the neutron flux incident on the surface S of the block in the $\vec{\Omega}$ direction. The number of neutrons absorbed per unit time in a black block is

$$p^0 = \int_S dS \int_\Omega |\Omega_n| \varphi^0(\vec{r}_0, \vec{\Omega})\, d\Omega, \tag{56.17}$$

where the index zero again denotes the corresponding quantities for a black block.

If $\varphi(\vec{r}_0, \vec{\Omega})$ and $\varphi^0(\vec{r}_0, \vec{\Omega})$ are exact solutions of the corresponding kinetic equations, Equations (56.16) and (56.17) can be used to find the total number of neutrons absorbed by the block per unit time. Consider the identity

$$p = \frac{\int_S dS \int_\Omega |\Omega_n| \varphi(\vec{r}_0, \vec{\Omega}) \frac{\Sigma_c}{\Sigma} (1 - e^{-\Sigma l})\, d\Omega}{\int_S dS \int_\Omega |\Omega_n| \varphi^0(\vec{r}_0, \vec{\Omega})\, d\Omega} p^0. \tag{56.18}$$

Let us express p and p^0 in terms of the asymptotic values of $\varphi_0(\vec{r})$ and $\varphi_0^0(\vec{r})$. To do this, consider a closed surface S_R surrounding a block and sufficiently far from it. We then arrive at

$$p = \frac{1}{3}\int_{S_R} \nabla_n \varphi_0 \, dS \tag{56.19}$$

and, similarly,

$$p^0 = \frac{1}{3}\int_{S_R} \nabla_n \varphi_0^0 \, dS. \tag{56.20}$$

These equations show that the diffusion flux* is compensated by absorption in the block.

Inserting these into (56.18), we arrive at

$$\int_{S_R} \nabla_n \varphi_0 \, dS = \frac{\int_S dS \int_\Omega |\Omega_n| \varphi(\vec{r}_0, \vec{\Omega}) \frac{\Sigma_c}{\Sigma}(1 - e^{-\Sigma l})\, d\Omega}{\int_S dS \int_\Omega |\Omega_n| \varphi^0(\vec{r}_0, \vec{\Omega})\, d\Omega} \int_{S_R} \nabla_n \varphi_0^0 \, dS. \tag{56.21}$$

If the total asymptotic fluxes $\varphi_0(\vec{r})$ and $\varphi_0^0(\vec{r})$ are again written

$$\left.\begin{aligned} \varphi_0(\vec{r}) &= 1 + q_\infty I(\vec{r}), \\ \varphi_0^0(\vec{r}) &= 1 + q_\infty^0 I(\vec{r}), \end{aligned}\right\} \tag{56.22}$$

the desired functions $\varphi(\vec{r}_0, \vec{\Omega})$ and $\varphi^0(\vec{r}_0, \vec{\Omega})$ are found by integrating along the $\vec{\Omega}$ ray emating from a point M on the surface of the block:

$$\left.\begin{aligned} \varphi(\vec{r}_0, \vec{\Omega}) &= \frac{1}{4\pi}\int_0^\infty \varphi_0(\vec{r})\, e^{-\xi}\, d\xi, \\ \varphi^0(\vec{r}_0, \vec{\Omega}) &= \frac{1}{4\pi}\int_0^\infty \varphi_0^0(\vec{r})\, e^{-\xi}\, d\xi. \end{aligned}\right\} \tag{56.23}$$

Inserting this into (56.22), we obtain

$$\left.\begin{aligned} \varphi(\vec{r}_0, \vec{\Omega}) &= \frac{1}{4\pi}\left[1 + q_\infty \int_0^\infty I(\vec{r})\, e^{-\xi} d\xi\right]. \\ \varphi^0(\vec{r}_0, \vec{\Omega}) &= \frac{1}{4\pi}\left[1 + q_\infty^0 \int_0^\infty I(\vec{r})\, e^{-\xi} d\xi\right]. \end{aligned}\right\} \tag{56.24}$$

Hence, (56.21) becomes

$$q_\infty = \frac{h\int_S dS \int_\Omega |\Omega_n| (1 - e^{-\Sigma l})\, d\Omega + h q_\infty \int_S dS \int_\Omega |\Omega_n| (1 - e^{-\Sigma l}) \int_0^\infty I(\vec{r}) e^{-\xi}\, d\xi\, d\Omega}{\int_S dS \int_\Omega |\Omega_n|\, d\Omega + q_\infty^0 \int_S dS \int_\Omega |\Omega_n| \int_0^\infty I(\vec{r})\, e^{-\xi}\, d\xi\, d\Omega}\, q_\infty^0, \tag{56.25}$$

*Far from the block the flux becomes almost isotropic, since diffusion theory gives a good description of the neutron field.

where

$$h = \Sigma_c / \Sigma.$$

Solving this for the unknown q_∞, we have

$$q_\infty = \frac{h \int\limits_S dS \int\limits_\Omega |\Omega_n| (1 - e^{-\Sigma l})\, d\Omega}{\int\limits_S dS \int\limits_\Omega |\Omega_n|\, d\Omega + q_\infty^0 \int\limits_S dS \int\limits_\Omega |\Omega_n| [1 - h(1 - e^{-\Sigma l})] \int\limits_0^\infty I(\vec{r})\, e^{-\xi}\, d\xi\, d\Omega}\, q_\infty^0. \qquad (56.26)$$

This is convenient to write in the form

$$q_\infty = \frac{hA}{1 + q_\infty^0 (1 - hF)\, \Phi}\, q_\infty^0, \qquad (56.27)$$

where

$$A = \frac{\int\limits_S dS \int\limits_\Omega |\Omega_n| (1 - e^{-\Sigma l})\, d\Omega}{\int\limits_S dS \int\limits_\Omega |\Omega_n|\, d\Omega}, \qquad (56.28)$$

$$\Phi = \frac{1}{\pi S} \int\limits_S dS \int\limits_\Omega |\Omega_n|\, d\Omega \int\limits_0^\infty I(\vec{r})\, e^{-\xi}\, d\xi, \qquad (56.29)$$

$$F = \frac{\int\limits_S dS \int\limits_\Omega |\Omega_n| (1 - e^{-\Sigma l})\, d\Omega \int\limits_0^\infty I(\vec{r})\, e^{-\xi}\, d\xi}{\int\limits_S dS \int\limits_\Omega |\Omega_n|\, d\Omega \int\limits_0^\infty I(\vec{r})\, e^{-\xi}\, d\xi}. \qquad (56.30)$$

For a plane layer, A, Φ and F are given by

$$\left.\begin{aligned} A(\beta) &= 1 - 2E_3(\beta) \\ \Phi(a) &= \frac{2}{3} \\ F(\beta) &= 1 - 3E_4(\beta). \end{aligned}\right\} \qquad (56.31)$$

For a cylindrical block,

$$\left.\begin{aligned} A(\beta) &= \frac{4}{\pi} \int\limits_0^{\pi/2} \cos\psi\, d\psi \int\limits_0^{\pi/2} \sin^2\vartheta \left(1 - e^{-\beta \frac{\cos\psi}{\sin\vartheta}}\right) d\vartheta, \\ \Phi(a) &= \frac{4}{\pi} \int\limits_0^{\pi/2} \cos\psi\, d\psi \int\limits_0^{\pi/2} \sin^2\vartheta\, T\left(\frac{2\sin\vartheta\cos\psi}{a};\ \frac{\sin^2\vartheta}{a^2}\right) d\vartheta, \\ F(a, \beta) &= \frac{4}{\pi\Phi(a)} \int\limits_0^{\pi/2} \cos\psi\, d\psi \int\limits_0^{\pi/2} \sin^2\vartheta \times \\ &\times T\left(\frac{2\sin\vartheta\cos\psi}{a};\ \frac{\sin^2\vartheta}{a^2}\right)\left(1 - e^{-\beta\frac{\cos\psi}{\sin\vartheta}}\right) d\vartheta, \end{aligned}\right\} \qquad (56.32)$$

where $T(\alpha;\ \beta)$ is given by (55.38). In obtaining these formulas we again use the relations

$$I(\vec{r}) = \begin{cases} x & \text{for the plane geometry} \\ \ln\frac{r}{a} & \text{for the cylindrical geometry} \end{cases}$$

$$\int_0^\infty I(\vec{r})\, e^{-\xi}\, d\xi = \begin{cases} \cos\vartheta & \text{for the plane geometry} \\ \int_0^\infty e^{-\xi} \ln \sqrt{1 + 2\sin\vartheta\cos\psi\,\frac{\xi}{a} + \sin^2\vartheta\,\frac{\xi^2}{a^2}}\, d\xi - & \\ & \text{for the cylindrical geometry} \end{cases}$$

Figures 23 and 25 give graphs of $A(\beta)$ and $\Phi(a)$ (see also Tables XI and XIII of Appendix F), while $F(\beta)$ and $F(\alpha, \beta)$ are given in Appendix F (Tables XIV and XV).

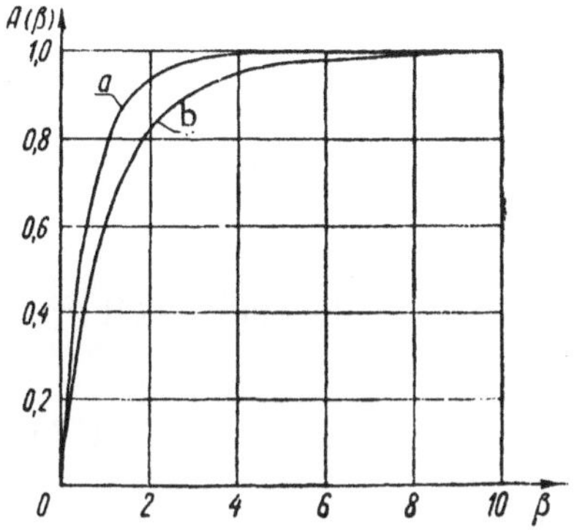

Fig. 25. The neutron transmission function $A(\beta)$ in the block $\beta = \Sigma d$; Σ is the total cross section in the block; d is the thickness of the block; a) plane layer; b) cylindrical block.

Since $F(\beta)$ and $F(\alpha, \beta)$ can be replaced with sufficient accuracy by $A(\beta)$, (56.27) can be written in the approximate form

$$q_\infty = \frac{hA(\beta)}{1 + q^0_\infty [1 - hA(\beta)]\,\Phi(a)}\, q^0_\infty, \tag{56.33}$$

where $\beta = \Sigma d$, $a = \Sigma^0_s r_0$, and r_0 is the geometric radius of the block. The structure of (56.33) is obviously the same as that of the Zaretskii-Odintsov Formula (56.15), except that $\ln\frac{1}{a}$ is now replaced by $\Phi(a)$. The fact that the present formula is an improvement is seen from Table XVI of Appendix F. The comparison given in this table shows that the Zaretskii-Odintsov theory can be used if $0 \leq a \leq 0.8$. Outside this interval their theory may lead to large errors.

In conclusion, we give the formula for γ, namely

$$\gamma = \frac{1 + \frac{1}{\gamma^0}[1 - hA(\beta)]\, a\Phi(a)}{hA(\beta)} \cdot \gamma^0. \tag{56.34}$$

The transition from dimensionless quantities a to the nondimensionless ones r_0 is performed according to (55.25').

57. The Wigner-Seitz Method

The Wigner-Seitz method was first developed to calculate wave functions in the periodic lattice of a crystal. This method is based on the observation that because the periodic lattice is symmetric, it is sufficient to solve the problem in a single cell. Further, if this cell has a high order of symmetry, it can be replaced with sufficient accuracy by a sphere of the same volume [64].

The Wigner-Seitz method has been found to be very effective also when the cell considered is one in a heterogeneous nuclear reactor. It may be formulated in the following way.

Consider a cylindrical cell of a heterogeneous instrument whose cross section is a square, a hexagon, etc. Let us replace this cell by an equivalent circular cylinder of equal volume on whose radius $r = R$ we assume the neutron flux to vanish. As for the condition on the surface of the uranium block ($r = a$) in the center of the cell, according to what we have already said it must be written in the form

$$\frac{\partial\varphi}{\partial r} = \frac{1}{\gamma}\varphi,$$

where γ is found by Peierls' integral equation using (56.34), and the index zero on the function φ_0 has been dropped. Thus, the calculation of the neutron density φ in a circular cell reduces to solving the problem

$$\left.\begin{aligned} &\nabla D^0 \nabla\varphi - \Sigma^0_c\varphi = -f(r) \\ &D^0 \frac{\partial\varphi}{\partial r} = \frac{D^0}{\gamma}\varphi \qquad (r = a), \\ &\frac{\partial\varphi}{\partial r} = 0 \qquad (r = R), \end{aligned}\right\} \tag{57.1}$$

where the index zero denotes quantities belonging to the moderator. The solution of (57.1) gives a function $\varphi(r)$ which is then used to obtain the effective constants averaged over the cell of the heterogeneous reactor. Proceeding, we first calculate the most important characteristic, the effective capture cross section Σ_c^{eff}. The total number of neutrons absorbed in the cell is found from the neutrons absorbed in the uranium block and those absorbed in the scatterer.*

The number of neutrons absorbed in the uranium block is found by integrating the second of Equations (57.1) over the surface S of the block, and may be written

$$p = \frac{D^0}{\gamma} \varphi(a) S, \tag{57.2}$$

and the number of neutrons absorbed by the scatterer is

$$p^0 = \int_{V^0} \Sigma_c^0 \varphi \, d\vec{r}, \tag{57.3}$$

where V^0 is the volume of the scatterer in the cell. Thus the total number of neutrons absorbed in the cell will be

$$p = \frac{D^0 S}{\gamma} \varphi(a) + \int_{V^0} \Sigma_c^0 \varphi \, d\vec{r}. \tag{57.4}$$

Let us now find the total number of neutrons in the stationary distribution over the cell of the reactor. It is also given by adding the number of neutrons in the block to the number in the rest of the cell.

The number of neutrons in the block can be found from the relation

$$\int_V \Sigma_c \varphi \, d\vec{r} = \frac{D^0 S}{\gamma} \varphi(a), \tag{57.5}$$

where V is the volume of the block. Since inside the block Σ_c = const, we obtain

$$\int_V \varphi \, d\vec{r} = \frac{D^0 S}{\gamma \Sigma_c} \varphi(a). \tag{57.6}$$

The number of neutrons in the cell outside the block is found simply by integrating (57.1):

$$\int_{V^0} \varphi \, d\vec{r}.$$

The total number of neutrons in the block is then given in the form of the sum

$$\frac{D^0 S}{\gamma \Sigma_c} \varphi(a) + \int_{V^0} \varphi \, d\vec{r}. \tag{57.7}$$

Dividing p by the number of neutrons as given by (57.7), we obtain the following expression for the effective capture cross section:

$$\Sigma_c^{eff} = \frac{\frac{D^0 S}{\gamma} \varphi(a) + \int_{V^0} \Sigma_c^0 \varphi \, d\vec{r}}{\frac{D^0 S}{\gamma \Sigma_c} \varphi(a) + \int_{V^0} \varphi \, d\vec{r}}. \tag{57.8}$$

*We assume that many more neutrons are captured in the uranium block than in the scatterer.

As for the effective fission cross section, it is obtained analogously, and is

$$\Sigma_f^{\text{eff}} = \frac{\Sigma_f}{\Sigma_c} \frac{\frac{D^0 S}{\gamma} \varphi(a)}{\frac{D^0 S}{\gamma \Sigma_c} \varphi(a) + \int\limits_{V_0} \varphi \, d\vec{r}} . \tag{57.9}$$

Although it is a relatively simple problem to calculate Σ_c^{eff} and Σ_f^{eff}, the effective diffusion coefficient is not so easily found. This is because the diffusion coefficient depends strongly on the direction of motion of the neutron, so that it is a tensor quantity. We shall not go into a rigorous determination of the effective diffusion coefficient, but shall merely refer the reader to the literature [137, 89]. Here we shall consider only the simplest case, in which the inhomogeneities of scattering in a cell are small. We then have the approximation

$$D^{\text{eff}} = \frac{D \frac{D^0 S}{\gamma \Sigma_c} \varphi(a) + \int\limits_{V} D^0 \varphi \, d\vec{r}}{\frac{D^0 S}{\gamma \Sigma_c} \varphi(a) + \int\limits_{V} \varphi \, d\vec{r}} , \tag{57.10}$$

where D is the diffusion coefficient in the block.

We note finally that the considerations of this section can be carried over formally to the case of neutron slowing-down. Usually this is done in terms of the so-called "adiabatic" approximation, which assumes that the solution $\varphi(\vec{r}, u)$ can be represented by a product of the form

$$\varphi(\vec{r}, u) = f(u) \, \varphi(\vec{r}),$$

which means that one makes the approximation that the variables can be separated. This approximation is usually good for thermal and intermediate reactors.

Thus, by performing the calculations for the cell of a heterogeneous reactor one is able to obtain all the necessary effective constants of the equivalent homogeneous reactor. Further computations can therefore be performed with these constants by the ordinary methods discussed in the previous chapters. For intermediate energies, at which the flux incident on the block is almost isotropic, the most useful method is that of Section 54.

Chapter XIII

FAST NEUTRON REACTORS

In 1949, A.I. Leipunskii pointed out the possibility of using a fast-neutron reactor for the simultaneous production of energy and a significant quantity of atomic fuel (see Blokhintsev [9]). His considerations were the following.*

In the first place, the multiplication factor in fast-neutron reactors, which contain U^{238}, is much larger than in thermal and intermediate reactors.

Second, in fast reactors the ratio between radiative capture and fission in the fuel is lower than in others, and for this reason there is less nonproductive neutron capture in the reactor.

Third, for high-energy neutrons the capture cross section of structural elements is much lower than that of uranium, so that this contribution to nonproductive absorption is also small.

Fourth, in fast-neutron reactors the cross section for capture by slags is also smaller than in other reactors, and there is no poisoning at all.

As a result, the breeding ratio in a fast reactor will be much larger than unity. This is the basis for the idea that a fast reactor can be used to breed nuclear fuel at the same time as it produces industrial energy.

The general kinetic equation for a fast-neutron reactor was first obtained in 1950 by D.I. Blokhintsev (see [9]), who also pointed out some methods for solving this equation. Later, L.N. Usachev contributed further to the development of this theory. In particular, he obtained the adjoint reactor equation in the most complete form, and worked out the application of perturbation theory to this case [81, 82]. American work on the theory and computational methods for fast-neutron reactors is given in the review paper by D. Okrent, R. Avery and H. Hummel [60]. In this chapter we will study the basic and adjoint reactor equations and reduce them to the set of multigroup equations, and will discuss some methods for solving the kinetic equations.

58. Kinetic Equation for a Fast-Neutron Reactor

In fast reactors, neutrons are slowed down primarily by inelastic scattering. The energy loss in inelastic collisions is determined by the energy levels of the given nucleus, and its probability is found from the quantum mechanical properties of the neutron-nucleus interaction. If we include elastic as well as inelastic scattering, the basic kinetic equation of the reactor can be written

$$\vec{\Omega}\nabla\varphi+\Sigma\varphi=\int d\Omega'\int_{-\infty}^{\infty} du'\,\varphi g(\mu_0,u,u'), \tag{58.1}$$

where $\varphi(\vec{r},\vec{\Omega},u)$ is the neutron flux per unit area perpendicular to $\vec{\Omega}$ in $(\vec{r},\vec{\Omega},u)$ phase space, $\Sigma=\Sigma_s+\Sigma_{in}+\Sigma_c$ is the total cross section, Σ_s is the elastic scattering cross section, Σ_{in} is the inelastic scattering cross section, Σ_c is the capture cross section, and $\mu_0=(\vec{\Omega}\vec{\Omega}')$.

The function $g(\mu_0, u, u')$ can be written in the form

$$\begin{aligned} g(\mu_0,u,u') &= \Sigma_s(u')W_s(\mu_0,u,u')+\Sigma_{in}(u')W_{in}(\mu_0,u,u')+\\ &+\nu_f\Sigma_f(u')W_f(\mu_0,u,u'). \end{aligned} \tag{58.2}$$

*In describing the physical content of fast-neutron reactor calculations, we will follow L.N. Usachev [81].

Here $W_s(\mu_0, u, u')$, $W_{in}(\mu_0, u, u')$ and $W_f(\mu_0, u, u')$ are the probability densities for elastic scattering, inelastic scattering, and fission, respectively. It should be noted that if the elastic scattering is isotropic in the center-of-mass system and if one ignores the thermal motion of the moderator nuclei, then

$$W_s(\mu_0, u, u') = f(\mu_0, u - u'),$$

where $f(\mu_0, u - u')$ is the function defined in Chapter I.

The $W_\nu(\mu_0, u, u')$ functions are normalized according to

$$\int d\Omega \int_{-\infty}^{\infty} du\, W_\nu(\mu_0, u, u') = 1. \qquad (58.3)$$

In this equation W_ν represents any of:

$$W_s,\ W_{in} \text{ and } W_f.$$

To the integro-differential equation (58.1) we must add the boundary condition

$$\varphi(\vec{r}, \vec{\Omega}, u) = 0 \quad \text{on} \quad S, \text{ for } \quad (\vec{\Omega}, \vec{n}^0) < 0. \qquad (58.4)$$

In solving the critical size problem for the reactor, it is convenient to introduce the parameter K_{eff}. Then $g(\mu_0, u, u')$ is given by

$$g(\mu_0, u, u') = \Sigma_s(u') W_s(\mu_0, u, u') + \Sigma_{in}(u') W_{in}(\mu_0, u, u') + \\ + \frac{\nu_f \Sigma_f(u')}{K_{eff}} W_f(\mu_0, u, u'). \qquad (58.5)$$

59. Adjoint Reactor Equation

We may write (58.1) in the form

$$\mathbf{L}\varphi = 0, \qquad (59.1)$$

where $\mathbf{L}$ is an integro-differential operator defined on a function set $\{f\}$. The elements of this set are defined on the space $\mathbf{G} \times \mathbf{\Omega} \times \mathbf{U}$, satisfy Condition (58.4), and are such that the operator $\mathbf{L}$ has meaning when acting on them.

We introduce the function set $\{f^*\}$ also defined on $\mathbf{G} \times \mathbf{\Omega} \times \mathbf{U}$, and define the scalar product

$$(f, f^*) = \int_G d\vec{r} \int_\Omega d\Omega \int_U du\, f f^*. \qquad (59.2)$$

We then use the functional relation

$$(f^*, \mathbf{L}f) = (f, \mathbf{L}^* f^*) \qquad (59.3)$$

to arrive at the operator adjoint in the sense of Lagrange and at the adjoint reactor equation.

Repeating the considerations of Section 6, we use Gauss' theorem and integration by parts to obtain the adjoint equation in the form

$$-\vec{\Omega}\nabla\varphi^* + \Sigma\varphi^* = \int d\Omega' \int_{-\infty}^{\infty} du'\, \varphi^* g(\mu_0, u', u), \qquad (59.4)$$

with the condition

$$\varphi^*(\vec{r}, \vec{\Omega}, u) = 0 \quad \text{on} \quad S, \quad \text{for} \quad (\vec{\Omega}, \vec{n}^0) > 0. \tag{59.5}$$

Thus, the adjoint problem is also defined and can be used to find the critical size and neutron importance in the reactor.

60. Multigroup Set of Basic Reactor Equations

Great computational difficulties arise in attempting to solve the kinetic equation of a fast-neutron reactor. The mathematical statement of the problem must therefore be simplified. This can be done by approximating the reactor equations in the form of a multigroup set of kinetic equations depending only on $\vec{r}$ and $\vec{\Omega}$.

In this section we will obtain the multigroup set of basic reactor equations.

We will do this using perturbation theory. Essentially what we shall do is replace our basic kinetic equation by a similar one with piecewise constant functions Σ and $g(\mu_0, u, u')$ of the lethargy and position in the reactor in such a way that K_{eff} remain unchanged under this replacement.

We shall call the original reactor equation the unperturbed equation, and that with the piecewise constant parameters the perturbed equation.

We will denote the quantities for the perturbed reactor by primes.

Let us, therefore, consider the basic reactor equation in the form

$$\mathbf{L}\varphi = 0, \tag{60.1}$$

where $\mathbf{L}$ is an operator defined on a function set $\{f\}$, and let us write the adjoint equation of the perturbed reactor in the form

$$\mathbf{L}^{*\prime}\varphi^{*\prime} = 0, \tag{60.2}$$

where $\mathbf{L}^{*\prime}$ is an operator defined on the function set $\{f^*\}$. We multiply (60.1) by $\varphi^{*\prime}$, and subtract. Then after some simple mathematical operations, we arrive at

$$\int d\vec{r} \int d\Omega \int_{-\infty}^{\infty} du \left\{ (\Sigma' - \Sigma)\, \varphi\varphi^{*\prime} - \right.$$

$$\left. - \varphi^{*\prime} \int d\Omega' \int_{-\infty}^{\infty} du'\, \varphi \left[g'(\mu_0, u, u') - g(\mu_0, u, u') \right] \right\} = 0. \tag{60.3}$$

Changing the order of integration, we obtain

$$\int d\vec{r} \int d\Omega \int_{-\infty}^{\infty} du \left\{ (\Sigma' - \Sigma)\, \varphi\varphi^{*\prime} - \right.$$

$$\left. - \varphi \int d\Omega' \int_{-\infty}^{\infty} du'\, \varphi^{*\prime} \left[g'(\mu_0, u', u) - g(\mu_0, u', u) \right] \right\} = 0. \tag{60.4}$$

Let us now divide the entire lethargy interval into separate energy groups.

Let this division be $u_{j-1} < u < u_j$. Within each group, we find an effective constant cross section which, by assumption, does not change K_{eff}.

We obtain these effective group constants from (60.4), which we write in the form

$$\sum_k \sum_j \int_{G_k} d\vec{r} \int d\Omega \int_{u_{j-1}}^{u_j} du \left\{ (\Sigma^j - \Sigma)\varphi\varphi^{*\prime} - \varphi \sum_l \int d\Omega' \int_{u_{l-1}}^{u_l} du' \varphi^{*\prime} [\overset{j\to l}{g}(\vec{\Omega}', \vec{\Omega}) - g(\mu_0, u', u)] \right\} = 0. \quad (60.5)$$

Here we have used the fact that within each group Σ' and g' are constant, and we have denoted them by Σ^j and $\overset{j\to l}{g}(\vec{\Omega}', \vec{\Omega})$.

Obviously (60.5) is satisfied if we write

$$\left.\begin{aligned} \Sigma^j &= \frac{\int_{G_h} d\vec{r} \int d\Omega \int_{u_{j-1}}^{u_j} du\, \Sigma\, \varphi\varphi^{*\prime}}{\int_{G_h} d\vec{r} \int d\Omega \int_{u_{j-1}}^{u_j} du\, \varphi\varphi^{*\prime}}, \\ \overset{j\to l}{g}(\vec{\Omega}', \vec{\Omega}) &= \frac{1}{\Delta u_j} [\overset{j\to l}{g_s}(\vec{\Omega}', \vec{\Omega}) + \overset{j\to l}{g_{in}}(\vec{\Omega}', \vec{\Omega}) + \overset{j\to l}{g_f}(\vec{\Omega}', \vec{\Omega})], \end{aligned}\right\} \quad (60.6)$$

where

$$\left.\begin{aligned} \overset{j\to l}{g_s}(\vec{\Omega}', \vec{\Omega}) &= \frac{\int_{G_k} d\vec{r} \int_{u_{j-1}}^{u_j} du\, \Sigma_s \varphi \int_{u_{l-1}}^{u_l} du' \varphi^{*\prime} W_s(\mu_0, u', u)}{\int_{G_k} d\vec{r} \int_{u_{j-1}}^{u_j} du\, \varphi \int_{u_{l-1}}^{u_l} du' \varphi^{*\prime}} \\ \overset{j\to l}{g_{in}}(\vec{\Omega}', \vec{\Omega}) &= \frac{\int_{G_k} d\vec{r} \int_{u_{j-1}}^{u_j} du\, \Sigma_{in} \varphi \int_{u_{l-1}}^{u_l} du' \varphi^{*\prime} W_{in}(\mu_0, u', u)}{\int_{G_k} d\vec{r} \int_{u_{j-1}}^{u_j} du\, \varphi \int_{u_{l-1}}^{u_l} du' \varphi^{*\prime}} \\ \overset{j\to l}{g_f}(\vec{\Omega}', \vec{\Omega}) &= \frac{\int_{G_k} d\vec{r} \int_{u_{j-1}}^{u_j} du\, \nu_f \Sigma_f \varphi \int_{u_{l-1}}^{u_l} du' \varphi^{*\prime} W_f(\mu_0, u', u)}{\int_{G_k} d\vec{r} \int_{u_{j-1}}^{u_j} du\, \varphi \int_{u_{l-1}}^{u_l} du' \varphi^{*\prime}} \end{aligned}\right\} \quad (60.7)$$

We now proceed to obtain the multigroup set of basic reactor equations. To do this, consider the basic equation with the perturbed parameters

$$\vec{\Omega}\nabla\varphi' + \Sigma'\varphi' = \int d\Omega' \int_{-\infty}^{\infty} du' \varphi' g'(\vec{\Omega}, \vec{\Omega}', u, u'), \quad (60.8)$$

where the piecewise constant group constants are obtained from (60.6) and (60.7).

Integrating (60.8) over u within a single group, we obtain

$$\vec{\Omega}\nabla\varphi^j + \Sigma^j\varphi^j = \int d\Omega' \int_{-\infty}^{\infty} du' \varphi' g^j(\vec{\Omega}, \vec{\Omega}' u'), \quad (60.9)$$

where

$$\varphi^j = \int_{u_{j-1}}^{u_j} \varphi' du, \quad g^j(\vec{\Omega}, \vec{\Omega}' u') = \int_{u_{j-1}}^{u_j} g'(\vec{\Omega}, \vec{\Omega}', u, u')\, du.$$

Since $g^j(\vec{\Omega}, \vec{\Omega}', u')$ is piecewise constant in u within the limits $u_{j-1} < u < u_j$ of the group, the integral on the right side of (60.9) can be written

$$\int d\Omega' \int_{-\infty}^{\infty} du'\varphi' g^j(\vec{\Omega}, \vec{\Omega}', u') = \sum_l \int d\Omega' \varphi^l \overset{l\to j}{g_0}(\vec{\Omega}, \vec{\Omega}'), \tag{60.10}$$

where

$$\overset{j\to l}{g_0}(\vec{\Omega}', \vec{\Omega}) = \Delta u_j \overset{j\to l}{g}(\vec{\Omega}', \vec{\Omega}).$$

We thus arrive at the multigroup set of reactor equations

$$\vec{\Omega}\nabla\varphi^j + \Sigma^j\varphi^j = \sum_l \int d\Omega'\varphi^l \overset{l\to j}{g_0}(\vec{\Omega}, \vec{\Omega}'). \tag{60.11}$$

The boundary conditions for (60.11) are

$$\varphi^j(\vec{r}, \vec{\Omega}) = 0 \text{ on } S, \text{ for } (\vec{\Omega}, \vec{n}^0) < 0. \tag{60.12}$$

If we set the lethargy interval (u_{j-1}, u_j) equal to the entire lethargy interval $(-\infty, \infty)$, we arrive at the one-group kinetic equation

$$\vec{\Omega}\nabla\varphi + \Sigma\varphi = \int d\Omega'\varphi g(\vec{\Omega}, \vec{\Omega}') \tag{60.13}$$

with the condition

$$\varphi(\vec{r}, \vec{\Omega}) = 0 \text{ on } S, \text{ for } (\vec{\Omega}, \vec{n}^0) < 0. \tag{60.14}$$

61. Multigroup Set of Adjoint Reactor Equations

We will obtain the multigroup set of adjoint reactor equations from the adjoint equation with piecewise constant parameters

$$-\vec{\Omega}\nabla\varphi^{*\prime} + \Sigma'\varphi^{*\prime} = \int d\Omega' \int_{-\infty}^{\infty} du'\varphi^* g'(\Omega', \Omega\ u', u), \tag{61.1}$$

with the condition

$$\varphi^{*\prime}(\vec{r}, \vec{\Omega}, u) = 0 \text{ on } S, \text{ for } (\vec{\Omega}, \vec{n}^0) > 0. \tag{61.2}$$

Taking into account the properties of Σ' and g', we arrive simply at the multigroup set of adjoint reactor equations

$$-\vec{\Omega}\nabla\varphi^{*j} + \Sigma^j\varphi^{*j} = \sum_l \int d\Omega'\varphi^{*l} \overset{j\to l}{g_0}(\vec{\Omega}', \vec{\Omega}) \tag{61.3}$$

with the condition

$$\varphi^{*j}(\vec{r}, \vec{\Omega}) = 0 \text{ on } S, \text{ for } (\vec{\Omega}, \vec{n}^0) > 0. \tag{61.4}$$

Here $\varphi^{*j}(\vec{r}, \vec{\Omega})$ are values of $\varphi^*(\vec{r}, \vec{\Omega}, u)$ in (u_{j-1}, u_j), independent of $\underline{u}$. If (u_{j-1}, u_j) is the entire lethargy interval, we obtain the one-velocity adjoint equation

$$-\vec{\Omega}\nabla\varphi^* + \Sigma\varphi^* = \int d\Omega'\varphi^* g(\vec{\Omega}', \vec{\Omega}) \tag{61.5}$$

with the condition

$$\varphi^*(\vec{r}, \vec{\Omega}) = 0 \text{ on } S, \text{ for } (\vec{\Omega}, \vec{n}^0) > 0. \tag{61.6}$$

62. Mean Group Constants

In the above sections the methods for averaging the group constants assumed knowledge of the exact solution. This is unnecessary only in one exceptional case, that of cross sections constant within the groups and transition probabilities within the groups depending only on the angle of collision. If the physical quantities are slowly varying functions of the lethargy, one would naturally suppose that the effective group constants can be obtained if one knows the neutron flux and importance spectra very roughly.

For critical mass problems one can usually calculate the effective group constants on the following assumptions. In Equations (60.6) and (60.7) the neutron flux and importance can be assumed spherically symmetrical, writing

$$\varphi(\vec{r}, \vec{\Omega}, u) \cong \frac{1}{4\pi}\varphi_0(\vec{r}, u),$$
$$\varphi^*(\vec{r}, \vec{\Omega}, u) \cong \frac{1}{4\pi}\varphi_0^*(\vec{r}, u),$$

where

$$\varphi_0 = \int \varphi\, d\Omega, \quad \varphi_0^* = \int \varphi^*\, d\Omega.$$

The total flux φ_0 and importance φ_0^* can be approximated by solving simpler problems such as, for instance, the diffusion approximation (or the diffusion-age approximation if there is little inelastic scattering).

With these assumptions, the group constants for a zone G_k in the reactor can be found from the equations

$$\left.\begin{aligned} \Sigma^j &= \frac{\int_{u_{j-1}}^{u_j} du\, \overline{\Sigma\varphi_0\varphi_0^{*\prime}}}{\int_{u_{j-1}}^{u_j} du\, \overline{\varphi_0\varphi_0^{*\prime}}}, \\ \overset{l\to j}{g}(\vec{\Omega}, \vec{\Omega}') &= \frac{1}{\Delta u_j}\left[\overset{l\to j}{g_s}(\vec{\Omega}, \vec{\Omega}') + \overset{l\to j}{g_{in}}(\vec{\Omega}, \vec{\Omega}') + \overset{l\to j}{g_f}(\vec{\Omega}, \vec{\Omega}')\right], \end{aligned}\right\} \tag{62.1}$$

where

$$\overset{l\to j}{g_\nu}(\vec{\Omega}, \vec{\Omega}') = \frac{\int_{u_{j-1}}^{u_j} du\, \Sigma_\nu \int_{u_{l-1}}^{u_l} du' \overline{W_\nu(\mu_0, u', u)\, \varphi_0(\vec{r}, u')\, \varphi_0^{*\prime}(\vec{r}, u)}}{\int_{u_{j-1}}^{u_j} du \int_{u_{l-1}}^{u_l} du' \overline{\varphi_0(\vec{r}, u')\, \varphi^{*\prime}(\vec{r}, u)}}, \tag{62.2}$$

and the index ν denotes the fact that this last equation is valid for elastic and inelastic scattering, as well as for fission. In these equations we have written

$$\overline{A(\vec{r})} = \int_{G_k} d\vec{r}\, A(\vec{r}).$$

An important fact in the averaging formulas obtained is that all the desired effective group constants are expressed by quotients of linear functionals. Therefore, even a large error in the determination of the weight functions will not cause a large error in the functionals.

After the group constants are found from Equations (62.1) and (62.2), one can write out the multigroup set of basic and adjoint reactor equations. If all the neutrons are included in a single group, we may use Equation (60.6) to obtain the effective constants for the one-group equation.

Nuclear reactor calculations are usually performed in the diffusion (or the diffusion-age) approximation. If the reactor is much larger than the neutron mean free path, the diffusion approximation can be used validly to describe the space and energy distribution of the neutron flux and importance, as well as to find the critical size of the reactor. In this case, however, it is desirable to obtain an accurate value for the critical mass, the basic characteristic of the reactor which depends on the kinetic effects. At the same time it should be emphasized that whereas the preliminary calculation is performed in the diffusion (or diffusion-age) approximation, the more accurate calculations should be based on a more rigorous treatment of the collision mechanism.

It is therefore desirable to use the kinetic equation to improve the preliminary results obtained by approximation methods. For this purpose we proceed as follows.

Let us assume that we have solved the problem in some approximation. We use this solution to calculate the effective constants in the equivalent one-group problem and attempt to solve the one-velocity kinetic equation. This solution can be used to correct the eigenvalue of the problem and therefore the critical mass. The correction obtained is better when more terms are maintained in the Legendre polynomial expansion of the indicatrix $g(\vec{\Omega}, \vec{\Omega}')$, which we write

$$g(\vec{\Omega}, \vec{\Omega}') = \frac{1}{4\pi}[g_0 + g_1P_1(\mu') + g_2P_2(\mu') + \dots]. \tag{62.3}$$

The above methods of averaging the basic and adjoint reactor equations and of correcting the approximate solutions by reducing the problems to equivalent one-group problems and then solving the kinetic equation are valid in general and can be used for reactors with any spectra.

Further, these methods can be used to obtain the multigroup reactor equations in the diffusion and diffusion-age approximations.* This has been discussed, however, in previous chapters from another point of view. We shall here indicate without sufficient proof some new algorithms which may in some cases be preferable.

Finally, let us turn to some simplifications that can be made in the basic and adjoint equations when treating fast-neutron reactors if one assumes that the neutrons are slowed down primarily by inelastic collisions and that elastic scattering can essentially be ignored. In this case $g(\mu_0, u, u')$ may be considered isotropic, so that we may write

$$g(\mu_0, u, u') = \frac{1}{4\pi}\left[\Sigma_{in}(u')W_{in}(u, u') + \frac{\nu_f\Sigma_f(u')}{K_{\text{eff}}}\chi(u)\right]. \tag{62.4}$$

Then the group constants $\overset{l\to j}{g_0}(\vec{\Omega}, \vec{\Omega}')$ are independent of the scattering angle, so that

$$\overset{l\to j}{g_0} = \frac{1}{4\pi}\overset{l\to j}{\Sigma}. \tag{62.5}$$

In this way the multigroup set of reactor equations becomes

$$\left.\begin{aligned} &\vec{\Omega}\nabla\varphi^j + \Sigma^j\varphi^j = \frac{1}{4\pi}\sum_{l=1}^{j}\overset{l\to j}{\Sigma}\varphi_0^l + \frac{\chi^j}{4\pi K_{\text{eff}}}Q(\vec{r}),\\ &\varphi^j(\vec{r}, \vec{\Omega}) = 0 \text{ on } S, \text{ for } (\vec{\Omega}, \vec{n}^0) < 0, \end{aligned}\right\} \tag{62.6}$$

*The method of reducing the multigroup equation in the diffusion approximation to the "equivalent" one-group fast-neutron reactor equation by using the neutron importance function was proposed independently by S.B. Shikhov. For the perturbed adjoint neutron importance function, Shikhov uses the solution of the unperturbed problem.

where

$$\left.\begin{aligned}\varphi_0^j(\vec{r}) = \int d\Omega\varphi^j(\vec{r}, \vec{\Omega}), \quad \chi^j = \int\limits_{u_{j-1}}^{u_j} \chi(u)\, du, \\ Q(\vec{r}) = \sum_{j=1}^{m} \nu_f^j \Sigma_f^j \varphi_0^j.\end{aligned}\right\} \qquad (62.7)$$

The perturbation theory formula (60.3) can be used to show that the quantity $(\nu_f \Sigma_f^j)$ in (62.7) should be averaged according to

$$\nu_f^j \Sigma_f^j = \frac{\int\limits_{G_k} d\vec{r} \int d\Omega \int\limits_{u_{j-1}}^{u_j} du\, \nu_f \Sigma_f \varphi S^*}{\int\limits_{G_k} d\vec{r} \int d\Omega \int\limits_{u_{j-1}}^{u_j} du\, \varphi S^*}, \qquad (62.8)$$

where

$$S^*(\vec{r}) = \int d\Omega' \int\limits_{-\infty}^{\infty} du' \varphi^{*\prime} \chi(u'), \qquad (62.9)$$

and $\varphi^{*\prime}$ is a solution of the adjoint equation obtained on the assumption that $\nu_f \Sigma_f$ is a piecewise constant function of the lethargy and that the fission spectrum is unperturbed.

The adjoint set of equations then becomes

$$\left.\begin{aligned}-\vec{\Omega}\nabla\varphi^{*j} + \Sigma^j \varphi^{*j} = \frac{1}{4\pi} \sum_{l=j}^{m} \Sigma^{j \to l} \varphi_0^{*l} + \frac{\nu_f^j \Sigma_f^j}{4\pi K_{\text{eff}}} Q^*(\vec{r}), \\ \varphi^{*j}(\vec{r}, \vec{\Omega}) = 0 \text{ on } S, \text{ for } (\vec{\Omega}, \vec{n}^0) > 0,\end{aligned}\right\} \qquad (62.10)$$

where

$$\varphi_0^{*l} = \int d\Omega \varphi^{*l}(\vec{r}, \vec{\Omega}), \quad Q^*(\vec{r}) = \sum_{j=1}^{m} \chi^j \varphi_0^{*j}. \qquad (62.11)$$

The solution of these equations can be attempted using the successive approximation method defined in Chapter I.

63. The Method of Spherical Harmonics

Outstanding among the approximation methods for solving the kinetic equations is the method of spherical harmonics, which is due to S. Chandrasekhar [87] and has been further developed by Marchak [118], Wang and Guth [140], Bethe, Tonks and Hurwitz [94], Davison [98], Usachev [81], and others. Essentially, this method involves solving the angular part of the kinetic equation in terms of a series of Legendre polynomials. The resulting unknown Fourier coefficients are given by an infinite set of ordinary differential equations in the geometric coordinate. If the Fourier series is cut off at the (2n + 1)-st term, we obtain a system of (2n + 1) equations whose solution, with the appropriate boundary conditions, gives the approximate solution of the kinetic equation.

Consider the one-velocity Boltzmann kinetic equation for a plane slab, namely

$$\mu \frac{\partial \varphi}{\partial z} + \Sigma \varphi = \frac{\Sigma_s}{2} \int_{-1}^{1} \varphi \, d\mu + f_0(z), \tag{63.1}$$

where $\Sigma_s + \Sigma_c = \Sigma$, and $f_0(z)$ is a given source function which, for simplicity, we shall assume symmetric about the center of the slab, so that

$$f_0(-z) = f_0(z)$$

for $z \in [-H/2, H/2]$, where H is the thickness of the slab.

We wish to find a solution of (63.1) in the form

$$\varphi(z, \mu) = \frac{1}{2} \sum_{l=0}^{\infty} (2l+1) \varphi_l(z) P_l(\mu), \tag{63.2}$$

where the $P_l(\mu)$ are Legendre polynomials and satisfy the conditions

$$\left.\begin{aligned} &\int_{-1}^{+1} P_l(\mu) P_m(\mu) \, d\mu = \begin{cases} 0 & (l \neq m), \\ \dfrac{2}{2l+1} & (l = m), \end{cases} \\ &\varphi_l(z) = \int_{-1}^{+1} \varphi(z, \mu) P_l(\mu) \, d\mu. \end{aligned}\right\} \tag{63.3}$$

Inserting (63.2) into (63.1) and using well-known properties of Legendre polynomials, we obtain

$$\mu P_l(\mu) = \frac{l+1}{2l+1} P_{l+1}(\mu) + \frac{l}{2l+1} P_{l-1}(\mu). \tag{63.4}$$

Multiplying this by $P_m(\mu)$ and integrating over μ in the range $-1 \leq \mu \leq 1$, we arrive at the ordinary differential equations

$$m \frac{d\varphi_{m-1}}{dz} + (m+1) \frac{d\varphi_{m+1}}{dz} + (2m+1) \Sigma \varphi_m = \delta_{0m} (\Sigma_s \varphi_0 + 2f_0)$$
$$(m = 0, 1, 2, \ldots), \tag{63.5}$$

where

$$\delta_{0m} = \begin{cases} 0 & (m \neq 0), \\ 1 & (m = 0). \end{cases}$$

In order to specify the problem completely, we must add to (63.5) the appropriate boundary conditions. We therefore assume that at the center $z = 0$ of the slab

$$\left.\begin{aligned} &\frac{d\varphi_m}{dz} = 0 \quad (m = 0, 2, \ldots 2n, \ldots), \\ &\varphi_m = 0 \quad (m = 1, 3, \ldots, 2n+1, \ldots). \end{aligned}\right\} \tag{63.6}$$

On the outer layer at $z = H/2$, the function $\varphi(z, \mu)$ must satisfy the condition

$$\varphi(z, \mu) = 0 \quad (-1 < \mu < 0). \tag{63.7}$$

Marchak has shown that in order to obtain the boundary conditions it is convenient to replace the exact expression by the set of integral expressions

$$\int_{-1}^{0} \mu^{2i+1} \varphi \, d\mu = 0 \quad \text{on} \quad S \qquad (63.8)$$

$$(i = 0, 1, 2, \ldots).$$

Inserting (63.2) into this, and performing the indicated integration, we arrive at the algebraic equations

$$\sum_{m=0}^{\infty} a_{im} \varphi_m = 0 \quad (i = 0, 1, 2, \ldots), \qquad (63.9)$$

where

$$a_{im} = (2m+1) \int_{-1}^{0} \mu^{2i+1} P_m(\mu) \, d\mu. \qquad (63.10)$$

The problem stated in Equations (63.5), (63.6) and (63.9) is thus completely defined. In practice, in order to find the approximate solution, we cut off the set of equations at a term with odd n.

Let us now proceed to treat spherically symmetric problems. The one-velocity Boltzmann kinetic equation is then

$$\mu \frac{\partial \varphi}{\partial r} + \frac{1-\mu^2}{r} \frac{\partial \varphi}{\partial \mu} + \Sigma \varphi = \frac{\Sigma_s}{2} \int_{-1}^{1} \varphi \, d\mu + f_0(r) \qquad (63.11)$$

with the condition that on the external reactor boundary S

$$\varphi(r, \mu) = 0 \quad (-1 < \mu < 0). \qquad (63.12)$$

We wish to find a solution to (63.11) in the form

$$\varphi(r, \mu) = \frac{1}{2} \sum_{l=0}^{\infty} (2l+1) \varphi_l(r) P_l(\mu). \qquad (63.13)$$

Inserting this into (63.11) and making use of (63.4) and

$$(1-\mu^2) \frac{dP_l}{d\mu} = (l+1)(\mu P_l - P_{l+1}), \qquad (63.14)$$

we arrive at the set of ordinary differential equations

$$m\left(\frac{d\varphi_{m-1}}{dr} - \frac{m-1}{r}\varphi_{m-1}\right) + (m+1)\left(\frac{d\varphi_{m+1}}{dr} + \frac{m+2}{r}\varphi_{m+1}\right) +$$
$$+ (2m+1)\Sigma\varphi_m = \delta_{0m}(\Sigma_s \varphi_0 + 2f_0(r)) \qquad (63.15)$$
$$(m = 0, 1, 2, \ldots).$$

To this we add the following boundary conditions. At the center of the sphere where r = 0, we have

$$\frac{d\varphi_0}{dr}=0,\quad \varphi_m=0 \quad (m=1,\ 2\ldots), \tag{63.16}$$

and on the external boundary where $r = R$ we have

$$\sum_{m=0}^{\infty} a_{im}\varphi_m=0 \quad (i=0,\ 1,\ 2,\ \ldots), \tag{63.17}$$

where the a_{im} are again given by (63.10).

We remark that the multigroup equations do not differ essentially from the one-velocity equation we are here discussing. The difference is only that in the multigroup equations the expression $(\Sigma_s\varphi_0 + 2f_0)$ on the right is replaced by

$$F^j=\sum_{l=1}^{j-1}\varphi_0^l\, g_0^{l\to j}\ . \tag{63.18}$$

Here F^j is the external source function for the j-th group.

64. The Spherical Harmonic Equations in Matrix Form

A set of linear equations is very conveniently written in matrix form. Let us thus use matrix calculus to write the equations of the method of spherical harmonics. We turn first to (63.5) and introduce the vectors

$$\boldsymbol{\varphi}=\begin{Vmatrix}\varphi_0\\ \varphi_2\\ \ldots\\ \varphi_{n-1}\end{Vmatrix},\quad \boldsymbol{I}=\begin{Vmatrix}\varphi_1\\ \varphi_3\\ \ldots\\ \varphi_n\end{Vmatrix},\quad \boldsymbol{F}=2\begin{Vmatrix}f_0\\ 0\\ \ldots\\ 0\end{Vmatrix} \qquad (n\ \ \text{odd})$$

and matrices

$$\boldsymbol{\alpha}=\begin{Vmatrix}1 & 0 & 0 & 0\ldots\\ 2 & 1+2 & 0 & 0\ldots\\ 0 & 2\cdot 2 & 1+2\cdot 2 & 0\ldots\\ 0 & 0 & 2\cdot 3 & 1+2\cdot 3\ldots\\ \ldots & \ldots & \ldots & \ldots\end{Vmatrix},$$

$$\boldsymbol{\beta}=\begin{Vmatrix}1 & 2 & 0 & 0\ldots\\ 0 & 1+2 & 2\cdot 2 & 0\ldots\\ 0 & 0 & 1+2\cdot 2 & 2\cdot 3\ldots\\ 0 & 0 & 0 & 1+2\cdot 3\ldots\\ \ldots & \ldots & \ldots & \ldots\end{Vmatrix},$$

$$\boldsymbol{a}=\begin{Vmatrix}\Sigma_c & 0 & 0 & 0\ \ldots\\ 0 & (1+4)\,\Sigma & 0 & 0\ \ldots\\ 0 & 0 & (1+2\cdot 4)\,\Sigma & 0\ \ldots\\ 0 & 0 & 0 & (1+3\cdot 4)\,\Sigma\ldots\\ \ldots & \ldots & \ldots & \ldots\end{Vmatrix},$$

$$\boldsymbol{b}=\begin{Vmatrix}3\Sigma & 0 & 0 & 0\ \ldots\\ 0 & (3+4)\,\Sigma & 0 & 0\ \ldots\\ 0 & 0 & (3+2\cdot 4)\,\Sigma & 0\ \ldots\\ 0 & 0 & 0 & (3+3\cdot 4)\,\Sigma\ldots\\ \ldots & \ldots & \ldots & \ldots\end{Vmatrix}.$$

Then Equations (63.5) can be written

$$\left.\begin{aligned}\boldsymbol{\alpha}\cdot\frac{d\boldsymbol{I}}{dz}+\boldsymbol{a}\varphi=\boldsymbol{F},\\ \boldsymbol{\beta}\cdot\frac{d\varphi}{dz}+\boldsymbol{b}\boldsymbol{I}=0.\end{aligned}\right\} \tag{64.1}$$

These equations can be reduced formally to a single second order diffusion equation.

To do this, we multiply the first of these on the left by $\boldsymbol{\alpha}^{-1}$ and the second by $\mathbf{b}^{-1}$, obtaining

$$\frac{d\boldsymbol{I}}{dz}+\Sigma\varphi=\boldsymbol{f}, \tag{64.2}$$

$$\boldsymbol{I}=-\boldsymbol{D}\frac{d\varphi}{dz}, \tag{64.3}$$

where

$$\boldsymbol{D}=\boldsymbol{b}^{-1}\boldsymbol{\beta},\quad \Sigma=\boldsymbol{\alpha}^{-1}\boldsymbol{a},\quad \boldsymbol{f}=\boldsymbol{\alpha}^{-1}\cdot\boldsymbol{F}.$$

Inserting (64.3) into (64.2), we have

$$\frac{d}{dz}\boldsymbol{D}\frac{d\varphi}{dz}-\Sigma\varphi=-\boldsymbol{f}. \tag{64.4}$$

The boundary condition at the center of a plane slab will be

$$\frac{d\varphi}{dz}=0 \qquad (z=0). \tag{64.5}$$

On the boundary z = H/2, we use (63.9), which we write in the form

$$\boldsymbol{A}\varphi+\boldsymbol{B}\boldsymbol{I}=0 \quad \left(z=\frac{H}{2}\right), \tag{64.6}$$

where

$$\boldsymbol{A}=\|a_{im}\|,\quad \begin{pmatrix} i=0,\ 1,\ 2,\ 3\ldots \\ m=0,\ 2,\ 4,\ 6,\ldots \end{pmatrix},$$

$$\boldsymbol{B}=\|a_{im}\|,\quad \begin{pmatrix} i=0,\ 1,\ 2,\ 3,\ldots \\ m=1,\ 3,\ 5,\ \ldots \end{pmatrix},$$

and the a_{im} are again given by (63.10).

Now (64.6) can be simplified somewhat by replacing **I** by φ using (64.3). We then arrive at

$$\boldsymbol{A}\varphi-\boldsymbol{B}\boldsymbol{D}\frac{d\varphi}{dz}=0 \quad \left(z=\frac{H}{2}\right), \tag{64.7}$$

or

$$\varphi-\Gamma\frac{d\varphi}{dz}=0 \quad \left(z=\frac{H}{2}\right),$$

where

$$\Gamma=\boldsymbol{A}^{-1}\boldsymbol{B}\boldsymbol{D}.$$

Thus, Boltzmann's kinetic equation for the plane slab has been reduced by the method of spherical harmonics to the boundary value problem

$$\left.\begin{aligned} \frac{d}{dz} D \frac{d\varphi}{dz} - \Sigma\varphi = -f, \\ \frac{d\varphi}{dz} = 0 \quad (z = 0), \\ \varphi - \Gamma \frac{d\varphi}{dz} = 0 \quad \left(z = \frac{H}{2}\right). \end{aligned}\right\} \tag{64.8}$$

In spherically symmetric regions, we arrive in the same way at

$$\left.\begin{aligned} \frac{dI}{dr} + \frac{\omega}{r} I + \Sigma\varphi = f, \\ I = -D\left(\frac{d\varphi}{dr} + \frac{\delta}{r}\varphi\right), \end{aligned}\right\} \tag{64.9}$$

where

$$\delta = \beta^{-1}\varkappa, \quad \omega = \alpha^{-1}\sigma.$$

$$\varkappa = 2\begin{Vmatrix} 0 & 1\cdot 3 & 0 & 0 & 0 & 0\ldots \\ 0 & -1\cdot 3 & 2\cdot 5 & 0 & 0 & 0\ldots \\ 0 & 0 & -2\cdot 5 & 3\cdot 7 & 0 & 0\ldots \\ 0 & 0 & 0 & -3\cdot 7 & 4\cdot 9 & 0\ldots \\ 0 & 0 & 0 & 0 & -4\cdot 9 & 5\cdot 11\ldots \\ \ldots & \ldots & \ldots & \ldots & \ldots & \ldots \end{Vmatrix},$$

$$\sigma = 2\begin{Vmatrix} 1 & 0 & 0 & 0 & 0\ldots \\ -1 & 3\cdot 2 & 0 & 0 & 0\ldots \\ 0 & -3\cdot 2 & 5\cdot 3 & 0 & 0\ldots \\ 0 & 0 & -5\cdot 3 & 7\cdot 4 & 0\ldots \\ 0 & 0 & 0 & -7\cdot 4 & 9\cdot 5\ldots \\ \ldots & \ldots & \ldots & \ldots & \ldots \end{Vmatrix}.$$

We must now establish the boundary conditions. To do this at the center of the sphere, let us analyze the behavior of the individual terms in (64.9). It is easily shown that asymptotically we have

$$\lim_{r\to 0} \omega \frac{I}{r} = \omega \frac{dI}{dr}, \quad \lim_{r\to 0} \delta \frac{\varphi}{r} = \delta \frac{d\varphi}{dr}. \tag{64.10}$$

This means that in the neighborhood of $r = 0$, Equations (64.9) become

$$\begin{aligned} (E + \omega)\frac{dI}{dr} + \Sigma\varphi = f, \\ I = -D(E + \delta)\frac{d\varphi}{dr}. \end{aligned} \tag{64.11}$$

The desired condition at the center of the sphere can now be obtained by using these expressions and the condition that φ is a vector symmetric about the origin $r = 0$.

On the outer surface of the sphere we require that (63.17) be satisfied, which can be written in the form

$$\varphi - \Gamma\left(\frac{d\varphi}{dr} + \frac{\delta}{R}\varphi\right) = 0. \tag{64.12}$$

Consider the matrix equation (64.8) which, again, is

$$\frac{d}{dz} \boldsymbol{D} \frac{d\varphi}{dz} - \Sigma\varphi = -\boldsymbol{f},$$ (65.1)

which we shall now rewrite in the form of the two equations

$$\frac{d\boldsymbol{I}}{dz} = \boldsymbol{f} - \Sigma\varphi,$$ (65.2)

$$\frac{d\varphi}{dz} = -\boldsymbol{D}^{-1}\boldsymbol{I}.$$ (65.3)

Let us break up the interval $0 \leq z \leq H/2$ into segments (z_k, z_{k+1}). We will write $\Delta z_{k+1/2} = z_{k+1} - z_k$.

We now integrate (65.2) in the interval $(z_{k-1/2}, z_{k+1/2})$, obtaining

$$\boldsymbol{I}_{k+1/2} - \boldsymbol{I}_{k-1/2} = \int_{z_{k-1/2}}^{z_{k+1/2}} (\boldsymbol{f} - \Sigma\varphi)\, dz.$$ (65.4)

Assuming that $\boldsymbol{f}$ and Σ are piecewise continuous with discontinuities at z_k, and that φ and $\boldsymbol{I}$ are continuous, we may write approximately

$$\boldsymbol{I}_{k+1/2} - \boldsymbol{I}_{k-1/2} = (\boldsymbol{f}\Delta z)_k - (\Sigma\Delta z)_k \varphi_k,$$ (65.5)

where

$$(\psi)_k = \frac{1}{2}(\psi_k^+ + \psi_k^-), \qquad \psi_k^+ = \lim_{\varepsilon\to 0}\psi(x_k + \varepsilon),$$

$$\psi_k^- = \lim_{\varepsilon\to 0}\psi(x_k - \varepsilon).$$

Let us now integrate (65.3) from z_k to z_{k+1}. We then have

$$\varphi_{k+1} - \varphi_k = -\boldsymbol{D}_{k+1/2}^{-1}\int_{z_k}^{z_{k+1}} \boldsymbol{I}\, dz.$$ (65.6)

We assume here that $\boldsymbol{D}^{-1}$ is constant within the range of integration. Since $\boldsymbol{I}$ is continuous, the integral in this last equation can be written approximately in the form

$$\int_{z_k}^{z_{k+1}} \boldsymbol{I}\, dz = \boldsymbol{I}_{k+1/2}\Delta z_{k+1/2}.$$

We thus arrive at the simplified expression

$$\varphi_{k+1} - \varphi_k = -\boldsymbol{D}_{k+1/2}^{-1}\Delta z_{k+1/2}\boldsymbol{I}_{k+1/2}$$ (65.7)

or

$$\boldsymbol{I}_{k+1/2} = -\frac{\boldsymbol{D}_{k+1/2}}{\Delta z_{k+1/2}}(\varphi_{k+1} - \varphi_k).$$ (65.8)

Inserting this into (65.5), we have

$$\frac{D_{k+1/2}}{\Delta z_{k+1/2}}(\varphi_{k+1}-\varphi_k)-\frac{D_{k-1/2}}{\Delta z_{k-1/2}}(\varphi_k-\varphi_{k-1})-(\Sigma\Delta z)_k\varphi_k=-(f\Delta z)_k. \tag{65.9}$$

We multiply this on the left by $\Delta z_{k+1/2}D^{-1}_{k+1/2}$, arriving at the finite-difference equation

$$\varphi_{k+1}-B_k\varphi_k+C_k\varphi_{k-1}=-F_k, \tag{65.10}$$

where

$$C_k=\frac{\Delta z_{k+1/2}}{\Delta z_{k-1/2}}D^{-1}_{k+1/2}D_{k-1/2},\quad B_k=E+C_k+D^{-1}_{k+1/2}(\Sigma\Delta z)_k\Delta z_{k+1/2},$$

$$F_k=D^{-1}_{k+1/2}(f\Delta z)_k\Delta z_{k+1/2}.$$

Let us now establish the boundary conditions for the finite-difference equations. We choose the finite-difference net so that the interpolation point with $k=0$ lies a half step into the negative $\underline{z}$ values and that the last point, with $k=n$, lies a half step from the boundary, i.e., so that

$$z_n=\frac{H}{2}+\frac{1}{2}\Delta z_{n+1/2}.$$

Thus, the origin and the boundary of the slab lie in the center of the first and last interpolation intervals, respectively.

We recall the last two equations of (64.8), namely

$$\left.\begin{aligned}\frac{d\varphi}{dz}&=0 \qquad (z=0),\\ \varphi-\Gamma\frac{d\varphi}{dz}&=0 \qquad \left(z=\frac{H}{2}\right),\end{aligned}\right\} \tag{65.11}$$

and integrate, to obtain the approximate boundary conditions

$$\begin{aligned}\varphi_0&=\varphi_1,\\ \varphi_{n-1}&=\gamma\varphi_n,\end{aligned} \tag{65.12}$$

where

$$\gamma=\gamma_1^{-1}\gamma_2,$$

$$\gamma_1=2\mu_{n-1/2}R+E;\qquad R=A^{-1}T;\qquad T=Bd$$

$$\gamma_2=2\mu_{n-1/2}R-E;\qquad d=3\Sigma D;$$

$$\mu_{k+1/2}=\frac{1}{(3\Sigma\Delta z)_{k+1/2}}.$$

Thus, the set of finite-difference equations for the method of spherical harmonics is completely specified.

Similarly, one can find the corresponding set of equations for one-dimensional spherically symmetric regions.

In conclusion, we will give the computational formulas for the components of the F_k vector and for the matrix coefficients B_k, C_k and γ in the P_n-approximation. The explicit form of these for the plane problem is*

* The formulas were reduced to this form by Zh. N. Bel'skaia, N. P. Kochubei, and I. P. Tiuterev.

$$F_k = \frac{(f_0\Delta z)_k}{\mu_{k+1/2}}\psi; \qquad \psi = |\psi_i|; \qquad (i = 0,\ 1,\ 2,\ \ldots,\ n)$$

$$B_k = \left(1 + \frac{\mu_{k-1/2}}{\mu_{k+1/2}}\right)E + \frac{(\Sigma\Delta z)_k}{\mu_{k+1/2}}P - \frac{(\Sigma_s\Delta z)_k}{\mu_{k+1/2}}Q; \qquad (65.13)$$

where

$$P = \| p_{im} \|, \qquad Q = \| q_{im} \| \qquad (i,\ m = 0,\ 1,\ 2,\ \ldots,\ n)$$

$$C_k = \frac{\mu_{k-1/2}}{\mu_{k+1/2}}E$$

$$A = \| a_{im} \| \qquad \begin{pmatrix} i = 0,\ 1,\ 2,\ \ldots,\ \frac{n-1}{2} \\ m = 0,\ 2,\ 4,\ \ldots,\ n-1 \end{pmatrix}$$

$$T = \| t_{im} \| \qquad \begin{pmatrix} i = 0,\ 1,\ 2,\ \ldots,\ \frac{n-1}{2} \\ m = 1,\ 3,\ 5,\ \ldots,\ n \end{pmatrix},$$

and where

$$\psi_i = (-1)^i \frac{2^{i+1}(2i-1)!!}{3(i!)} \sum_{l=i}^{n} (4l+3)\left[\frac{2^{l-i}\, l!}{(2l+1)!!}\right]^2 \quad \left(i = 0,\ 1,\ 2,\ \ldots,\ \frac{n-1}{2}\right).$$

$$p_{im} = (-1)^{i+m} 2^{|i-m|} \frac{4m+1}{3} \cdot \frac{[(2i-1)!!][(2m-1)!!]}{i!\, m!} \times$$

$$\times \sum_{l=\max(i,\,m)}^{n} (4l+3)\left[\frac{2^{l-\max(i,\,m)}\, l!}{(2l+1)!!}\right]^2 \quad \left(i,\ m = 0,\ 1,\ 2,\ \ldots,\ \frac{n-1}{2}\right).$$

$$q_{im} = \begin{cases} (-1)^i \cdot \dfrac{2^i(2i-1)!!}{3(i!)} \displaystyle\sum_{l=i}^{n} (4l+3)\left[\frac{2^{l-i}\, l!}{(2l+1)!!}\right]^2, & m = 0, \\ 0, & m \neq 0, \end{cases}$$

$$a_{im} = \begin{cases} -\dfrac{1}{2(i+1)}, & m = 0, \\ -(2m+1)\left[\prod_{l=2}^{m}(2i-l+3)\right] \Big/ \left[\prod_{l=0}^{m}(2i+l+2)\right], & m = 2, 4, \ldots, n-1 \\ & i = 0,\ 1,\ 2,\ \ldots,\ \frac{n-1}{2} \end{cases}$$

and

$$t_{im} = \begin{cases} \dfrac{3}{2i+3} & m = 1, \\ \dfrac{6}{2i+3} + \dfrac{9 \cdot 2i}{(2i+3)(2i+5)} & m = 3, \\ 3(m-1) \cdot \dfrac{\prod_{l=5}^{m}(2i-l+5)}{\prod_{l=3}^{m}(2i+l)} + 3m \cdot \dfrac{\prod_{l=3}^{m}(2i-l+3)}{\prod_{l=1}^{m}(2i+l+2)}, & m = 5,\ 7,\ \ldots,\ n, \\ & i = 0,\ 1,\ 2,\ \ldots,\ \frac{n-1}{2}. \end{cases}$$

In these equations we set $(-1)!! = 1$.

Proceeding to spherically symmetric regions, we start with the equations

$$\left.\begin{aligned}\frac{dI}{dr}+\frac{\omega}{r}I&=f-\Sigma\varphi,\\ \frac{d\varphi}{dr}+\frac{\delta}{r}\varphi&=-D^{-1}I.\end{aligned}\right\}\qquad(65.14)$$

We integrate the first of these within the limits ($r_{k-1/2}$, $r_{k+1/2}$). This gives

$$I_{k+1/2}-I_{k-1/2}+\omega\int\limits_{r_{k-1/2}}^{r_{k+1/2}}\frac{I}{r}\,dr=\int\limits_{r_{k-1/2}}^{r_{k+1/2}}(f-\Sigma\varphi)\,dr.\qquad(65.15)$$

Setting, as an approximation,*

$$\int\limits_{r_{k-1/2}}^{r_{k+1/2}}\frac{I}{r}\,dr=\frac{\Delta r_k}{2}\left(\frac{I_{k+1/2}}{r_{k+1/2}}+\frac{I_{k-1/2}}{r_{k-1/2}}\right),$$

$$\int\limits_{r_{k-1/2}}^{r_{k+1/2}}(f-\Sigma\varphi)\,dr=(f\Delta r)_k-(\Sigma\Delta r)_k\,\varphi_k,$$

where $\Delta r_k=r_{k+1/2}-r_{k-1/2}$, we rewrite (65.15) in the form

$$\left(E+\frac{\Delta r_k}{2r_{k+1/2}}\omega\right)I_{k+1/2}-\left(E-\frac{\Delta r_k}{2r_{k-1/2}}\omega\right)I_{k-1/2}+(\Sigma\Delta r)_k\,\varphi_k=(f\Delta r)_k.\qquad(65.16)$$

We now integrate the second of (65.14) within the limits (r_k, r_{k+1}), obtaining

$$\varphi_{k+1}-\varphi_k+\delta\int\limits_{r_k}^{r_{k+1}}\frac{\varphi}{r}\,dr=-\int\limits_{r_k}^{r_{k+1}}D^{-1}I\,dr.\qquad(65.17)$$

We may approximate the integrals by *

$$\int\limits_{r_k}^{r_{k+1}}\frac{\varphi}{r}\,dr=\frac{\Delta r_{k+1/2}}{2r_{k+1/2}}(\varphi_{k+1}+\varphi_k),$$

$$\int\limits_{r_k}^{r_{k+1}}D^{-1}I\,dr=D^{-1}_{k+1/2}\Delta r_{k+1/2}I_{k+1/2}.$$

In this way we rewrite (65.17) in the form

$$\left(E+\frac{\Delta r_{k+1/2}}{2r_{k+1/2}}\delta\right)\varphi_{k+1}-\left(E-\frac{\Delta r_{k+1/2}}{2r_{k+1/2}}\delta\right)\varphi_k=-D^{-1}_{k+1/2}\Delta r_{k+1/2}I_{k+1/2}.\qquad(65.18)$$

We solve this for $I_{k+1/2}$, obtaining

$$I_{k+1/2}=-\frac{D_{k+1/2}}{\Delta r_{k+1/2}}\left[\left(E+\frac{\Delta r_{k+1/2}}{2r_{k+1/2}}\delta\right)\varphi_{k+1}-\left(E-\frac{\Delta r_{k+1/2}}{2r_{k+1/2}}\delta\right)\varphi_k\right].\qquad(65.19)$$

This equation is now used to eliminate $I_{k+1/2}$ and $I_{k-1/2}$ from (65.16). We then obtain the finite-difference equation

* Other approximation formulas may also be applied here.

$$a_k\varphi_{k+1} - b_k\varphi_k + c_k\varphi_{k-1} = -(f\Delta r)_k, \tag{65.20}$$

where

$$\left.\begin{aligned}
a_k &= \left(E + \frac{\Delta r_k}{2r_{k+1/2}}\,\omega\right)\frac{D_{k+1/2}}{\Delta r_{k+1/2}}\left(E + \frac{\Delta r_{k+1/2}}{2r_{k+1/2}}\,\delta\right),\\
b_k &= \left(E + \frac{\Delta r_k}{2r_{k+1/2}}\,\omega\right)\frac{D_{k+1/2}}{\Delta r_{k+1/2}}\left(E - \frac{\Delta r_{k+1/2}}{2r_{k+1/2}}\,\delta\right) +\\
&+ \left(E - \frac{\Delta r_k}{2r_{k-1/2}}\,\omega\right)\frac{D_{k-1/2}}{\Delta r_{k-1/2}}\left(E + \frac{\Delta r_{k-1/2}}{2r_{k-1/2}}\,\delta\right) + (\Sigma\Delta r)_k,\\
c_k &= \left(E - \frac{\Delta r_k}{2r_{k-1/2}}\,\omega\right)\frac{D_{k-1/2}}{\Delta r_{k-1/2}}\left(E - \frac{\Delta r_{k-1/2}}{2r_{k-1/2}}\,\delta\right).
\end{aligned}\right\} \tag{65.21}$$

We now multiply (65.20) on the left by a_k^{-1}, obtaining

$$\varphi_{k+1} - B_k\varphi_k + C_k\varphi_{k-1} = -F_k, \tag{65.22}$$

where

$$B_k = a_k^{-1}b_k, \qquad C_k = a_k^{-1}c_k, \qquad F_k = a_k^{-1}(f\Delta r)_k. \tag{65.22'}$$

We must now find the appropriate boundary conditions. That for the center of the sphere is found by using (64.11), which is valid in the neighborhood of $r = 0$. We will assume that the region of definition of φ extends a half of a Δr interval in the negative r direction, so that the center $r = 0$ of the region lies on the interpolation point with $k = 0$. Then integration of the first of (64.11) over the range $-\Delta r/2 \le r \le \Delta r/2$, and the second over the range $-\Delta r \le r \le 0$ and $0 \le r \le \Delta r$, will lead to the finite-difference equation (if we take into account the behavior of the integrands in the appropriate regions)

$$(E+\omega)\,D\,(E+\delta)\,(\varphi_1 - 2\varphi_0 + \varphi_{-1}) - \Sigma\Delta r^2\varphi_0 = -f_0\Delta r^2. \tag{65.23}$$

Recalling the symmetry of φ, that is, that $\varphi_1 = \varphi_{-1}$, we can write this in the form

$$2\,(E+\omega)\,D\,(E+\delta)\,\varphi_1 - [2\,(E+\omega)\,D\,(E+\delta) + \Sigma\Delta r^2]\,\varphi_0 = -f_0\Delta r^2 \tag{65.24}$$

or

$$\varphi_0 = \eta\varphi_1 + \xi f_0, \tag{65.25}$$

where

$$\eta = 2\,[2\,(E+\omega)\,D\,(E+\delta) + \Sigma\Delta r^2]^{-1}\,[(E+\omega)\,D\,(E+\delta)],$$
$$\xi = \Delta r^2\,[2\,(E+\omega)\,D\,(E+\delta) + \Sigma\Delta r^2]^{-1}.$$

The condition on the outer surface of the sphere is obtained by integrating (64.12):

$$\left[\Gamma\left(E + \delta\frac{\Delta r}{2R}\right) - \frac{\Delta r}{2}\,E\right]\varphi_n = \left[\Gamma\left(E - \delta\frac{\Delta r}{2R}\right) + \frac{\Delta r}{2}\,E\right]\varphi_{n-1}. \tag{65.26}$$

We can write this in the form

$$\varphi_{n-1} = \gamma\varphi_n, \tag{65.27}$$

where

$$\gamma = \left[\Gamma\left(E - \delta\frac{\Delta r}{2R}\right) + \frac{\Delta r}{2}\,E\right]^{-1}\cdot\left[\Gamma\left(E + \delta\frac{\Delta r}{2R}\right) - \frac{\Delta r}{2}\,E\right].$$

Thus, the sets of finite difference equations for plane and spherical regions have been completely specified. All else reduces to finding the solutions of these sets. Thus, to the finite-difference equations (65.22) we must add the relations

$$\left.\begin{aligned}\varphi_0 &= \eta\varphi_1 + \xi f_0,\\ \varphi_{n-1} &= \gamma\varphi_n.\end{aligned}\right\} \tag{65.28}$$

The problem then becomes complete.

It should be noted in conclusion that for spherically symmetric regions it is impossible to obtain final formulas for **B**, **C**, η, ξ and γ in simple form, as was done for the case of plane geometry. In the P_3 and P_5-approximations, however, such formulas are easily obtained. When one recalls that the large majority of problems can be solved quite satisfactorily in these approximations, it becomes clear that this method can be used to solve a large category of problems.

66. Solution of the Finite-Difference Spherical Harmonic Equations

In this section we shall apply the methods of matrix factorization and successive approximations to the solution of the finite-difference equations for the method of spherical harmonics. We first treat matrix factorization. As was mentioned in Section 36, this method reduces the finite-difference matrix equation

$$\varphi_{k+1} - B_k\varphi_k + C_k\varphi_{k-1} = -F_k \tag{66.1}$$

to the very simple system

$$\left.\begin{aligned}\beta_{k+1} &= C_{k+1}(B_k - \beta_k)^{-1},\\ z_{k+1} &= \beta_{k+1}(z_k + F_k),\\ \varphi_k &= C_{k+1}^{-1}(\beta_{k+1}\varphi_{k+1} + z_{k+1}).\end{aligned}\right\} \tag{66.2}$$

To solve (66.2) we must have initial conditions, and these can be obtained from the boundary conditions. For plane regions we have, according to (65.12),

$$\varphi_0 = \varphi_1.$$

$$\varphi_{n-1} = \gamma\varphi_n.$$

We then obtain

$$\begin{aligned}\beta_1 &= C_1, \quad z_1 = 0,\\ \varphi_n &= (C_n\gamma - \beta_n)^{-1} z_n.\end{aligned} \tag{66.3}$$

For spherically symmetric regions, we use (65.28), obtaining

$$\begin{aligned}\beta_1 &= C_1\eta, \quad z_1 = C_1\xi f_0,\\ \varphi_n &= (C_n\gamma - \beta_n)^{-1} z_n.\end{aligned} \tag{66.4}$$

Let us now turn to the solution of (66.1) by means of successive approximations. In that equation $\mathbf{B}_k$ and $\mathbf{C}_k$ are the matrices defined by (65.22'), and $\boldsymbol{\varphi}_k$ and F_k are vectors of the form

$$\varphi_k = \begin{vmatrix}\varphi_{0,k}\\ \varphi_{2,k}\\ \cdot\\ \cdot\\ \cdot\\ \varphi_{n-1,k}\end{vmatrix}, \qquad F_k = \begin{vmatrix}F_{0,k}\\ F_{2,k}\\ \cdot\\ \cdot\\ \cdot\\ F_{n-1,k}\end{vmatrix}. \tag{66.5}$$

Now (66.1) is a vector equation with (n + 1)/2 components, for each of which we may write

$$\varphi_{i,k+1} - B_k^{i,i}\varphi_{i,k} + C_k^{i,i}\varphi_{i,k-1} = -f_{ik}, \tag{66.6}$$

where

$$f_{ik} = F_{ik} - \sum_{j \neq i} (B_k^{i,j}\varphi_{jk} - C_k^{i,j}\varphi_{j,k-1}), \quad (i = 0, 2, \ldots, n-1).$$

The boundary condition at the center may be written in the form

$$\varphi_{i,0} = \varphi_{i,1} \quad (i = 0, 2, \ldots, n-1). \tag{66.7}$$

On the outer surface we have

$$\varphi_{i,n-1} - \gamma^{i,i}\varphi_{i,n} = \sum_{j \neq i} \gamma^{i,j}\varphi_{j,n}, \quad (i = 0, 2, \ldots, n-1), \tag{66.8}$$

where the $\gamma^{i,j}$ are the matrix elements of $\boldsymbol{\gamma}$.

We find the solution of the problem by assuming that the sums in (66.6) and (66.8) vanish. It is clear that this solution is an approximation. We use it to calculate the sums in (66.6) and (66.8), and then obtain a new approximation. We repeat this process until two successive approximate solutions agree with each other to the required accuracy. When the functions on the right are given, all of Equations (66.6) are of the diffusion type, so that they can be solved by the method of difference factorization.

The rate of convergence of the successive approximations can be somewhat increased by using the approximate values for the components $\varphi_{0,k}, \varphi_{2,k}, \ldots, \varphi_{l-2,k}$ already obtained in any given cycle of iteration when solving the diffusion equation for the l-th component of the vector $\boldsymbol{\varphi}_k$.

67. Vladimirov's Numerical Method for Solving the Kinetic Equations

In 1952, V.S. Vladimirov suggested a method for solving the kinetic equations for spherically symmetric regions by using the properties of Boltzmann's integro-differential equation [11, 15]. This method is easily programmed for electronic computers in solving both one-group and multigroup problems. Let us consider Vladimirov's method for the approximate solution of the one-velocity equation

$$\mu\frac{\partial\varphi}{\partial r} + \frac{1-\mu^2}{r}\frac{\partial\varphi}{\partial\mu} + \Sigma\varphi = \frac{\Sigma_s}{2}\int_{-1}^{+1}\varphi(r,\mu')\,d\mu' + f(r) \tag{67.1}$$

with the boundary conditions

$$\varphi(R,\mu) = 0 \quad (\mu \leq 0). \tag{67.2}$$

The extension of the computational algorithm for the numerical solution of the multigroup problem presents no difficulty. We shall assume that the functions Σ, Σ_s and $f(r)$ are piecewise continuous on [0, R], possibly having points of discontinuity of the first kind. In Equation (67.1) we perform the substitution

$$x = r\mu, \quad y = r\sqrt{1-\mu^2} \tag{67.3}$$

transforming it into the form

$$\frac{\partial\psi}{\partial x} + \Sigma\left(\sqrt{x^2+y^2}\right)\psi(x,y) = = {}^1/_2\Sigma_s\left(\sqrt{x^2+y^2}\right)\varphi_0\left(\sqrt{x^2+y^2}\right) + f\left(\sqrt{x^2+y^2}\right), \tag{67.4}$$

where

$$\psi(x, y) = \varphi\left(\sqrt{x^2+y^2}, \cos \operatorname{arctg} \frac{y}{x}\right), \quad \varphi_0 = \int_{-1}^{+1} \varphi \, d\mu. \tag{67.5}$$

In this transformation the rectangle ($0 \leq r \leq R$, $-1 \leq \mu \leq 1$) is transformed continuously into the semicircle ($x^2 + y^2 < R^2$, $y > 0$), and the lines $r\sqrt{1-\mu^2} = \text{const}$, which are characteristics of the differential part of the operator in (67.1), are transformed into the straight lines $y = \text{const}$. The boundary condition for $\psi(x, y)$ becomes

$$\psi\left(-\sqrt{R^2-y^2}, y\right) = 0 \quad 0 \leqslant y \leqslant R. \tag{67.6}$$

We now solve (67.4) as an ordinary first order differential equation, obtaining

$$\psi(x, y) = \int_{-\sqrt{R^2-y^2}}^{x} \left[\frac{1}{2}\Sigma_s\left(\sqrt{x'^2+y^2}\right)\varphi_0\left(\sqrt{x'^2+y^2}\right) + f\left(\sqrt{x'^2+y^2}\right)\right] e^{-\int_{x'}^{x} \Sigma\left(\sqrt{x''^2+y^2}\right) dx''} dx'. \tag{67.7}$$

According to (67.3), the net in the (r, μ) coordinates becomes

$$r = r_k, \quad \mu = \pm \mu_{ki}, \quad \mu_{ki} = \frac{x_{ki}}{r_k} = \sqrt{1 - \frac{y_i^2}{r_k^2}}. \quad \begin{matrix} (i = 0, 1, \ldots, k) \\ (k = 0, 1, \ldots, s) \end{matrix}$$

Setting

$$y = y_i, \quad x = \pm x_{ki}$$

in (67.7) and writing

$$\left.\begin{aligned} \varphi(r_k, -\mu_{ki}) &= \varphi_{ki}^- = \psi(-x_{ki}, y_i), \\ \varphi(r_k, \mu_{ki}) &= \varphi_{ki}^+ = \psi(x_{ki}, y_i), \\ \varphi_0(r_k) = \varphi_{0k}, \quad &(k = i, i+1, \ldots, s; \ i = 0, 1, \ldots, s) \end{aligned}\right\} \tag{67.8}$$

we obtain

$$\left.\begin{aligned} \varphi_{k-1,i}^- &= \varphi_{ki}^- \exp\left[-\int_{-x_{ki}}^{-x_{k-1,i}} \Sigma\left(\sqrt{x''^2+y_i^2}\right) dx''\right] + \\ &+ \int_{-x_{ki}}^{-x_{k-1,i}} Q\left(\sqrt{x'^2+y_i^2}\right) \exp\left[-\int_{x'}^{-x_{k-1,i}} \Sigma\left(\sqrt{x''^2+y_i^2}\right) dx''\right] dx', \\ &(k = s, s-1, \ldots, i+1; \ i = 0, 1, \ldots, s) \\ \varphi_{ki}^+ &= \varphi_{k-1,i}^+ \exp\left[-\int_{x_{k-1,i}}^{x_{ki}} \Sigma\left(\sqrt{x''^2+y_i^2}\right) dx''\right] + \\ &+ \int_{x_{k-1,i}}^{x_{ki}} Q\left(\sqrt{x'^2+y_i^2}\right) \exp\left[-\int_{x'}^{x_{ki}} \Sigma\left(\sqrt{x''^2+y_i^2}\right) dx''\right] dx' \\ &(k = i+1, i+2, \ldots, s; \ i = 0, 1, \ldots, s), \end{aligned}\right\} \tag{67.9}$$

where

$$Q(r)=\frac{\Sigma_s}{2}\int_{-1}^{+1}\varphi(r,\mu)\,d\mu+f(r).$$

Let us now assume that on the (r_{k-1}, r_k) intervals the functions Σ_s and Σ are constant and equal to Σ_{sk} and Σ_k. Then the integrals in the exponentials can be calculated, and we use (67.8) to obtain

$$\left.\begin{aligned}
&\varphi^-_{k-1,i}=\varphi^-_{ki}\exp(-\Sigma_k\Delta x_{ki})+\\
&+\int_{x_{k-1,i}}^{x_{ki}}Q\left(\sqrt{x'^2+y_i^2}\right)\exp[-\Sigma_k(x'-x_{k-1,i})]\,dx',\\
&\varphi^+_{ki}=\varphi^+_{k-1,i}\exp(-\Sigma_k\Delta x_{ki})+\\
&+\int_{x_{k-1,i}}^{x_{ki}}Q\left(\sqrt{x'^2+y_i^2}\right)\exp[-\Sigma_k(x_{ki}-x')]\,dx',
\end{aligned}\right\}\qquad(67.10)$$

where

$$\Delta x_{ki}=x_{ki}-x_{k-1,i}.$$

Assuming that the function $Q\left(\sqrt{x^2+y_i^2}\right)$ in the integrand is linear, we obtain the approximate equations

$$\left.\begin{aligned}
&\varphi^-_{k-1,i}=\varphi^-_{ki}\exp(-\Sigma_k\Delta x_{ki})+A^-_{ki}\lambda(r_k-0)\,Q_k+\\
&+[1-\exp(-\Sigma_k\Delta x_{ki})-A^-_{ki}]\,\lambda(r_{k-1}+0)\,Q_{k-1}\\
&(k=s,\ s-1,\ \ldots,\ i+1;\quad i=0,\ 1,\ 2,\ \ldots,\ s),\\
&\varphi^+_{ki}=\varphi^+_{k-1,i}\exp(-\Sigma_k\Delta x_{ki})+A^+_{ki}\lambda(r_{k-1}+0)\,Q_{k-1}+\\
&+[1-\exp(-\Sigma_k\Delta x_{ki})-A^+_{ki}]\,\lambda(r_k-0)\,Q_k\\
&(k=i+1,\ i+2,\ \ldots,\ s;\quad i=0,\ 1,\ 2,\ \ldots,\ s),
\end{aligned}\right\}\qquad(67.11)$$

where

$$A^-_{ki}=A^+_{ki}=A(\Sigma_k\Delta x_{ki}),\quad\text{a}\quad A(u)=\frac{1-e^{-u}}{\Delta u}-e^{-u},$$

$$\lambda=\frac{1}{\Sigma}.$$

The recursion formulas (67.11) and the boundary conditions

$$\varphi^-_{si}=0,\ \varphi^-_{ii}=\varphi^+_{ii},\ (i=0,\ 1,\ \ldots,\ s)$$

completely define the $\varphi^{\pm}_{ki}$ from the known values of the Q_k. What remains is to define the method of successive approximations for finding an approximate solution to the kinetic equation.

We do this in the following way. From the given initial $Q(r)$ function and (67.11), we find $\varphi^{\pm}_{ki}$ at all interpolation points of the net (Fig. 26). We now integrate this function over the variable μ. This gives a new approximation expression for $Q(r)$, namely

$$Q(r)=\frac{\Sigma_s}{2}\int_{-1}^{+1}\varphi(r,\mu)\,d\mu+f(r).\qquad(67.12)$$

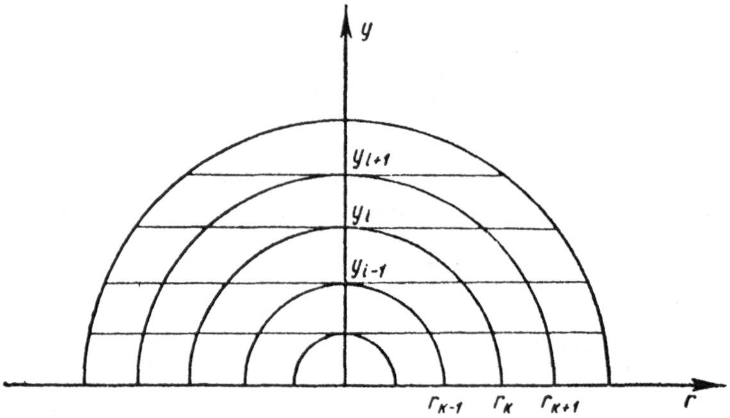

Fig. 26. Interpolation points of the net for a spherically symmetric region.

We use some quadrature formula to write

$$Q_k = \sum_i \alpha_{ki}\varphi_{ki} + f_k, \tag{67.13}$$

where the α_{ki} are the coefficients of the quadrature formula for the points of the given circle $x^2 + y^2 = r_k^2$. This process is continued until the ratio between two successive approximations is sufficiently close to unity. The values of $\varphi(r, \mu)$ obtained are then the solutions of the problem. E.S. Kuznetsov [38] and V.S. Vladimirov [11, 12] have shown that this method of successive approximations converges to the exact solution of the problem.

If in (67.1) we write $f(r) = \frac{1}{2}\nu_f \Sigma_f \varphi_0$, we arrive at the eigenvalue problem

$$\mu\frac{\partial\varphi}{\partial r} + \frac{1-\mu^2}{r}\frac{\partial\varphi}{\partial\mu} + \Sigma\varphi = \frac{1}{2}(\Sigma_s + \nu_f\Sigma_f)\,\varphi_0(r), \tag{67.14}$$

$$\varphi(R, \mu) = 0 \quad \text{for} \quad \mu \leqslant 0,$$

where, as before,

$$\varphi_0 = \int_{-1}^{+1} \varphi\, d\mu.$$

It has been shown [11] that this method of successive approximations converges to the exact solution of the problem. The approximate value of K_{eff} is then given by

$$K_{eff} \simeq K_{eff}^{(p)} = \frac{\|\varphi_0^{(p)}\|}{\|\varphi_0^{(p-1)}\|}, \tag{67.15}$$

where

$$\|\varphi\| = \sqrt{(\varphi, \varphi)}, \quad (\varphi, \varphi) = \int_0^R (\Sigma_s + \nu_f\Sigma_f)\,\varphi^2 r^2 dr.$$

Further, it has been shown that $K_{eff}^{(p)}$ increases monotonically in p and that it converges to K_{eff} as q^{2p}, where q is some number less than unity.

In conclusion, it is useful to note that in spherically symmetric problems Boltzmann's integro-differential equation is very easily reduced to the Peierls integral equation. Indeed, returning to the original variables r and μ in (67.7), we obtain

$$
\varphi(r,\mu)=\begin{cases}
\displaystyle\int\limits_r^R \frac{\left[\frac{1}{2}\Sigma_s(\rho)\varphi_0(\rho)+f(\rho)\right]\rho\,d\rho}{\sqrt{\rho^2+r^2(1-\mu^2)}}\,e^{-\int\limits_r^{\rho}\frac{\Sigma(u)u\,du}{\sqrt{u^2-r^2(1-\mu^2)}}} & \mu\leqslant 0,\\[2ex]
\displaystyle\varphi(r,-\mu)\,e^{-2\int\limits_{r\sqrt{1-\mu^2}}^{r}\frac{\Sigma(u)u\,du}{\sqrt{u^2-r^2(1-\mu^2)}}}+ & \\
\displaystyle+2e^{-\int\limits_{r\sqrt{1-\mu^2}}^{r}\frac{\Sigma(u)u\,du}{\sqrt{u^2-r^2(1-\mu^2)}}}\times & \\
\displaystyle\times\int\limits_{r\sqrt{1-\mu^2}}^{r}\frac{\left[\frac{1}{2}\Sigma_s(\rho)\varphi_0(\rho)+f(\rho)\right]\rho\,d\rho}{\sqrt{\rho^2-r^2(1-\mu^2)}}\times & \\
\displaystyle\times\,\mathrm{ch}\left(\int\limits_{r\sqrt{1-\mu^2}}^{\rho}\frac{\Sigma(u)\,u\,du}{\sqrt{u^2-r^2(1-\mu^2)}}\right) & \mu\geqslant 0.
\end{cases}
\tag{67.16}
$$

This formula was first obtained by E.S. Kuznetsov [38]. If it is inserted into the expression

$$
\varphi_0=\int\limits_{-1}^{+1}\varphi\,d\mu,
$$

we obtain the integral equation for spherically symmetric regions

$$
\varphi_0(r)=\int\limits_0^R\frac{\rho}{r}K(r,\rho)\,\Sigma_s(\rho)\,\varphi_0(\rho)\,d\rho+F(r),
\tag{67.17}
$$

where

$$
K(r,\rho)=K(\rho,r)=\frac{1}{2}\int\limits_{|r-\rho|}^{r+\rho}\frac{dt}{t}\,e^{-\int\limits_0^t\Sigma\left(\sqrt{s^2-\frac{s}{t}(t^2+\rho^2-r^2)+\rho^2}\right)ds},
\tag{67.18}
$$

and

$$
F(r)=\int\limits_0^R\frac{\rho}{r}f(\rho)\,K(r,\rho)\,d\rho.
\tag{67.19}
$$

68. Carlson's Numerical Method for Solving the Kinetic Equations

Simple and convenient is another numerical method, suggested by Carlson (see [60]) for solving the kinetic equations. We will first illustrate this method on the one-group Boltzmann kinetic equation.

Consider a spherically symmetric region $0 \leq r \leq R$, with a given neutron source function $f(r)$. The neutron flux φ is then given by

$$
\mu\frac{\partial\varphi}{\partial r}+\frac{1-\mu^2}{r}\frac{\partial\varphi}{\partial\mu}+\Sigma\varphi=Q(r)
\tag{68.1}
$$

where

$$Q(r) = \frac{\Sigma_s}{2} \int_{-1}^{+1} \varphi \, d\mu + f(r), \tag{68.2}$$

with the condition

$$\varphi(R, \mu) = 0, \quad \mu \leqslant 0. \tag{68.3}$$

In solving these equations Carlson also uses an iterated source method. In accordance with the concept of successive approximations, we will assume Q(r) given. The problem is then defined $\varphi(r, \mu)$.

For an approximate solution of the partial differential Equation (68.1), we break up the interval

$$-1 \leq \mu \leq 1$$

into 2n equal parts and integrate over μ within each of these subintervals

$$\mu_{j-1} \leq \mu \leq \mu_j$$

We then obtain

$$\int_{\mu_{j-1}}^{\mu_j} \mu \frac{\partial \varphi}{\partial r} d\mu + \frac{1}{r} \int_{\mu_{j-1}}^{\mu_j} (1 - \mu^2) \frac{\partial \varphi}{\partial \mu} d\mu + \Sigma \int_{\mu_{j-1}}^{\mu_j} \varphi d\mu = Q(r) \Delta\mu, \tag{68.4}$$

where

$$\Delta\mu = \mu_j - \mu_{j-1}.$$

Within each interval of integration, we shall consider φ linear in μ, so that

$$\varphi(r, \mu) = \frac{\mu_j - \mu}{\Delta\mu} \varphi^{j-1} + \frac{\mu - \mu_{j-1}}{\Delta\mu} \varphi^j, \tag{68.5}$$

where

$$\varphi^j = \varphi(r, \mu_j).$$

We now insert this into (68.4) and again integrate over μ. We then obtain the set of ordinary differential equations

$$\frac{d\varphi^j}{dr} + \frac{\Sigma + \frac{2}{r}\beta^j}{\alpha^j} \varphi^j = f^j \quad (j = 1, 2, \ldots, 2n), \tag{68.6}$$

where

$$f^j = \frac{1}{\alpha^j} \left[2Q - \frac{\gamma^j}{2} \frac{d\varphi^{j-1}}{dr} - \left(\Sigma - \frac{2\beta^j}{r} \right) \varphi^{j-1} \right], \tag{68.7}$$

and the α^j, β^j and γ^j depend only on the μ_j, and are defined by

$$\alpha^j = \mu_{j-1} + \frac{2}{3}\Delta\mu, \quad \beta^j = \frac{1}{\Delta\mu}\left(1 - \mu_j\mu_{j-1} - \frac{1}{3}\Delta\mu^2\right), \tag{68.8}$$

$$\gamma^j = 2\left(\mu_j - \frac{2}{3}\Delta\mu\right).$$

To (68.6) we must add another equation for φ^0. This equation is easily obtained by setting $\mu = -1$ in (68.1). We then obtain

$$-\frac{d\varphi^0}{dr} + \Sigma\varphi^0 = Q. \tag{68.9}$$

Equations (68.6) and (68.9) now form a closed set. In order that its solution be completely specified, we must find the appropriate boundary conditions. We shall use the following:

$$\left.\begin{aligned} &\varphi^j(R) = 0 \quad (j = 0,1,2\ \ldots,\ n). \\ &\varphi^j(0) = \varphi^0(0) \quad (j = n+1,\ n+2,\ \ldots,\ 2n). \end{aligned}\right\} \tag{68.10}$$

Let us now solve these equations. To do this we break up the whole $0 \le r \le R$ interval into subintervals of width $\Delta r_{k+1/2} = r_{k+1} - r_k$. We integrate (68.6) in the interval $r_k \le r \le r_{k+1}$, obtaining

$$\varphi^j_{k+1} - \varphi^j_k + \frac{1}{\alpha^j}\int\limits_{r_k}^{r_{k+1}} \left(\Sigma + \frac{2\beta^j}{r}\right)\varphi^j dr = \int\limits_{r_k}^{r_{k+1}} f^j dr. \tag{68.11}$$

We shall assume as an approximation that

$$\frac{1}{\alpha^j}\int\limits_{r_k}^{r_{k+1}} \left(\Sigma + \frac{2\beta^j}{r}\right)\varphi^j dr = \frac{1}{2\alpha^j}\left[(\Sigma\Delta r)_{k+1/2} + 2\beta^j\frac{\Delta r_{k+1/2}}{r_{k+1/2}}\right](\varphi^j_{k+1} + \varphi^j_k),$$

$$\int\limits_{r_k}^{r_{k+1}} f^j dr = \frac{2}{\alpha^j}(Q\Delta r)_{k+1/2} - c^j_k\varphi^{j-1}_{k+1} + d^j_k\varphi^{j-1}_k,$$

where

$$c^j_k = \frac{1}{2\alpha^j}\left[\gamma^j + (\Sigma\Delta r)_{k+1/2} - 2\beta^j\frac{\Delta r_{k+1/2}}{r_{k+1/2}}\right],$$

$$d^j_k = \frac{1}{2\alpha^j}\left[\gamma^j - (\Sigma\Delta r)_{k+1/2} + 2\beta^j\frac{\Delta r_{k+1/2}}{r_{k+1/2}}\right].$$

Here we have used the notation

$$(\Phi)_{k+1/2} = \frac{1}{2}(\Phi_{k+1} + \Phi_k).$$

Then (68.11) becomes

$$a^j_k\varphi^j_{k+1} - b^j_k\varphi^j_k = F^j_k, \tag{68.12}$$

where

$$a^j_k = 1 + \frac{1}{2\alpha^j}\left[(\Sigma\Delta r)_{k+1/2} + 2\beta^j\frac{\Delta r_{k+1/2}}{r_{k+1/2}}\right],$$

$$b^j_k = 1 - \frac{1}{2\alpha^j}\left[(\Sigma\Delta r)_{k+1/2} + 2\beta^j\frac{\Delta r_{k+1/2}}{r_{k+1/2}}\right],$$

$$F^j_k = \frac{2}{\alpha^j}(Q\Delta r)_{k+1/2} - C^j_k\varphi^{j-1}_{k+1} + d^j_k\varphi^{j-1}_k.$$

Using (68.9), we again arrive at (68.12) for $j = 0$, where

$$a_k = 1 - \frac{1}{2}(\Sigma\Delta r)_{k+1/2}, \quad b_k^0 = 1 + \frac{1}{2}(\Sigma\Delta r)_{k+1/2},$$

$$F_k^0 = -(Q\Delta r)_{k+1/2}.$$

We shall attempt to solve (68.12) by means of the recursion formulas

$$\left.\begin{aligned} \varphi_k^j &= \frac{a_k^j}{b_k^j}\varphi_{k+1}^j - \frac{F_k^j}{b_k^j} \quad (j = 0, 1, 2, \ldots, n), \\ \varphi_{k+1}^j &= \frac{b_k^j}{a_k^j}\varphi_k^j + \frac{F_k^j}{a_k^j} \quad (j = n+1,\ n+2,\ \ldots,\ 2n). \end{aligned}\right\} \qquad (68.13)$$

Numerical computations based on (68.13) are not difficult. The method is the following. We first find the function φ_k^0 with the condition that $\varphi^0(R) = 0$. The other unknown functions φ_k^j are found by means of the successive solution of (68.13) with Condition (68.10).

After all the φ_k^j are found, we must calculate a new approximate value for Q(r) using (68.2) by means of the simple trapeze formula

$$Q_k = \Delta\mu \frac{\Sigma_{sk}}{2} \sum_{j=0}^{2n}{}' \xi^j \varphi_k^j + f_k, \qquad (68.14)$$

where

$$\xi^j = \begin{cases} \frac{1}{2} & (j = 0,\ j = 2n), \\ 1 & (j = 1, 2, \ldots, 2n-1). \end{cases}$$

In conclusion,it should be noted that in most cases of practical interest the critical mass of a reactor can be found with sufficient accuracy by breaking up the interval $(-1 \leq \mu \leq 1)$ into four subintervals of length $\Delta\mu = {}= 1/2$.

This approximation is even more accurate than the P_3-approximation using spherical harmonics.

69. The Equations of a Fast-Neutron Reactor in the Diffusion Approximation

In performing fast-neutron reactor calculations, for which the dimensions of the reactor zones are much larger than the scattering length, the ordinary diffusion approximation is quite valid.

In view of the fact the method for obtaining the diffusion equations for fast-neutron reactors by means of Equations (60.11) and (61.3) does not differ in principle from the method we have analyzed in Chapter II, we will refrain from repeating the details, and shall merely state the final results.

Thus, after the necessary transformations, the multigroup set of basic reactor equations (60.11) becomes

$$\left.\begin{aligned} \nabla D^j \nabla \varphi^j - \Sigma^j \varphi^j &= -\sum_{l=1}^{j} \overset{l\to j}{\Sigma} \varphi_0^l - \frac{\chi^j}{K_{\text{eff}}} Q(\vec{r}), \\ Q(\vec{r}) &= \sum_{l=1}^{m} \nu_f^l \cdot \Sigma_f^l \varphi_0^l \ (j = 1, 2, \ldots, n), \end{aligned}\right\} \qquad (69.1)$$

where

$$D^j = \frac{1}{3\Sigma_j}, \quad \Sigma^j = \Sigma_{in}^j + \Sigma_c^j$$

with the condition that on the boundary S of the reactor we have

$$2D^j\nabla_{\vec{n}0}\varphi^j + \varphi^j = 0 \quad (j = 1, 2, \ldots, m). \tag{69.2}$$

The adjoint reactor equations are

$$\left.\begin{aligned} \nabla D^j\nabla\varphi^{*j} - \Sigma^j\varphi^{*j} &= -\sum_{l=j}^{m} \overset{j\to l}{\Sigma}\varphi_0^{*l} - \frac{\nu_f^j\Sigma_f^j}{K_{\text{eff}}} Q^*(\vec{r}), \\ Q^*(\vec{r}) &= \sum_{l=1}^{m} \chi^l\varphi^{*l} \quad (j = m, m-1, \ldots, 1). \end{aligned}\right\} \tag{69.3}$$

The boundary condition for these is

$$2D^j\nabla_{\vec{n}0}\varphi^{*j} + \varphi^{*j} = 0 \quad (j = m, m-1, \ldots, 1). \tag{69.4}$$

The basic and adjoint equations are solved in the diffusion approximation as previously described.

70. Elastic Scattering

The methods for performing fast-neutron reactor calculations described in the present chapter have assumed that the neutrons are slowed down by means of inelastic scattering. This assumption is well justified under a vanishingly small concentration of elements whose elastic moderating power is high. If elastic scattering cannot be ignored, the theory and method of calculation must be altered.

Consider Boltzmann's kinetic equation

$$\vec{\Omega}\nabla\varphi + \Sigma\varphi = \int d\Omega' \int_{-\infty}^{\infty} du'\varphi\left[g_s(\mu_0, u, u') + g_{in}(\mu_0, u, u')\right] + \frac{\chi(u)}{4\pi K_{\text{eff}}} Q(\vec{r}), \tag{70.1}$$

where

$$Q(r) = \int_{-\infty}^{\infty} \nu_f\Sigma_f\varphi_0 du.$$

In many cases it is convenient to replace g_s by an approximate function of simpler form. It is necessary to require only that some of the moments of this function be the same as some of the scattering moments.

Let us consider two important practical applications. 1. If slowing down by elastic scattering is only a small effect, $g_s(\mu_0, u, u')$ can be approximated by an isotropic function, and the anisotropy can be taken into account in the so-called transport approximation. Then in (70.1) Σ_s should be replaced by $\Sigma_{tr} = \Sigma_s(1 - \bar{\mu}_0)$, and we write

$$g_s(\mu_0, u, u') = \frac{1}{4\pi}\Sigma_{tr}W_s(u, u'), \tag{70.2}$$

where

$$W_s(u, u') = \begin{cases} \dfrac{1}{\xi} e^{-\frac{u-u'}{\xi}}, & \text{for } u < u' \\ 0, & \text{for } u > u' \end{cases}$$

It is easily shown that the exact moments of $\frac{1}{4\pi}W_s$ are

$$\frac{1}{4\pi}\int d\Omega \int_{-\infty}^{\infty} duW_s(u, u') = 1, \frac{1}{4\pi}\int d\Omega \int_{-\infty}^{\infty} du\,(u - u')\,W_s(u, u') = \xi. \tag{70.3}$$

Using the isotropy of the scattering function, we arrive at the reactor equation

$$\vec{\Omega}\nabla\varphi+\Sigma\varphi=\frac{1}{4\pi}\int_{-\infty}^{\infty}du'\varphi_0 g_0(u,u')+\frac{\chi(u)}{4\pi K_{eff}}Q(\vec{r}), \tag{70.4}$$

where

$$g_0(u,u')=\Sigma_{tr}(u')W_s(u,u')+\Sigma_{in}(u')W_{in}(u,u')$$

$$\Sigma=\Sigma_{tr}+\Sigma_{in}+\Sigma_c.$$

2. If elastic scattering is of importance, $g(\mu_0, u, u')$ can be approximated by the function

$$g(\mu_0,u,u')=\frac{\Sigma_s}{4\pi}[W_{s0}(u,u')+3\mu_0 W_{s1}(u,u')], \tag{70.5}$$

where

$$\left.\begin{aligned} W_{s0}(u,u')&=\left(1-\frac{\xi}{\gamma}\right)\delta(u-u')+\frac{\xi}{\gamma^2}e^{-\frac{u-u'}{\gamma}},\\ W_{s1}(u,u')&=e^{-\frac{u-u'}{\bar{\mu}_0}}.\end{aligned}\right\} \tag{70.6}$$

The function $W(\mu_0, u, u')=\frac{1}{4\pi}(W_{s0}+3\mu_0 W_{s1})$ has the following exact moments:

$$\int d\Omega\int_{-\infty}^{\infty}du W(\mu_0 u,u')=1,\quad \int d\Omega\int_{-\infty}^{\infty}du(u-u')W(\mu_0,u,u')=\xi,$$

$$\int d\Omega\int_{-\infty}^{\infty}du(u-u')^2 W(\mu_0,u,u')=\bar{\xi}^2, \tag{70.7}$$

$$\int d\Omega\int_{-\infty}^{\infty}du\mu_0 W(\mu_0,u,u')=\bar{\mu}_0.$$

With these assumptions, the basic reactor equation becomes

$$\vec{\Omega}\nabla\varphi+\Sigma\varphi=\frac{1}{4\pi}\int_{-\infty}^{\infty}du'\varphi_0 g_0(u,u')+\frac{3\mu}{4\pi}\int_{-\infty}^{\infty}du'\varphi_1 g_1(u,u')+\frac{\chi(u)}{4\pi K_{eff}}Q(\vec{r}), \tag{70.8}$$

where

$$\left.\begin{aligned} \varphi_0(\vec{r},u)&=\int d\Omega\varphi(\vec{r},\vec{\Omega},u),\quad \varphi_1(\vec{r},u)=\int d\Omega\mu\varphi(\vec{r},\vec{\Omega},u)\\ g_0(u,u')&=\Sigma_s(u')W_{s0}(u,u')+\Sigma_{in}W_{in}(u,u'),\\ g_1(u,u')&=\Sigma_s(u')W_{s1}(u,u').\end{aligned}\right\} \tag{70.9}$$

The corresponding adjoint equations are easily obtained.

In conclusion we note that the multigroup representations of Equations (70.1) and (70.8) are obtained in the same way as indicated in the previous sections of this chapter.

Chapter XIV

REACTORS WITH HYDROGENOUS MODERATORS*

A specific property of numerical reactor calculations with hydrogenous moderators is that the mechanism of elastic scattering of neutrons on hydrogen nuclei cannot be described sufficiently well by the model of continuous slowing down which is used in age theory. It is necessary in this case to use more accurate kinetic slowing-down equations.

It is found that the critical mass and neutron spectrum of reactors with hydrogenous moderators can, as a rule, be found with sufficient accuracy in the diffusion approximation. Even in this approximation, however, the numerical calculation involves many important mathematical difficulties. These are particularly significant in the case of intermediate reactors, for which the neutron spectrum varies strongly in the region of U^{235} resonance energies.

Thermal reactors are calculated by various approximate methods based primarily on experimental values of the mean square slowing-down length. In the final analysis the problem reduces to solving a multigroup set of diffusion equations. In this set of equations the diffusion lengths are chosen so as to describe satisfactorily the spectrum of neutrons from a point source in an infinite medium [23, 54, 118, 146].

More accurate methods for hydrogenous moderators have been developed by Solenguth and D. Goertzel (see [146]) and by Shirkov [90].

In the present chapter we shall discuss numerical methods for intermediate and thermal reactors with hydrogenous moderators.

71. The Neutron Slowing-Down Equations in the Reactor

Consider a reactor whose moderator is a one-component medium of hydrogen nuclei. If the reactor is sufficiently large compared with the neutron mean free path, the calculations may be performed in the diffusion approximation. The slowing-down equations are then written

$$\left.\begin{aligned} \nabla\vec{\varphi}_1 + \Sigma\varphi_0 &= \int\limits_{-\infty}^{u} \Sigma_s\varphi_0 f_0(u-u')\,du' + S, \\ \frac{1}{3}\nabla\varphi_0 + \Sigma\vec{\varphi}_1 &= \int\limits_{-\infty}^{u} \Sigma_s\vec{\varphi}_1 f_1(u-u')\,du', \end{aligned}\right\} \tag{71.1}$$

where $f_0(u)$ and $f_1(u)$ are of the form

$$f_0(u) = e^{-u}, \quad f_1(u) = e^{-\frac{3}{2}u},$$

for hydrogen, and where $S(\vec{r}, u) = \chi(u)Q(\vec{r})$ describes the fission neutron sources.

By differentiating with respect to u and then eliminating the integral terms, Equations (71.1) can be written as the differential equations

*The present chapter was added to the book after the manuscript had been prepared for publication. The author apologizes for the lack of proper order which results in placing this chapter at the end of the book.

$$\left.\begin{aligned}\nabla\vec{\varphi}_1+\Sigma_c\varphi_0+\frac{\partial}{\partial u}(\nabla\vec{\varphi}_1+\Sigma_c\varphi_0)&=-\frac{\partial\Sigma_s\varphi_0}{\partial u}+S+\frac{\partial S}{\partial u},\\ \frac{1}{3}\nabla\varphi_0+\Sigma_{tr}\vec{\varphi}_1+\frac{2}{3}\frac{\partial}{\partial u}\left(\frac{1}{3}\nabla\varphi_0+\Sigma\vec{\varphi}_1\right)&=0,\end{aligned}\right\} \tag{71.2}$$

where

$$\Sigma_{tr}=\Sigma_s(1-\bar{\mu}_0)+\Sigma_c,\quad \bar{\mu}_0=\frac{2}{3}.$$

These equations are equivalent to (71.1) and can be used to calculate the neutron slowing-down spectrum in the reactor.

Let us now go on to study slowing down by a mixture of hydrogen and other elements. We shall assume that most of the slowing down is due to scattering by hydrogen, i.e., that

$$\Sigma_{s0}\gg\sum_{k\neq 0}\xi_k\Sigma_{sk},$$

where Σ_{s0} is the cross section for scattering by hydrogen nuclei, and Σ_{sk} (with $k\neq 0$) is the cross section for scattering by other elements in the mixture.

With this assumption we can formulate a set of approximate slowing-down equations which in the limit of a one-component hydrogen medium becomes the exact set of equations given by (71.2). We will start with the integro-differential slowing-down equations

$$\left.\begin{aligned}\nabla\vec{\varphi}_1+\Sigma\varphi_0&=\int_{-\infty}^{u}\Sigma_s\varphi_0 f_0(u,u')\,du'+S,\\ \frac{1}{3}\nabla\varphi_0+\Sigma\vec{\varphi}_1&=\int_{-\infty}^{u}\Sigma_s\vec{\varphi}_1 f_1(u,u')\,du',\end{aligned}\right\}. \tag{71.3}$$

where

$$f_0(u,u')=\frac{1}{\Sigma_s}\sum_k\Sigma_{sk}f_{0k}(u-u'),$$

$$f_1(u,u')=\frac{1}{\Sigma_s}\sum_k\Sigma_{sk}f_{1k}(u-u'),$$

and f_{0k} and f_{1k} are the functions defined in (8.7). We will approximate the exact f_0 and f_1 functions by

$$\left.\begin{aligned}W_{s0}(u,u')&=\left(1-\frac{\xi}{\gamma}\right)\delta(u-u')+\frac{\xi}{\gamma^2}e^{-\frac{u-u'}{\gamma}},\\ W_{s1}(u,u')&=e^{-\frac{u-u'}{\bar{\mu}_0}}\end{aligned}\right\} \tag{71.4}$$

where

$$\xi=\frac{1}{\Sigma_s}\sum_k\xi_k\Sigma_{sk},\quad \gamma=\frac{\sum_k\gamma_k\xi_k\Sigma_{sk}}{\sum_k\xi_k\Sigma_{sk}},\quad \bar{\mu}_0=\frac{1}{\Sigma_s}\sum_k\bar{\mu}_{0k}\Sigma_{sk}$$

ξ_k is the mean lethargy loss in a collision between a neutron and nucleus, $\gamma_k=\overline{\xi_k^2}/2\xi_k$, where $\overline{\xi_k^2}$ is the mean square lethargy loss, $\bar{\mu}_{0k}$ is the mean cosine of the scattering angle, and W_{s0} and W_{s1} are defined for $0\leq u<\infty$. The W_{s0} function was constructed by E. Greuling and D. Goertzel and for a one-component moderator it has the correct zeroth, first and second energy moments, while W_{s1} has the correct zeroth moment.

These functions are identical with the exact f_0 and f_1 functions for a hydrogen moderator. Replacing f_0 and f_1 in (71.3) by their approximations given by (71.4), we have

$$\left.\begin{aligned} \nabla\vec{\varphi}_1 + \Sigma_c\varphi_0 &= -\frac{\xi\Sigma_s}{\gamma}\varphi_0 + \int_{-\infty}^{u} \varphi_0 \frac{\xi\Sigma_s}{\gamma^2} e^{-\frac{u-u'}{\gamma}} du' + S, \\ \frac{1}{3}\nabla\varphi_0 + \Sigma\vec{\varphi}_1 &= \int_{-\infty}^{u} \vec{\varphi}_1 \Sigma_s e^{-\frac{u-u'}{\bar{\mu}_0}} du'. \end{aligned}\right\} \tag{71.5}$$

We now differentiate with respect to u, obtaining

$$\left.\begin{aligned} \frac{\partial}{\partial u}(\nabla\vec{\varphi}_1 + \Sigma_c\varphi_0) &= -\frac{\partial}{\partial u}\left(\varphi_0 \frac{\xi\Sigma_s}{\gamma}\right) + \varphi_0 \frac{\xi\Sigma_s}{\gamma^2} - \\ &- \int_{-\infty}^{u} \varphi_0 \frac{\xi\Sigma_s}{\gamma^3} e^{-\frac{u-u'}{\gamma}} du' + \frac{\partial S}{\partial u}, \\ \frac{\partial}{\partial u}\left(\frac{1}{3}\nabla\varphi_0 + \Sigma\vec{\varphi}_1\right) &= \Sigma_s\vec{\varphi}_1 - \int_{-\infty}^{u} \vec{\varphi}_1 \frac{\Sigma_s}{\bar{\mu}_0} e^{-\frac{u-u'}{\bar{\mu}_0}} du'. \end{aligned}\right\} \tag{71.6}$$

Let us write

$$\left.\begin{aligned} \gamma' &= \frac{\int_{-\infty}^{u} \varphi_0 \frac{\xi\Sigma_s}{\gamma^2} e^{-\frac{u-u'}{\gamma}} du'}{\int_{-\infty}^{u} \varphi_0 \frac{\xi\Sigma_s}{\gamma^3} e^{-\frac{u-u'}{\gamma}} du'}, \\ \bar{\mu}_0' &= \frac{\int_{-\infty}^{u} \vec{\varphi}_1 \Sigma_s e^{-\frac{u-u'}{\bar{\mu}_0}} du'}{\int_{-\infty}^{u} \vec{\varphi}_1 \frac{\Sigma_s}{\bar{\mu}_0} e^{-\frac{u-u'}{\bar{\mu}_0}} du'}. \end{aligned}\right\} \tag{71.7}$$

We now multiply the first of Equations (71.6) by γ', adding the result to the first of (71.5); we multiply the second of (71.6) by $\bar{\mu}_0'$, adding this to the second of (71.5). The resulting equations are

$$\left.\begin{aligned} \nabla\vec{\varphi}_1 + \Sigma_c\varphi_0 + \gamma'\frac{\partial}{\partial u}(\nabla\vec{\varphi}_1 + \Sigma_c\varphi_0) &= -\frac{\partial \xi\Sigma_s'\varphi_0}{\partial u} + S + \gamma'\frac{\partial S}{\partial u} - \varepsilon, \\ \frac{1}{3}\nabla\varphi_0 + \Sigma_{tr}\vec{\varphi}_1 + \bar{\mu}_0'\frac{\partial}{\partial u}\left(\frac{1}{3}\nabla\varphi_0 + \Sigma\vec{\varphi}_1\right) &= 0, \end{aligned}\right\} \tag{71.8}$$

where

$$\left.\begin{aligned} \xi\Sigma_s' = \frac{\gamma'}{\gamma}\xi\Sigma_s, \quad \Sigma_{tr} &= \Sigma_s(1-\bar{\mu}_0') + \Sigma_c, \\ \varepsilon &= \left(1-\frac{\gamma'}{\gamma}\right)\frac{\xi\Sigma_s}{\gamma}\varphi_0. \end{aligned}\right\} \tag{71.9}$$

Now ϵ is usually quite small and can be neglected. Indeed, if the moderator consists of a single element with atomic weight M_k, Equations (71.7) and (71.9) imply that

$$\gamma' = \gamma_k, \quad \bar{\mu}_0' = \bar{\mu}_{0k}, \quad \xi\Sigma_s' = \xi_k\Sigma_{sk}, \quad \varepsilon = 0. \tag{71.10}$$

This is true, in particular, for a one-component hydrogen medium.

When the moderator is a mixture of nuclei with different atomic weights, we make the approximation of taking $1/\gamma^2$, $1/\gamma^3$ and $1/\bar{\mu}_0$ outside the integral sign in (71.7) assigning them their values at the lethargy $u' = u$. We then have

$$\left.\begin{aligned} \gamma' = \gamma &= \frac{1}{\xi\Sigma_s}\sum_k \gamma_k\xi_k\Sigma_{sk}. \\ \bar{\mu}_0' = \bar{\mu}_0 &= \frac{1}{\Sigma_s}\sum_k \bar{\mu}_{0k}\Sigma_{sk}, \quad \varepsilon = 0, \\ \xi\Sigma_s' = \xi\Sigma_s &= \sum_k \xi_k\Sigma_{sk}. \end{aligned}\right\} \tag{71.11}$$

From the limiting relations given by (71.10) and (71.11) one may expect that ϵ will be small also when the moderator consists of a mixture of hydrogen with a low concentration of other nuclei.

We thus finally arrive at the following set of equations for the spectrum of slowing-down neutrons:

$$\left.\begin{aligned} \nabla\vec{\varphi}_1 + \Sigma_c\varphi_0 + \gamma\frac{\partial}{\partial u}(\nabla\vec{\varphi}_1 + \Sigma_c\varphi_0) &= -\frac{\partial\xi\Sigma_s\varphi_0}{\partial u} + S + \gamma\frac{\partial S}{\partial u}, \\ \frac{1}{3}\nabla\varphi_0 + \Sigma_{tr}\vec{\varphi}_1 + \bar{\mu}_0\frac{\partial}{\partial u}\left(\frac{1}{3}\nabla\varphi_0 + \Sigma\vec{\varphi}_1\right) &= 0. \end{aligned}\right\} \tag{71.12}$$

This is the same as (71.2) when the moderator consists of hydrogen alone. In this case, as is well known, $\xi = 1$, $\gamma = 1$ and $\mu_0 = 2/3$. We note in conclusion that for a mixture in which most of the nuclei are hydrogen, $\xi\Sigma_s$, γ and $\bar{\mu}_0$ can, according to (71.9-71.11), be written

$$\left.\begin{aligned} \xi\Sigma_s = \sum_k \xi_k\Sigma_{sk}, \quad \gamma &= \frac{1}{\xi\Sigma_s}\sum_k \gamma_k\xi_k\Sigma_{sk}, \\ \bar{\mu}_0 &= \frac{1}{\Sigma_s}\sum_k \bar{\mu}_{0k}\Sigma_{sk}. \end{aligned}\right\} \tag{71.13}$$

72. The Diffusion Equation for Thermal Neutrons

Consider a reactor with a hydrogen moderator. Then, uniting all the thermal neutrons into a single group, the space distribution of neutrons over the reactor volume is described in the diffusion approximation by the equations

$$\left.\begin{aligned} \nabla\vec{\Phi}_1 + \Sigma_{cT}\Phi_0 &= \int_{-\infty}^{u_T} \Sigma_s\varphi_0 e^{-(u_T-u)}\,du, \\ \frac{1}{3}\nabla\Phi_0 + \Sigma_{tr\,T}\vec{\Phi}_1 &= \frac{2}{3}\int_{-\infty}^{u_T} \Sigma_s\vec{\varphi}_1 e^{-\frac{3}{2}(u_T-u)}\,du. \end{aligned}\right\} \tag{72.1}$$

We have here made use of the relations

$$\left.\begin{aligned} F_0(u) &= \int_{u_T}^{\infty} f_0(u'-u)\,du' = e^{-(u_T-u)}, \\ F_1(u) &= \int_{u_T}^{\infty} f_1(u'-u)\,du' = \frac{2}{3}e^{-\frac{3}{2}(u_T-u)}. \end{aligned}\right\} \tag{72.2}$$

Let us now turn to (71.1), in which we set $u = u_T$ and $S = 0$. We then obtain

$$\left.\begin{aligned} \int_{-\infty}^{u_T} \Sigma_s \varphi_0 e^{-(u_T-u)} du &= \nabla \vec{\varphi}_1^m + \Sigma^m \varphi_0^m, \\ \int_{-\infty}^{u_T} \Sigma_s \vec{\varphi}_1 e^{-\frac{3}{2}(u_T-u)} du &= \frac{1}{3} \nabla \varphi_0^m + \Sigma^m \vec{\varphi}_1^m, \end{aligned}\right\} \tag{72.3}$$

where the index $\underline{m}$ refers to quantities with energy $E = E_C$. Further, let us use (72.3) to eliminate the integral terms from (72.1). We then obtain

$$\left.\begin{aligned} \nabla \vec{\Phi}_1 + \Sigma_{cT} \Phi_0 &= \nabla \vec{\varphi}_1^m + \Sigma^m \varphi_0^m, \\ \frac{1}{3} \nabla \Phi_0 + \Sigma_{tr\,T} \vec{\Phi}_1 &= \frac{2}{3} \left(\frac{1}{3} \nabla \varphi_0^m + \Sigma^m \vec{\varphi}_1^m \right). \end{aligned}\right\} \tag{72.4}$$

If the moderator is a mixture of several elements, the equations for the thermal neutron group become

$$\left.\begin{aligned} \nabla \vec{\Phi}_1 + \Sigma_{cT} \Phi_0 &= \int_{-\infty}^{u_T} \varphi_0 \frac{\xi \Sigma_s}{\gamma} e^{-\frac{(u_T-u)}{\gamma}} du, \\ \frac{1}{3} \nabla \Phi_0 + \Sigma_{tr\,T} \vec{\Phi}_1 &= \int_{-\infty}^{u_T} \vec{\varphi}_1 \bar{\mu}_0 \Sigma_s e^{-\frac{u_T-u}{\mu_0}} du. \end{aligned}\right\} \tag{72.5}$$

We have here used (72.2), in which f_0 and f_1 are replaced by their approximations given by (71.4).

Performing an identity transformation on (72.5), we arrive at

$$\left.\begin{aligned} \nabla \vec{\Phi}_1 + \Sigma_{cT} \Phi_0 &= \gamma_T' \int_{-\infty}^{u_T} \varphi_0 \frac{\xi \Sigma_s}{\gamma^2} e^{-\frac{u_T-u}{\gamma}} du, \\ \frac{1}{3} \nabla \Phi_0 + \Sigma_{trT} \vec{\Phi}_1 &= \bar{\mu}_{0T}' \int_{-\infty}^{u_T} \vec{\varphi}_1 \Sigma_s e^{-\frac{u_T-u}{\mu_0}} du, \end{aligned}\right\} \tag{72.6}$$

where

$$\left.\begin{aligned} \gamma_T' &= \frac{\displaystyle\int_{-\infty}^{u_T} \varphi_0 \frac{\xi \Sigma_s}{\gamma} e^{-\frac{u_T-u}{\gamma}} du}{\displaystyle\int_{-\infty}^{u_T} \varphi_0 \frac{\xi \Sigma_s}{\gamma^2} e^{-\frac{u_T-u}{\gamma}} du}, \\ \bar{\mu}_{0T}' &= \frac{\displaystyle\int_{-\infty}^{u_T} \vec{\varphi}_1 \bar{\mu}_0 \Sigma_s e^{-\frac{u_T-u}{\mu_0}} du}{\displaystyle\int_{-\infty}^{u_T} \vec{\varphi}_1 \Sigma_s e^{-\frac{u_T-u}{\mu_0}} du}. \end{aligned}\right\} \tag{72.7}$$

If the moderator consists of nuclei of a single type, we have

$$\gamma'_T = \gamma_h, \quad \bar{\mu}'_{0T} = \bar{\mu}_{0h}.$$

These relations are true, in particular, for a one-component hydrogen medium.

If the medium consists of several types of nuclei, we take the functions $1/\gamma$, $1/\gamma^2$ and $\bar{\mu}_0$ out from under the integral signs in (72.2) and assign them their values at $u = u_T$, obtaining

$$\left.\begin{aligned} \gamma'_T = \gamma_T &= \frac{1}{\xi\Sigma_{sT}} \sum_h \gamma_h \xi_h \Sigma_{shT}, \\ \bar{\mu}'_{0T} = \bar{\mu}_{0T} &= \frac{1}{\Sigma_{sT}} \sum_h \bar{\mu}_{0h} \Sigma_{shT}\,. \end{aligned}\right\} \tag{72.8}$$

This equation can be used to calculate the effective constants γ_T and $\bar{\mu}_{0T}$ for a moderator in which most of the nuclei are hydrogen. Let us now consider (71.5). We set $u = u_T$ and $S = 0$. We use the relations obtained to eliminate the integral terms from (72.6). This gives

$$\left.\begin{aligned} \nabla\vec{\Phi}_1 + \Sigma_{cT}\Phi_0 &= \gamma_T \nabla\vec{\varphi}_1^m + (\xi_T \Sigma_s^m + \gamma_T \Sigma_c^m)\,\varphi_0^m, \\ \frac{1}{3}\nabla\Phi_0 + \Sigma_{trT}\vec{\Phi}_1 &= \bar{\mu}_{0T}\left(\frac{1}{3}\nabla\varphi_0^m + \Sigma^m \vec{\varphi}_1^m\right). \end{aligned}\right\} \tag{72.9}$$

73. Basic Reactor Equations

The results of the previous sections of this chapter can be used to obtain the complete set of reactor equations. For a moderator of hydrogen nuclei this set is

$$\left.\begin{aligned} \nabla\vec{\varphi}_1 + \Sigma_c\varphi_0 + \frac{\partial}{\partial u}(\nabla\vec{\varphi}_1 + \Sigma_c\varphi_0) &= -\frac{\partial\Sigma_s\varphi_0}{\partial u} + \left(\chi + \frac{d\chi}{du}\right) Q(\vec{r}), \\ \frac{1}{3}\nabla\varphi_0 + \Sigma_{tr}\vec{\varphi}_1 + \frac{2}{3}\frac{\partial}{\partial u}\left(\frac{1}{3}\nabla\varphi_0 + \Sigma\vec{\varphi}_1\right) &= 0, \\ \nabla\vec{\Phi}_1 + \Sigma_{cT}\,\Phi_0 &= \nabla\vec{\varphi}_1^m + \Sigma^m\varphi_0^m, \\ \frac{1}{3}\nabla\Phi_0 + \Sigma_{trT}\,\vec{\Phi}_1 &= \frac{2}{3}\left(\frac{1}{3}\nabla\varphi_0^m + \Sigma^m\vec{\varphi}_1^m\right), \\ Q(\vec{r}) &= \nu_f\left(\int_{-\infty}^{u_T} \Sigma_f\varphi_0\,du + \Sigma_{fT}\,\Phi_0\right). \end{aligned}\right\} \tag{73.1}$$

If the moderator is a mixture of several different nuclei of which hydrogen makes up the majority, the set of basic reactor equations becomes

$$\left.\begin{aligned} \nabla\vec{\varphi}_1 + \Sigma_c\varphi_0 + \gamma\frac{\partial}{\partial u}(\nabla\vec{\varphi}_1 + \Sigma_c\varphi_0) &= -\frac{\partial\xi\Sigma_s\varphi_0}{\partial u} + \left(\chi + \gamma\frac{d\chi}{du}\right) Q(\vec{r}), \\ \frac{1}{3}\nabla\varphi_0 + \Sigma_{tr}\vec{\varphi}_1 + \bar{\mu}_0\frac{\partial}{\partial u}\left(\frac{1}{3}\nabla\varphi_0 + \Sigma\vec{\varphi}_1\right) &= 0, \\ \nabla\vec{\Phi}_1 + \Sigma_{cT}\,\Phi_0 &= \gamma_T\nabla\vec{\varphi}_1^m + (\xi_T\Sigma_s^m + \gamma_T\Sigma_c^m)\,\varphi_0^m, \\ \frac{1}{3}\nabla\Phi_0 + \Sigma_{trT}\,\vec{\Phi}_1 &= \bar{\mu}_{0T}\left(\frac{1}{3}\nabla\varphi_0^m + \Sigma^m\vec{\varphi}_1^m\right), \\ Q(\vec{r}) &= \nu_f\left(\int_{-\infty}^{u_T} \Sigma_f\varphi_0\,du + \Sigma_{fT}\,\Phi_0\right). \end{aligned}\right\} \tag{73.2}$$

We wish to find solutions of these sets among the class of functions $\{\varphi_0, \vec{\varphi}_1, \Phi_0, \vec{\Phi}_1\}$, continuous over the entire reactor volume and satisfying the following condition on the reactor boundary S:

$$\left.\begin{aligned} 2(\vec{\varphi}_1)_{\vec{n}_0} - \varphi_0 &= 0, \\ 2(\vec{\Phi}_1)_{\vec{n}_0} - \Phi_0 &= 0. \end{aligned}\right\} \tag{73.3}$$

The parameter K_{eff} has been eliminated from (73.1) and (73.2) as being insignificant (see the footnote on page 73).

74. Slowing Down in Media With Different Hydrogen Concentrations

Let us consider Equations (73.2). They can be used to perform calculations for reactors with any concentration of hydrogen nuclei. Indeed, when the moderator consists entirely of hydrogen nuclei, these equations become the same as (73.1). When the moderator consists of nuclei whose atomic mass M is large, the quantities γ and $\bar{\mu}_0$ approach zero and these equations then become those of the diffusion-age approximation. If we keep the terms containing the parameters γ and $\bar{\mu}_0$, we arrive at the diffusion-age approximation corrected to take account of the second moment of energy. Thus, for one-component moderators, Equations (73.2) give the correct expressions in the limits as $M \to 1$ and $M \to \infty$, and in addition they correct the diffusion approximation for all elements with atomic number greater than one. This means that these equations can be used to perform reactor calculations for any one-component moderators, including hydrogen.

It is easily shown that when the moderator is a mixture of different kinds of heavy elements, all of the assumptions made in deriving Equation (73.2) are quite valid. One can therefore assert that in the limit of a mixture of heavy elements these equations are still correct. This is obviously true also for a mixture of heavy elements and hydrogen nuclei for vanishing concentrations of the latter. It can be shown that if the mixture consists of hydrogen with small impurities of other elements, Equations (73.2) again describe the neutron spectrum in the reactor with sufficient accuracy.

It is natural to suppose, from these considerations, that (73.2) can be used successfully in reactor calculations for any hydrogen concentration in the moderator. It should, however, be borne in mind that in general these equations are not exact. They can be used only as the basis of an interpolation procedure for performing reactor calculations over the whole range of hydrogen concentrations in moderator mixtures. Thus, Equations (73.2) are of the required generality and can be used to perform reactor calculations for any hydrogen concentration in the mixture. Henceforth we will deal only with reactor equations in the form of (73.2), whose effective constants are given by (71.13) and (72.8).

75. The Multigroup Set of Reactor Equations for Weak Absorption of Neutrons

There are several ways to represent the basic reactor equation (73.2) in the form of a multigroup set of diffusion equations. When the solution varies slowly with lethargy, this can be done by means of linear interpolation of the solution within each group. This approach is equivalent to representing the reactor equations in the form of a set of multigroup equations for weak absorption of slowing-down neutrons.

This method may, however, lead to significant errors in the case of strong absorption. It is therefore necessary to develop multigroup methods analogous to those already considered in Chapter VII. In the present section we shall consider reducing the basic reactor equations to a set of multigroup equations for weak absorption of the slowing-down neutrons.

Let us thus consider the slowing-down equations

$$\left.\begin{aligned} \nabla\vec{\varphi}_1 + \Sigma_c\varphi_0 + \gamma\frac{\partial}{\partial u}(\nabla\vec{\varphi}_1 + \Sigma_c\varphi_0) &= -\frac{\partial\xi\Sigma_s\varphi_0}{\partial u} + \left(\chi + \gamma\frac{d\chi}{du}\right)Q(\vec{r}), \\ \frac{1}{3}\nabla\varphi_0 + \Sigma_{tr}\vec{\varphi}_1 + \bar{\mu}_0\frac{\partial}{\partial u}\left(\frac{1}{3}\nabla\varphi_0 + \Sigma\vec{\varphi}_1\right) &= 0 \end{aligned}\right\} \tag{75.1}$$

and integrate them over a group given by $u_{j-1} \leq u \leq u_j$. We then have

$$\left.\begin{aligned}&\int\limits_{u_{j-1}}^{u_j}(\nabla\vec{\varphi}_1+\Sigma_c\varphi_0)\,du+\Big[\gamma\nabla\vec{\varphi}_1+(\xi\Sigma_s+\gamma\Sigma_c)\,\varphi_0\Big]_{u=u_{j-1}}^{u=u_j}=\eta^j Q(\vec{r}),\\&\int\limits_{u_{j-1}}^{u_j}\Big(\frac{1}{3}\nabla\varphi_0+\Sigma_{tr}\vec{\varphi}_1\Big)\,du+\bar{\mu}_0\Big[\frac{1}{3}\nabla\varphi_0+\Sigma\vec{\varphi}_1\Big]_{u=u_{j-1}}^{u=u_j}=0,\end{aligned}\right\}\qquad(75.2)$$

where

$$\eta^j=\gamma\,(\chi^j-\chi^{j-1})+\int\limits_{u_{j-1}}^{u_j}\chi\,du.$$

We calculate the integrals on the left side using the simplest interpolation formulas. Since, in general, φ_0 and $\vec{\varphi}_1$ do not change monotonically within a given lethargy interval, it is reasonable first to go to new functions q_0 and $\vec{q}_1$, which are monotonic and slowly varying, according to

$$\varphi_0=\frac{q_0}{\xi\Sigma_s+\gamma\Sigma_c},\quad \vec{\varphi}_1=\frac{\vec{q}_1}{\Sigma_{tr}}.\Big\}\qquad(75.3)$$

For q_0 and $\vec{q}_1$ we will use a linear interpolation procedure. As a result we obtain

$$\left.\begin{aligned}\varphi_0&=v_0^j(u)\,\varphi_0^j+w_0^j(u)\,\varphi_0^{j-1},\\ \vec{\varphi}_1&=v_1^j(u)\,\vec{\varphi}_1^j+w_1^j(u)\,\vec{\varphi}_1^{j-1},\end{aligned}\right\}\qquad(75.4)$$

where

$$\left.\begin{aligned}&v_0^j(u)=\frac{\xi\Sigma_s^j+\gamma\Sigma_c^j}{\xi\Sigma_s+\gamma\Sigma_c}\,\frac{u-u_{j-1}}{u_j-u_{j-1}},\quad w_0^j(u)=\frac{\xi\Sigma_s^{j-1}+\gamma\Sigma_c^{j-1}}{\xi\Sigma_s+\gamma\Sigma_c}\,\frac{u_j-u}{u_j-u_{j-1}},\\&v_1^j(u)=\frac{\Sigma_{tr}^j}{\Sigma_{tr}}\,\frac{u-u_{j-1}}{u_j-u_{j-1}},\quad w_1^j(u)=\frac{\Sigma_{tr}^{j-1}}{\Sigma_{tr}}\,\frac{u_j-u}{u_j-u_{j-1}}.\end{aligned}\right\}\qquad(75.5)$$

Using (75.4), Equations (75.2) become

$$\left.\begin{aligned}&A_0^j\nabla\vec{\varphi}_1^j+B_0^j\nabla\vec{\varphi}_1^{j-1}+C_0^j\varphi_0^j-D_0^j\varphi_0^{j-1}=\eta^jQ(\vec{r}),\\&A_1^j\frac{1}{3}\nabla\varphi_0^j+B_1^j\frac{1}{3}\nabla\varphi_0^{j-1}+C_1^j\vec{\varphi}_1^j-D_1^j\vec{\varphi}_1^{j-1}=0,\end{aligned}\right\}\qquad(75.6)$$

where

$$\left.\begin{aligned}&A_0^j=\int\limits_{u_{j-1}}^{u_j}v_1^j(u)\,du+\gamma,\quad B_0^j=\int\limits_{u_{j-1}}^{u_j}w_1^j(u)\,du-\gamma,\\&C_0^j=\int\limits_{u_{j-1}}^{u_j}\Sigma_cv_0^j(u)\,du+(\xi\Sigma_s^j+\gamma\Sigma_c^j),\\&D_0^j=-\int\limits_{u_{j-1}}^{u_j}\Sigma_cw_0^j(u)\,du+(\xi\Sigma_s^{j-1}+\gamma\Sigma_c^{j-1}),\\&A_1^j=\int\limits_{u_{j-1}}^{u_j}v_0^j(u)\,du+\bar{\mu}_0,\qquad B_1^j=\int\limits_{u_{j-1}}^{u_j}w_0^j(u)\,du-\bar{\mu}_0,\\&C_1^j=\int\limits_{u_{j-1}}^{u_j}\Sigma_{tr}v_1^j(u)\,du+\bar{\mu}_0\Sigma^j,\quad D_1^j=-\int\limits_{u_{j-1}}^{u_j}\Sigma_{tr}w_1^j(u)+\bar{\mu}_0\Sigma^{j-1}.\end{aligned}\right\}\qquad(75.7)$$

We now solve (75.6) for the unknown functions at lethargy $u = u_j$, obtaining

$$\left.\begin{aligned} \nabla\vec{\varphi}_1^j + \Sigma_0^j \varphi_0^j = f_0^j, \\ \frac{1}{3}\nabla\varphi_0^j + \Sigma_1^j \vec{\varphi}_1^j = \vec{f}_1^j, \end{aligned}\right\} \quad (75.8)$$

where

$$\left.\begin{aligned} \Sigma_0^j = \frac{C_0^j}{A_0^j}, \quad \Sigma_1^j = \frac{C_1^j}{A_1^j}, \\ f_0^j = \frac{D_0^j}{A_0^j}\varphi_0^{j-1} - \frac{B_0^j}{A_0^j}\nabla\vec{\varphi}_1^{j-1} + \frac{\eta^j}{A_0^j}Q(\vec{r}), \\ \vec{f}_1^j = \frac{D_1^j}{A_1^j}\vec{\varphi}_1^{j-1} - \frac{B_1^j}{A_1^j}\frac{1}{3}\nabla\varphi_0^{j-1}. \end{aligned}\right\} \quad (75.9)$$

The functions f_0^j and $\vec{f}_1^j$, according to (75.8), satisfy the computationally convenient recursion relation

$$\left.\begin{aligned} f_0^j = -K_0^j f_0^{j-1} + L_0^j \varphi_0^{j-1} + N^j Q, \\ \vec{f}_1^j = -K_1^j \vec{f}_1^{j-1} + L_1^j \vec{\varphi}_1^{j-1}, \end{aligned}\right\} \quad (75.10)$$

where

$$\left.\begin{aligned} K_\nu^j = \frac{B_\nu^j}{A_\nu^j}, \\ L_\nu^j = K_\nu^j \frac{C_\nu^{j-1}}{A_\nu^{j-1}} + \frac{D_\nu^j}{A_\nu^j}, \quad (\nu = 0;\ 1) \\ N^j = \frac{\eta^j}{A_0^j}. \end{aligned}\right\} \quad (75.11)$$

Thus the multigroup set of slowing-down equations is completely specified. Let us now go on to a consideration of the equations for the thermal neutron group

$$\left.\begin{aligned} \nabla\vec{\Phi}_1 + \Sigma_{cT}\Phi_0 = \gamma_T \nabla\vec{\varphi}_1^m + (\xi_T \Sigma_s^m + \gamma_T \Sigma_c^m)\varphi_0^m, \\ \frac{1}{3}\nabla\Phi_0 + \Sigma_{trT}\vec{\Phi}_1 = \bar{\mu}_{0T}\left(\frac{1}{3}\nabla\varphi_0^m + \Sigma^m \vec{\varphi}_1^m\right). \end{aligned}\right\} \quad (75.12)$$

From these equations we eliminate $\nabla\vec{\varphi}_1^m$ and $\nabla\varphi_0^m$ using (75.8), according to which

$$\left.\begin{aligned} \nabla\vec{\varphi}_1^m = f_0^m - \Sigma_0^m \varphi_0^m, \\ \frac{1}{3}\nabla\varphi_0^m = \vec{f}_1^m - \Sigma_1^m \vec{\varphi}_1^m. \end{aligned}\right\} \quad (75.13)$$

We then obtain

$$\left.\begin{aligned} \nabla\vec{\Phi}_1 + \Sigma_{cT}\Phi_0 = f_{0T}, \\ \frac{1}{3}\nabla\Phi_0 + \Sigma_{trT}\vec{\Phi}_1 = \vec{f}_{1T}, \end{aligned}\right\} \quad (75.14)$$

wnere

$$\left.\begin{aligned} f_{0T} &= \gamma_T f_0^m + (\Sigma_{0T} - \gamma_T \Sigma_0^m)\varphi_0^m, \\ \vec{f}_{1T} &= \bar{\mu}_{0T} \vec{f}_1^m + (\Sigma_{1T} - \bar{\mu}_{0T}\Sigma_1^m)\vec{\varphi}_0^m \end{aligned}\right\} \tag{75.15}$$

and

$$\left.\begin{aligned} \Sigma_{0T} &= \xi_T \Sigma_s^m + \gamma_T \Sigma_c^m, \\ \Sigma_{1T} &= \bar{\mu}_{0T}(\Sigma_s^m + \Sigma_c^m). \end{aligned}\right\} \tag{75.16}$$

To conclude, we must consider the expression for the number of secondary neutrons $Q(\vec{r})$, namely

$$Q(\vec{r}) = \nu_f \left(\int_{-\infty}^{u_T} \Sigma_f \varphi_0 \, du + \Sigma_{fT} \Phi_0 \right). \tag{75.17}$$

We represent the integral on the right side as a sum of integrals over the individual groups and then use the interpolation formulas, obtaining

$$Q(\vec{r}) = \nu_f \left(\sum_{j=1}^{m} a^j \varphi_0^j + \Sigma_{fT} \Phi_0 \right), \tag{75.18}$$

where

$$\left.\begin{aligned} a^j &= \begin{cases} s^j + t^{j+1} & (j = 1, 2, \ldots, m-1), \\ s^m & (j = m), \end{cases} \\ s^j &= \int_{u_{j-1}}^{u_j} \Sigma_f v_0^j(u)\, du, \\ t^j &= \int_{u_{j-1}}^{u_j} \Sigma_f w_0^j(u)\, du. \end{aligned}\right\} \tag{75.19}$$

Thus, the multigroup set of reactor equations becomes

$$\left.\begin{aligned} &\nabla\vec{\varphi}_1^j + \Sigma_0^j \varphi_0^j = f_0^j, \\ &\frac{1}{3}\nabla\varphi_0^j + \Sigma_1^j \vec{\varphi}_1^j = \vec{f}_1^j, \qquad (j = 1, 2, \ldots, m) \\ &\nabla\vec{\Phi}_1 + \Sigma_{cT}\Phi_0 = f_{0T} \\ &\frac{1}{3}\nabla\Phi_0 + \Sigma_{trT}\vec{\Phi}_1 = \vec{f}_{1T}. \end{aligned}\right\} \tag{75.20}$$

The solution of these equations lies in the class $\{\varphi_0^j, \vec{\varphi}_1^j, \Phi_0, \vec{\Phi}_1\}$ of functions continuous over the entire reactor volume G and satisfy the following boundary conditions on the external boundary S:

$$\left.\begin{aligned} 2(\vec{\varphi}_1^j)_{\vec{n}0} - \varphi_0^j &= 0, \\ 2(\vec{\Phi}_1)_{\vec{n}0} - \Phi_0 &= 0. \end{aligned}\right\} \tag{75.21}$$

76. The Multigroup Set of Reactor Equations for Strong Absorption of Slowing-Down Neutrons

We now proceed to derive the multigroup set of reactor equations for strong absorption of the slowing-down neutrons. To do this we again turn to (75.1), which we rewrite in the form

$$\left.\begin{aligned} &\frac{\partial q_0}{\partial u}+\frac{\Sigma_c}{\xi\Sigma_s+\gamma\Sigma_c}q_0=-\nabla\vec{\varphi}_1-\gamma\frac{\partial\nabla\vec{\varphi}_1}{\partial u}+S+\gamma\frac{\partial S}{\partial u},\\ &\frac{\partial\vec{q}_1}{\partial u}+\frac{1}{\bar{\mu}_0}\frac{\Sigma_{tr}}{\Sigma_s+\Sigma_c}\vec{q}_1=-\frac{1}{\bar{\mu}_0}\cdot\frac{1}{3}\nabla\varphi_0-\frac{1}{3}\frac{\partial\nabla\varphi_0}{\partial u},\end{aligned}\right\} \tag{76.1}$$

where

$$\begin{aligned} q_0&=(\xi\Sigma_s+\gamma\Sigma_c)\varphi_0,\\ \vec{q}_1&=(\Sigma_s+\Sigma_c)\vec{\varphi}_1.\end{aligned} \tag{76.2}$$

Let us break up the whole lethargy interval into subintervals (u_{j-1}, u_j). We shall solve the linear inhomogeneous Equation (76.1) for the functions q_0 and $\vec{q}_1$ within each group. We then have

$$\left.\begin{aligned} &q_0(\vec{r}, u)=q_0^{j-1}p_0^j(u)-\\ &-\int\limits_{u_{j-1}}^{u}\left(\nabla\vec{\varphi}_1+\gamma\frac{\partial\nabla\vec{\varphi}_1}{\partial u}-S-\gamma\frac{\partial S}{\partial u}\right)\frac{p_0^j(u)}{p_0^j(u')}du',\\ &\vec{q}_1(\vec{r}, u)=\vec{q}_1^{j-1}p_1^j(u)-\\ &-\frac{1}{3}\int\limits_{u_{j-1}}^{u}\left(\frac{1}{\bar{\mu}_0}\nabla\varphi_0+\frac{\partial\nabla\varphi_0}{\partial u}\right)\frac{p_1^j(u)}{p_1^j(u')}du',\end{aligned}\right\} \tag{76.3}$$

where

$$\left.\begin{aligned} &p_0^j(u)=e^{-\int\limits_{u_{j-1}}^{u}\frac{\Sigma_c}{\xi\Sigma_s+\gamma\Sigma_c}du}, &&p_1^j(u)=e^{-\frac{1}{\bar{\mu}_0}\int\limits_{u_{j-1}}^{u}\frac{\Sigma_{tr}}{\Sigma_s+\Sigma_c}du}\\ &q_0^{j-1}=q_0(\vec{r}, u_{j-1}), &&\vec{q}_1^{j-1}=\vec{q}_1(\vec{r}, u_{j-1}).\end{aligned}\right\} \tag{76.4}$$

Integrating by parts, we rewrite (76.3) in the form

$$\begin{aligned} &q_0(\vec{r}, u)=q_0^{j-1}p_0^j(u)-\gamma\nabla\vec{\varphi}_1+\gamma\nabla\vec{\varphi}_1^{j-1}p_0^j(u)-\\ &-\int\limits_{u_{j-1}}^{u}\nabla\vec{\varphi}_1\frac{\xi\Sigma_s}{\xi\Sigma_s+\gamma\Sigma_c}\frac{p_0^j(u)}{p_0^j(u')}du'+\\ &+\gamma S-\gamma S^{j-1}p_0^j(u)+\int\limits_{u_{j-1}}^{u}S\frac{\xi\Sigma_s}{\xi\Sigma_s+\gamma\Sigma_c}\frac{p_0^j(u)}{p^j_0(u')}du',\\ &\vec{q}_1(\vec{r}, u)=\vec{q}_1^{j-1}p_1^j(u)+\frac{1}{3}\nabla\varphi_0^{j-1}p_1^j(u)-\frac{1}{3}\nabla\varphi_0-\\ &-\frac{1}{3}\int\limits_{u_{j-1}}^{u}\nabla\varphi_0\frac{\Sigma_s}{\Sigma_s+\Sigma_c}\frac{p_1^j(u)}{p_1^j(u')}du'.\end{aligned} \tag{76.5}$$

We now express $\nabla\varphi_0$ and $\nabla\vec{\varphi}_1$ approximately by means of interpolation formulas relating these functions to their values on the boundaries of the (u_{j-1}, u_j) subintervals. Because φ_0 and $\vec{\varphi}_1$ do not vary monotonically within a given interval, we first go over to the new slowly varying monotonic functions q_0 and $\vec{q}_1$ according to (76.2). We then have

$$\varphi_0 = \frac{q_0}{\xi\Sigma_s + \gamma\Sigma_c}, \qquad \vec{\varphi}_1 = \frac{\vec{q}_1}{\Sigma_c + \Sigma_s}. \tag{76.6}$$

We may use linear interpolation for ∇q_0 and $\nabla\vec{q}_1$. As a result we have

$$\left.\begin{aligned} \nabla\varphi_0 &= v_0^j(u)\nabla\varphi_0^j + w_0^j(u)\nabla\varphi_0^{j-1}, \\ \nabla\vec{\varphi}_1 &= v_1^j(u)\nabla\vec{\varphi}_1^j + w_1^j(u)\nabla\vec{\varphi}_1^{j-1}, \end{aligned}\right\} \tag{76.7}$$

where

$$\left.\begin{aligned} v_0^j(u) &= \frac{\xi\Sigma_s^j + \gamma\Sigma_c^j}{\xi\Sigma_s + \gamma\Sigma_c}\,\frac{u - u_{j-1}}{u_j - u_{j-1}}, & w_0^j(u) &= \frac{\xi\Sigma_s^{j-1} + \gamma\Sigma_c^{j-1}}{\xi\Sigma_s + \gamma\Sigma_c}\,\frac{u_j - u}{u_j - u_{j-1}}, \\ v_1^j(u) &= \frac{\Sigma_s^j + \Sigma_c^j}{\Sigma_s + \Sigma_c}\,\frac{u - u_{j-1}}{u_j - u_{j-1}}, & w_1^j(u) &= \frac{\Sigma_s^{j-1} + \Sigma_c^{j-1}}{\Sigma_s + \Sigma_c}\,\frac{u_j - u}{u_j - u_{j-1}}. \end{aligned}\right\} \tag{76.8}$$

Inserting (76.7) into the right side of (76.5) we arrive at

$$\left.\begin{aligned} (\xi\Sigma_s + \gamma\Sigma_c)\varphi_0 &= (\xi\Sigma_s^{j-1} + \gamma\Sigma_c^{j-1})\,p_0^j(u)\,\varphi_0^{j-1} - \\ &\quad - A_0^j(u)\nabla\vec{\varphi}_1^j - B_0^j(u)\nabla\vec{\varphi}_1^{j-1} + \eta^j(u)\,Q(\vec{r}). \\ (\Sigma_s + \Sigma_c)\vec{\varphi}_1 &= (\Sigma_s^{j-1} + \Sigma_c^{j-1})\cdot p_1^j(u)\,\vec{\varphi}_1^{j-1} - \\ &\quad - A_1^j(u)\frac{1}{3}\nabla\varphi_0^j - B_1^j(u)\frac{1}{3}\nabla\varphi_0^{j-1}, \end{aligned}\right\} \tag{76.9}$$

where

$$\left.\begin{aligned} A_0^j(u) &= \gamma v_1^j(u) + \int_{u_{j-1}}^{u} v_1^j(u')\,\frac{\xi\Sigma_s}{\xi\Sigma_s + \gamma\Sigma_c}\,\frac{p_0^j(u)}{p_0^j(u')}\,du', \\ B_0^j(u) &= \gamma\,[w_1^j(u) - p_0^j(u)] + \int_{u_{j-1}}^{u} w_1^j(u')\,\frac{\xi\Sigma_s}{\xi\Sigma_s + \gamma\Sigma_c}\,\frac{p_0^j(u)}{p_0^j(u')}\,du', \\ A_1^j(u) &= v_0^j(u) + \int_{u_{j-1}}^{u} v_0^j(u')\,\frac{\Sigma_s}{\Sigma_s + \Sigma_c}\cdot\frac{p_1^j(u)}{p_1^j(u')}\,du', \\ B_1^j(u) &= w_0^j(u) - p_1^j(u) + \int_{u_{j-1}}^{u} w_0^j(u')\,\frac{\Sigma_s}{\Sigma_s + \Sigma_c}\cdot\frac{p_1^j(u)}{p_1^j(u')}\,du', \\ \eta^j(u) &= \gamma\chi - \gamma\chi^{j-1}p_0^j(u) + \int_{u_{j-1}}^{u} \chi(u')\,\frac{\xi\Sigma_s}{\xi\Sigma_s + \gamma\Sigma_c}\,\frac{p_0^j(u)}{p_0^j(u')}\,du'. \end{aligned}\right\} \tag{76.10}$$

If we set $u = u_j$ in (76.9) and solve for the desired functions at lethargy u_j, we arrive at the set of differential equations

$$\left.\begin{aligned} \nabla\vec{\varphi}_1^j + \Sigma_0^j\varphi_0^j &= f_0^j, \\ \frac{1}{3}\nabla\varphi_0^j + \Sigma_1^j\vec{\varphi}_1^j &= \vec{f}_1^j, \end{aligned}\right\} \tag{76.11}$$

where

$$\left.\begin{aligned}\Sigma_0^j &= \frac{\xi\Sigma_s^j+\gamma\Sigma_c^j}{A_0^j}, \qquad \Sigma_1^j = \frac{\Sigma_s^j+\Sigma_c^j}{A_1^j},\\ f_0^j &= \frac{\xi\Sigma_s^{j-1}+\gamma\Sigma_c^{j-1}}{A_0^j}\, p_0^j\varphi_0^{j-1} - \frac{B_0^j}{A_0^j}\nabla\vec{\varphi}_1^{j-1} + \frac{\eta^j}{A_0^j}Q,\\ \vec{f}_1^j &= \frac{\Sigma_s^{j-1}+\Sigma_c^{j-1}}{A_1^j}\, p_1^j\vec{\varphi}_1^{j-1} - \frac{B_1^j}{A_1^j}\frac{1}{3}\nabla\varphi_0^{j-1}.\end{aligned}\right\} \tag{76.12}$$

Here we have used the notation

$$\psi^j = \psi^j(u_j). \tag{76.13}$$

We may use (76.11) to obtain the following simple recursion formulas for f_0^j and $\vec{f}_1^j$:

$$\left.\begin{aligned} f_0^j &= -K_0^j f_0^{j-1} + L_0^j\varphi_0^{j-1} + N^jQ,\\ \vec{f}_1^j &= -K_1^j\vec{f}_1^{j-1} + L_1^j\vec{\varphi}_1^{j-1},\end{aligned}\right\} \tag{76.14}$$

where

$$\left.\begin{aligned} K_\nu^j &= \frac{B_\nu^j}{A_\nu^j}, \qquad L_\nu^j = M_\nu^j + K_\nu^j\Sigma_\nu^{j-1}, \qquad (\nu = 0;\ 1)\\ M_0^j &= \frac{\xi\Sigma_s^{j-1}+\gamma\Sigma_c^{j-1}}{A_0^j}\, p_0^j, \qquad M_1^j = \frac{\Sigma_s^{j-1}+\Sigma_c^{j-1}}{A_1^j}\cdot p_1^j,\\ N^j &= \frac{\eta^j}{A_0^j}.\end{aligned}\right\} \tag{76.15}$$

Let us consider further the equations for the thermal group. As in the previous section, this set of equations can be written

$$\left.\begin{aligned} \nabla\vec{\Phi}_1 + \Sigma_{cT}\Phi_0 &= f_{0T},\\ \frac{1}{3}\nabla\Phi_0 + \Sigma_{trT}\vec{\Phi}_1 &= \vec{f}_{1T},\end{aligned}\right\} \tag{76.16}$$

where

$$\left.\begin{aligned} f_{0T} &= \gamma_T f_0^m + (\Sigma_{0T} - \gamma_T\Sigma_0^m)\varphi_0^m,\\ \vec{f}_{1T} &= \bar{\mu}_{0T}\vec{f}_1^m + (\Sigma_{1T} - \bar{\mu}_{0T}\Sigma_1^m)\vec{\varphi}_1^m.\end{aligned}\right\} \tag{76.17}$$

We now go on to a consideration of the interpolation formulas for the solutions φ_0 and $\vec{\varphi}_1$ within each (u_{j-1}, u_j) group. For this purpose we bring (76.9) into the form

$$\left.\begin{aligned} \varphi_0 &= a_0^j(u)\varphi_0^j + b_0^j(u)\varphi_0^{j-1} - c_0^j(u)\nabla\vec{\varphi}_1^j + d^j(u)Q,\\ \vec{\varphi}_1 &= a_1^j(u)\vec{\varphi}_1^j + b_1^j(u)\vec{\varphi}_1^{j-1} - c_1^j(u)\frac{1}{3}\nabla\varphi_0^j,\end{aligned}\right\} \tag{76.18}$$

where

$$
\begin{aligned}
a_0^j(u) &= \frac{\xi\Sigma_s^j + \gamma\Sigma_c^j}{\xi\Sigma_s + \gamma\Sigma_c} \cdot \frac{B_0^j(u)}{B_0^j}, \\
b_0^j(u) &= \frac{\xi\Sigma_s^{j-1} + \gamma\Sigma_c^{j-1}}{\xi\Sigma_s + \gamma\Sigma_c} \cdot \left[p_0^j(u) - p_0^j \frac{B_0^j(u)}{B_0^j} \right], \\
c_0^j(u) &= \frac{1}{\xi\Sigma_s + \gamma\Sigma_c} \cdot \left[A_0^j(u) - A_0^j \frac{B_0^j(u)}{B_0^j} \right], \\
d^j(u) &= \frac{1}{\xi\Sigma_s + \gamma\Sigma_c} \cdot \left[\eta^j(u) - \eta^j \frac{B_0^j(u)}{B_0^j} \right], \\
a_1^j(u) &= \frac{\Sigma_s^j + \Sigma_c^j}{\Sigma_s + \Sigma_c} \cdot \frac{B_1^j(u)}{B_1^j}, \\
b_1^j(u) &= \frac{\Sigma_s^{j-1} + \Sigma_c^{j-1}}{\Sigma_s + \Sigma_c} \cdot \left[p_1^j(u) - p_1^j \frac{B_1^j(u)}{B_1^j} \right], \\
c_1^j(u) &= \frac{1}{\Sigma_s + \Sigma_c} \cdot \left[A_1^j(u) - A_1^j \frac{B_1^j(u)}{B_1^j} \right].
\end{aligned}
\tag{76.19}
$$

In deriving (76.18), we made use also of (76.9) in which we first set $u = u_j$. In conclusion, let us calculate the number of secondary neutrons

$$
Q = \nu_f \left(\int_{-\infty}^{u_T} \Sigma_f \varphi \, du + \Sigma_{fT} \Phi \right). \tag{76.20}
$$

As before, let us write the integral on the right as a sum of integrals over the (u_{j-1}, u_j) subintervals and make use of (76.18) in each of these. We then arrive at

$$
Q = \frac{\nu_f}{1 - \varepsilon\nu_f} \left[\sum_{j=1}^{m} (\alpha^j \varphi_0^j - \beta^j \nabla \vec{\varphi}_1^j) + \Sigma_{fT} \Phi_0 \right], \tag{76.21}
$$

where

$$
\left.
\begin{aligned}
&\alpha^j = \begin{cases} s^j + t^{j+1} & \text{for} \quad j = 1, 2, \ldots, m-1 \\ s^m & \text{for} \quad j = m \end{cases} \\
&s^j = \int_{u_{j-1}}^{u_j} \Sigma_f a_0^j(u)\, du, \qquad t^j = \int_{u_{j-1}}^{u_j} \Sigma_f b_0^j(u)\, du, \\
&\beta^j = \int_{u_{j-1}}^{u_j} \Sigma_f c_0^j(u)\, du, \qquad \varepsilon = \sum_{j=1}^{m} \int_{u_{j-1}}^{u_j} \Sigma_f d^j(u)\, du.
\end{aligned}
\right\} \tag{76.22}
$$

We now use (76.11) to eliminate $\nabla \vec{\varphi}_1^j$ from (76.21), which gives the computationally convenient expression

$$
Q = \frac{\nu_f}{1 - \varepsilon\nu_f} \left\{ \sum_{j=1}^{m} [(\alpha^j + \beta^j \Sigma_0^j) \varphi_0^j - \beta^j f_0^j] + \Sigma_{fT} \Phi_0 \right\}.
$$

As a result, the multigroup set of reactor equations can be written in the form

$$\left.\begin{aligned}&\nabla\vec{\varphi}_1^j+\Sigma_0^j\varphi_0^j=f_0^j, \qquad (j=1,\ 2,\ \ldots,\ m)\\&\frac{1}{3}\nabla\varphi_0^j+\Sigma_1^j\vec{\varphi}_1^j=\vec{f}_1^j,\\&\nabla\vec{\Phi}_1+\Sigma_{cT}\Phi_0=f_{0T},\\&\frac{1}{3}\nabla\Phi_0+\Sigma_{trT}\vec{\Phi}_1=\vec{f}_{1T}.\end{aligned}\right\} \tag{76.23}$$

The solution is found in the class of functions continuous over the entire reactor volume G and satisfying the boundary conditions

$$\left.\begin{aligned}&2(\vec{\varphi}_1^j)_{\vec{n}_0}-\varphi_0^j=0\\&2(\vec{\Phi}_1^j)_{\vec{n}_0}-\Phi_0^j=0.\end{aligned}\right\} \tag{76.24}$$

In conclusion, it is convenient to direct our attention to a possible simplification in the numerical procedure when the hydrogen concentration in the moderator is low. This simplification involves the use of the so-called transport approximation which consists, essentially, of replacing the second of Equations (75.8) by*

$$\frac{1}{3}\nabla\varphi_0^j+\Sigma_{tr}^j\vec{\varphi}_1^j=0,$$

where Σ_{tr}^j is an averaged transport cross section.

The quantity Σ_{tr}^j can be averaged over the group in several different ways. Seemingly the most reasonable is

$$\frac{1}{3\Sigma_{tr}^j}=\frac{\displaystyle\int\limits_{u_{j-1}}^{u_j}\frac{1}{3\xi\Sigma_s\Sigma_{tr}}\cdot\frac{p_0^j(u)}{p_0^j(u')}du'}{\displaystyle\int\limits_{u_{j-1}}^{u_j}\frac{1}{\xi\Sigma_s}\cdot\frac{p_0^j(u)}{p_0^j(u')}du'}.$$

We note finally that if a hydrogen-containing zone of the reactor has a common boundary with a moderator containing no hydrogen, the calculations can be performed as before using Equations (75.20) and (76.23) and the transport approximation for those zones which do not contain hydrogen.

77. Adjoint Equations

In addition to (76.23), let us derive the multigroup set of adjoint reactor equations. To do this we introduce the vector functions $\boldsymbol{f}$ and $\boldsymbol{f}^*$

$$\boldsymbol{f}=\left\|\begin{matrix}\varphi_0^j\\\vec{\varphi}_1^j\\\Phi_0\\\vec{\Phi}_1\end{matrix}\right\| \quad\text{and}\quad \boldsymbol{f}^*=\left\|\begin{matrix}\varphi_0^{*j}\\\vec{\varphi}_1^{*j}\\\Phi_0^*\\\vec{\Phi}_1^*\end{matrix}\right\|, \tag{77.1}$$

where $\boldsymbol{f}$ is a solution of (76.23), and $\boldsymbol{f}^*$ is a solution of the adjoint set of equations, which we have yet to construct.

We define the scalar product of the vector functions by

---

*Such a replacement is entirely permissible because the order of magnitude of $\vec{f}_1^j$ is the same as that of $\bar{\mu}_0$.

$$(f,\ f^*) = \int\limits_G d\vec{r}\left[\sum_{j=1}^{m}(\varphi_0^j\varphi_0^{*j} + \vec{\varphi}_1^j\vec{\varphi}_1^{*j}) + \Phi_0\Phi_0^* + \vec{\Phi}_1\vec{\Phi}_1^*\right]. \tag{77.2}$$

Consider the operator **L**. We will assume that when **L** acts on f we arrive at a new vector whose components are those given by Equation (76.23). If the functions f_0, $\vec{f}_1$ and f_{0T}, $\vec{f}_{1T}$, as well as Q, are replaced by their explicit expression (76.12), (76.17) and (76.21) and all of the terms are taken over to the left side of the equation, (76.23) can be written in the form

$$\mathbf{L}f = \left|\begin{array}{c} \nabla\vec{\varphi}_1^j + \Sigma_0^j\varphi_0^j - M_0^j\varphi_0^{j-1} + K_0^j\nabla\varphi_1^{j-1} - \\ -\dfrac{\nu_f N^j}{1-\varepsilon\nu_f}\cdot\left[\sum_{l=1}^{m}(\alpha^l\varphi_0^l - \beta^l\nabla\vec{\varphi}_1^l) + \Sigma_{fT}\Phi_0\right] \\ \dfrac{1}{3}\nabla\varphi_0^j + \Sigma_1^j\vec{\varphi}_1^j - M_1^j\vec{\varphi}_1^{j-1} + K_1^j\dfrac{1}{3}\nabla\varphi_0^{j-1} \\ \nabla\vec{\Phi}_1 + \Sigma_{cT}\Phi_0 - \gamma_T\nabla\vec{\varphi}_1^m - \Sigma_{0T}\varphi_0^m \\ \dfrac{1}{3}\nabla\Phi_0 + \Sigma_{trT}\vec{\Phi}_1 - \bar{\mu}_{0T}\dfrac{1}{3}\nabla\varphi_0^m - \Sigma_{1T}\vec{\varphi}_1^m \end{array}\right| = 0 \tag{77.3}$$

To these we must add (76.24) and boundary conditions.

We now consider the basic functional relation

$$(f^*,\ \mathbf{L}f) = (f,\ \mathbf{L}^*f^*). \tag{77.4}$$

Using the definition (77.2) of the scalar product, we can transform the functional on the left to the form of that on the right. We then arrive at the adjoint set of reactor equations

$$\mathbf{L}^*f^* = \left|\begin{array}{l} \dfrac{1}{3}\nabla\vec{\varphi}_1^{*j} + \Sigma_0^j\varphi_0^{*j} - M_0^{j+1}\varphi_0^{*j+1} + K_1^{j+1}\dfrac{1}{3}\nabla\vec{\varphi}_1^{*j+1} - \\ \qquad - \delta_{jm}(\Sigma_{0T}\Phi_0^* - \bar{\mu}_{0T}\dfrac{1}{3}\nabla\vec{\Phi}_1^*) - \alpha^j Q_0^* \\ -\nabla\varphi_0^{*j} + \Sigma_1^j\vec{\varphi}_1^{*j} - M_1^{j+1}\vec{\varphi}_1^{*j+1} + K_0^{j+1}\nabla\varphi_0^{*j+1} - \\ -\delta_{jm}(\Sigma_{1T}\vec{\Phi}_1^* - \gamma_T\nabla\Phi_0^*) - \beta^j\vec{Q}_1^* \\ -\dfrac{1}{3}\nabla\vec{\Phi}_1^* + \Sigma_{cT}\Phi_0^* - \Sigma_{fT}Q_0^*, \\ -\nabla\Phi_0^* + \Sigma_{trT}\vec{\Phi}_1^*, \end{array}\right| = 0 \tag{77.5}$$

where

$$Q_0^* = \frac{\nu_f}{1-\varepsilon\nu_f}\sum_{l=1}^{m}N^l\varphi_0^{*l}, \qquad \vec{Q}_1^* = \frac{\nu_f}{1-\varepsilon\nu_f}\sum_{l=1}^{m}N^l\nabla\varphi_0^{*l},$$

and

$$\delta_{jm} = \begin{cases} 1 & (j = m), \\ 0 & (j \neq m). \end{cases}$$

In constructing the vector $\mathbf{L}^*f^*$, we assume that f^* is a vector function continuous in the entire reactor volume G and that on the external boundary of the reactor

$$\left.\begin{array}{l} 2(\vec{\varphi}_1^{*j})_{\vec{n}_0} + \varphi_0^{*j} = 0, \\ 2(\vec{\Phi}_1^*)_{\vec{n}_0} + \Phi_0^* = 0. \end{array}\right\} \tag{77.6}$$

The adjoint reactor equations can be written in the form

$$\left.\begin{aligned}
&-\frac{1}{3}\nabla\vec{\varphi}_1^{*j}+\Sigma_0^j\varphi_0^{*j}=f_0^{*j},\\
&-\nabla\varphi_0^{*j}+\Sigma_1^j\vec{\varphi}_1^{*j}=\vec{f}_1^{*j},\\
&-\frac{1}{3}\nabla\vec{\Phi}_1^*+\Sigma_{cT}\Phi_0^*=f_{0T}^*,\\
&-\nabla\Phi_0^*+\Sigma_{trT}\vec{\Phi}_1^*=0,
\end{aligned}\right\} \tag{77.7}$$

where

$$\left.\begin{aligned}
&f_0^{*j}=K_0^{*j}f_0^{*j+1}+L_0^{*j}\varphi_0^{*j+1}+\\
&\qquad+\delta_{jm}\left[(\Sigma_{0T}-\bar{\mu}_{0T}\Sigma_{cT})\Phi_0^*+\bar{\mu}_{0T}f_{0T}^*\right]+\alpha^jQ_0^*,\\
&\vec{f}_1^{\,*j}=K_1^{*j}\vec{f}_1^{*j+1}+L_1^{*j}\vec{\varphi}_1^{*j+1}+\delta_{jm}(\Sigma_{1T}-\gamma_T\Sigma_{trT})\vec{\Phi}_1^*+\beta^j\vec{Q}_1^*,\\
&f_{0T}^*=\Sigma_{fT}Q_0^*,\\
&K_\nu^{*j}=K_{1-\nu}^{j+1},\quad L_\nu^{*j}=M_\nu^{j+1}-\Sigma_\nu^{j+1}K_{1-\nu}^{j+1},\qquad(\nu=0;\ 1)\\
&Q_0^*=\frac{\nu_f}{1-\epsilon\nu_f}\sum_{l=1}^{m}N^l\varphi_0^{*l},\\
&\vec{Q}_1^*=\frac{\nu_f^*}{1-\epsilon\nu_f}\sum_{l=1}^{m}N^l(\Sigma_1^l\vec{\varphi}_1^{*l}-\vec{f}_1^{*l}),
\end{aligned}\right\} \tag{77.8}$$

and the quantities K_ν^j, Σ_ν^j, L_ν^j, N^j, α^j, β^j, and $\Sigma_{\nu T}$ are defined by means of (76.12), (76.15), (76.22), and (75.16).

In conclusion it should be noted that if we set $\beta^j=0$ and $\epsilon=0$ in (77.8) and choose K_ν^j, Σ_ν^j, L_ν^j, N^j, and α^j according to (75.9), (75.11) and (75.19), we arrive at the set of adjoint reactor equations for the case of weak absorption of slowing-down neutrons.

78. Finite-Difference Equations

In the preceding sections of this chapter it was shown that the solution of the multigroup set of basic and adjoint reactor equations reduces eventually to integrating the set of differential equations

$$\left.\begin{aligned}
&\nabla\vec{\varphi}_1+\Sigma_0\varphi_0=f_0,\\
&\frac{1}{3}\nabla\varphi_0+\Sigma_1\vec{\varphi}_1=\vec{f}_1,
\end{aligned}\right\} \tag{78.1}$$

where Σ_0, Σ_1 and f_0, $\vec{f}_1$ are piecewise continuous functions of position in the reactor, together with the boundary conditions on S

$$2(\vec{\varphi}_1)_{\vec{n}0}-\varphi_0=0. \tag{78.2}$$

In the one-dimensional cases of plane, cylindrical, and spherical geometries, Equations (78.1) can be written

$$\begin{aligned}
&\frac{1}{r^\alpha}\frac{d}{dr}r^\alpha I+\Sigma_0\varphi=f_0,\\
&\frac{1}{3}\frac{d\varphi}{dr}+\Sigma_1 I=f_1,
\end{aligned} \tag{78.3}$$

where

$$\alpha=\begin{cases}0 \text{ for the plane geometry}\\ 1 \text{ for the cylindrical geometry}\\ 2 \text{ for the spherical geometry.}\end{cases}$$

We have here made use of symmetry considerations to project the second of (78.1) onto the normal to S, thus transforming from vectors to scalars according to $\varphi_1 = \mathrm{I}n^0$, and $f_1 = f_1 n^0$; in addition we have set $\varphi_0 = \varphi$ for simplicity.

The boundary conditions (78.2) can now be rewritten

$$2\mathrm{I} - \varphi = 0 \text{ on S.} \tag{78.4}$$

We shall attempt to find (78.3) with these boundary conditions by finite-difference methods.

Restricting our considerations to the most simple finite-difference procedures, we proceed to replace the differential equation (78.3) by a set of finite-difference equations. To do this we multiply the first of the equations by r^α and integrate over the limits $r_{k-1/2} \leq r \leq r_{k+1/2}$, and then integrate the second over the limits $r_k \leq r \leq r_{k+1}$. We then arrive at

$$\left.\begin{aligned} r^\alpha_{k+1}\,\mathrm{I}_{k+1/2} - r^\alpha_{k-1/2}\mathrm{I}_{k-1/2} + \int\limits_{r_{k-1/2}}^{r_{k+1/2}} (\Sigma_0\varphi - f_0)\, r^\alpha\, dr = 0, \\ \frac{1}{3}(\varphi_{k+1} - \varphi_k) + \int\limits_{r_k}^{r_{k+1}} (\Sigma_1 \mathrm{I} - f_1)\, dr = 0. \end{aligned}\right\} \tag{78.5}$$

In view of the fact that φ, I and r^α are continuous over the entire reactor volume, we can take their approximate values out from under the integral sign. We then have

$$\left.\begin{aligned} r^\alpha_{k+1/2}\mathrm{I}_{k+1/2} - r^\alpha_{k-1/2}\mathrm{I}_{k-1/2} + (\Sigma_0\Delta r)_k\varphi_k r^\alpha_k = (f_0\Delta r)_k r^\alpha_k, \\ \frac{1}{3}(\varphi_{k+1} - \varphi_k) + \mathrm{I}_{k+1/2}(\Sigma_1\Delta r)_{k+1/2} = (f_1\Delta r)_{k+1/2}, \end{aligned}\right\} \tag{78.6}$$

where we have written

$$(\Phi)_k = \frac{1}{2}\left[\Phi(r_k + 0) + \Phi(r_k - 0)\right].$$

We can now write this result in the form

$$\left.\begin{aligned} \frac{r^\alpha_{k+1/2}}{r^\alpha_k}\cdot \mathrm{I}_{k+1/2} - \frac{r^\alpha_{k-1/2}}{r^\alpha_k}\mathrm{I}_{k-1/2} + (\Sigma_0\Delta r)_k\varphi_k = (f_0\Delta r)_k, \\ \mathrm{I}_{k+1/2} = -\frac{\varphi_{k+1} - \varphi_k}{(3\Sigma_1\Delta r)_{k+1/2}} + \frac{(f_1\Delta r)_{k+1/2}}{(\Sigma_1\Delta r)_{k+1/2}}. \end{aligned}\right\} \tag{78.7}$$

Eliminating I from these equations, we arrive at the finite-difference equation

$$a_k\varphi_{k+1} - b_k\varphi_k + c_k\varphi_{k-1} = -F_k, \tag{78.8}$$

where

$$\left.\begin{aligned} a_k = \frac{\left(\frac{r_{k+1/2}}{r_k}\right)^\alpha}{(3\Sigma_1\Delta r)_{k+1/2}}, \qquad c_k = \frac{\left(\frac{r_{k-1/2}}{r_k}\right)^\alpha}{(3\Sigma_1\Delta r)_{k-1/2}}, \\ b_k = a_k + c_k + (\Sigma_0\Delta r)_k, \\ F_k = (f_0\Delta r)_k - 3a_k(f_1\Delta r)_{k+1/2} + 3c_k(f_1\Delta r)_{k-1/2}. \end{aligned}\right\} \tag{78.9}$$

The function φ is symmetric about the center of the reactor, so that

$$d\varphi/dr = 0, \quad (r = 0). \tag{78.10}$$

Now, if the first interpolation point ($k = 1$) is located a half of an interpolation interval from $r = 0$, we may write this equation in the finite-difference form

$$\varphi_0 = \varphi_1, \tag{78.11}$$

where

$$\varphi_0 = \varphi\left(-\frac{\Delta r}{2}\right), \quad \text{and} \quad \varphi_1 = \varphi\left(\frac{\Delta r}{2}\right).$$

If we assume further that the external reactor boundary is in the center of the last interpolation interval, we may use (78.7) to rewrite (78.4) in the form

$$\varphi_n - \Gamma\varphi_{n-1} = g, \tag{78.12}$$

where

$$\Gamma = \frac{4 - (3\Sigma_1 \Delta r)_{n-1/2}}{4 + (3\Sigma_1 \Delta r)_{n-1/2}}, \qquad g = \frac{12\,(f_1 \Delta r)_{n-1/2}}{4 + (3\Sigma_1 \Delta r)_{n-1/2}},$$

and the fictitious point r_n lies a half of an interpolation interval outside S.

Equations (78.8), (78.11) and (78.12) form a complete set of finite-difference equations. The solution of this set is very easily found by difference factorization, as described in Chapter IX.

To derive the two-dimensional finite-difference reactor equations is also quite simple, and it can be done as described in Chapter VIII.

APPENDICES

A. NEUTRON AGE

In nuclear reactor calculations, significant importance is assigned to the mean square distance from the source for neutrons slowing down to thermal energies, namely

$$\overline{r^2} = \frac{\int r^2 \varphi(r, u_T)\, d\vec{r}}{\int \varphi(r, u_T)\, d\vec{r}},$$

where $\varphi(r, u_T)$ is the epithermal neutron spectrum from the source in an infinite medium.

Rather than $\overline{r^2}$ itself, the literature usually considers the quantity τ, called the neutron age, given [13, 54] by

$$\tau = \begin{cases} \frac{1}{2}\overline{r^2} & \text{for the plane geometry} \\ \frac{1}{4}\overline{r^2} & \text{for the cylindrical geometry} \\ \frac{1}{6}\overline{r^2} & \text{for the spherical geometry} \end{cases}$$

In a medium of finite volume, the neutron age τ, as well as $\overline{r^2}$, can be related to the total leakage of epithermal neutrons from the given volume. This means that reactor calculations must be preceded by a calculation of τ, which is a most important characteristic of the physical properties of the moderator and is related to the critical mass of the reactor.

The neutron age τ can be calculated by the following formulas.

a) Fermi's diffusion-age theory, which takes into account the fission neutron spectrum, gives

$$\tau = \int_{-\infty}^{u_T} \frac{du}{3\xi\Sigma_s\Sigma_{tr}} \int_{-\infty}^{u} \chi(u')\, du'; \tag{A.1}$$

b) The diffusion-age approximation with the first collision taken into account (see Appendix B) gives

$$\tau = \frac{1}{3\Sigma_s^2(0)} + \int_0^{u_T} \frac{du}{3\xi\Sigma_s\Sigma_{tr}}; \tag{A.2}$$

c) Diffusion-age approximation with the second energy moment taken into account gives

$$\tau = \varepsilon\left(\frac{1}{3\Sigma_s\Sigma_{tr}}\bigg|_{u_T} + \int_0^{u_T} \chi\frac{du}{3\Sigma_s\Sigma_{tr}}\right) + \int_{-\infty}^{u_T} \frac{du}{3\xi\Sigma_s\Sigma_{tr}} \int_{-\infty}^{u} \chi(u')\, du'; \tag{A.3}$$

d) For hydrogenous moderators, the expression for τ can be obtained by using (71.12) as described by Marchak [120]. After some transformations one can obtain the formula*

*This formula was obtained at the author's request by V.P. Kochergin.

$$\tau(u) = \frac{\Phi_{02}(u)}{\Phi_{00}(u)}, \tag{A.4}$$

where

$$\Phi_{00}(u) = \frac{1}{\xi\Sigma_s + \gamma\Sigma_c}\left[\int\limits_{-\infty}^{u} \chi(u')\frac{\xi\Sigma_s}{\xi\Sigma_s+\gamma\Sigma_c} e^{-\int\limits_{u'}^{u}\frac{\Sigma_c}{\xi\Sigma_s+\gamma\Sigma_c}du''} du' + \gamma\chi(u)\right],$$

$$\Phi_{02}(u) = \frac{1}{\xi\Sigma_s + \gamma\Sigma_c}\left[\int\limits_{-\infty}^{u} \Phi_{11}(u')\frac{\xi\Sigma_s}{\xi\Sigma_s+\gamma\Sigma_c} e^{-\int\limits_{u'}^{u}\frac{\Sigma_c}{\xi\Sigma_s+\gamma\Sigma_c}du''} du' + \gamma\Phi_{11}(u)\right],$$

$$\Phi_{11}(u) = \frac{1}{3(\Sigma_s + \Sigma_c)}\left[\int\limits_{-\infty}^{u} \frac{\Sigma_s}{\Sigma_s+\Sigma_c}\Phi_{00}(u') e^{-\int\limits_{u'}^{u}\frac{1}{\bar{\mu}_0}\frac{\Sigma_r}{\Sigma_s+\Sigma_c}du''} au' + \Phi_{00}(u)\right].$$

When water is used as the moderator, Equation (A.4) goes over asymptotically into Orlov's [62] formula with $\Sigma_C = 0$.

Comparison of (A.1-A.4) shows that the different methods of calculation lead to different values of τ. It is therefore reasonable to compare reactor calculations by different methods only when all of the results are reduced to the same value of τ. This can always be done by changing the computed physical constants.

We note another important situation. In view of the fact that the experimental value of τ is known for most substances used as basic moderators, one must take care, in using one or another calculational procedure, to make sure that it gives a value of τ close to the experimental one. If a large deviation occurs, it must be liquidated by the proper choice of constants. When this test has been performed, the reactor calculation may proceed.

B. REACTOR EQUATIONS FOR A MONOENERGETIC FISSION SOURCE

In actually performing the calculations it is sometimes convenient to describe the chain reaction in the reactor by a model which assumes that all the fission neutrons are produced at the same energy. With this assumption the diffusion-age approximation for the set of basic and adjoint reactor equations can be formulated as follows. We will start with Equations (8.19), the neutron slowing-down equations in the diffusion approximation, namely*

$$\left.\begin{aligned} \nabla\vec{\varphi}_1+\Sigma\varphi_0 &= \int\limits_{u-r}^{u} du'\,\Sigma_s\varphi_0 f_0(u-u')+S(\vec{r},u), \\ \frac{1}{3}\nabla\varphi_0+\Sigma\vec{\varphi}_1 &= \int\limits_{u-r}^{u} du'\,\Sigma_s\vec{\varphi}_1 f_1(u-u'), \end{aligned}\right\} \tag{B.1}$$

where

$$S(\vec{r},u)=\nu_f\chi(u)\left(\int\limits_{-\infty}^{u_T}\Sigma_f\varphi_0 du+\Sigma_{fT}\Phi_0\right).$$

The source function $S(\vec{r}, u)$ is by assumption monochromatic and corresponds to lethargy $u = 0$, so that

$$S(\vec{r},u)=Q(\vec{r})\cdot\delta(u), \tag{B.2}$$

where

$$Q(\vec{r})=\nu_f\left(\int\limits_{-\infty}^{u_T}\Sigma_f\varphi_0 du+\Sigma_{fT}\Phi_0\right).$$

We write the solution of (B.1) in the form

$$\left.\begin{aligned} \varphi_0(\vec{r},u) &= \psi_0(\vec{r})\,\delta(u)+\varphi_0^0(\vec{r},u), \\ \vec{\varphi}_1(\vec{r},u) &= \vec{\psi}_1(\vec{r})\,\delta(u)+\vec{\varphi}_1^0(\vec{r},u). \end{aligned}\right\} \tag{B.3}$$

Inserting this into (B.1) we arrive at

$$\left.\begin{aligned} \nabla\vec{\psi}_1+\Sigma_0\psi_0 &= Q(\vec{r}), \\ \frac{1}{3}\nabla\psi_0+\Sigma\vec{\psi}_1 &= 0 \end{aligned}\right\} \tag{B.4}$$

and

$$\left.\begin{aligned} \nabla\vec{\varphi}_1^0+\Sigma\varphi_0^0 &= \int\limits_{u-r}^{u} du\Sigma_s(u')\,f_0(u-u')\,\varphi_0^0(\vec{r},u')+\Sigma_{s0}f_0(u)\,\psi_0, \\ \frac{1}{3}\nabla\varphi_0^0+\Sigma\vec{\varphi}_1^0 &= \int\limits_{u-r}^{u} du'\,\Sigma_s(u')\,f_1(u-u')\,\vec{\varphi}_1^0(\vec{r},u')+\Sigma_{s0}f_1(u)\,\vec{\psi}_1. \end{aligned}\right\} \tag{B.5}$$

*In Equations (B.1) we eliminate the parameter K_{eff} in accordance with the footnote on page 73.

Here and in what follows the index zero denotes quantities evaluated at u = 0.

Equations (B.4) imply that

$$\nabla D_0 \nabla \psi_0 - \Sigma_0 \psi_0 = -Q(\vec{r}), \tag{B.6}$$

where

$$D_0 = \frac{1}{3\Sigma_{s0}}.$$

We assume in (B.5) that $\Sigma_s \varphi_0^0$ and $\Sigma_s \vec{\varphi}_1^0$ are slowly varying functions of u, so that in performing the integration we may use the approximations

$$\left.\begin{aligned} \Sigma_s(u')\varphi_0^0(\vec{r}, u') &= \Sigma_s(u)\varphi_0^0(\vec{r}, u) + \frac{\partial \Sigma_s \varphi_0^0}{\partial u}(u' - u) + \dots, \\ \Sigma_s(u')\vec{\varphi}_1^0(\vec{r}, u') &= \Sigma_s(u)\vec{\varphi}_1^0(\vec{r}, u) + \dots . \end{aligned}\right\} \tag{B.7}$$

Inserting this into the right side of (B.5), we arrive at

$$\left.\begin{aligned} \nabla \vec{\varphi}_1^0 + \Sigma_c \varphi_0^0 &= -\frac{\partial \xi \Sigma_s \varphi_0^0}{\partial u} + \Sigma_{s0} f_0(u)\psi_0, \\ \frac{1}{3}\nabla \varphi_0^0 + \Sigma_{tr}\vec{\varphi}_1^0 &= -\Sigma_{s0} f_1(u)\vec{\psi}_1, \end{aligned}\right\} \tag{B.8}$$

where

$$\left.\begin{aligned} \Sigma_{tr} &= \Sigma_s(1 - \bar{\mu}_0) + \Sigma_c, \\ \xi &= \int_0^r t f_0(t)\,dt; \quad \bar{\mu}_0 = \int_0^r f_1(t)\,dt. \end{aligned}\right\} \tag{B.9}$$

We may neglect absorption for the primary neutrons (those with u = 0), so that

$$\Sigma_0 = \Sigma_{s0}.$$

According to (B.4) the second of Equations (B.8) can be written

$$\vec{\varphi}_1^0 = -D\nabla \varphi_0^0 - D f_1(u)\nabla \psi_0, \tag{B.10}$$

where we have written

$$D = \frac{1}{3\Sigma_{tr}}.$$

We now insert this into the first of (B.8). We then have

$$\nabla D \nabla \varphi - \Sigma_c \varphi = \frac{\partial q}{\partial u} - F(\vec{r}, u), \tag{B.11}$$

where

$$\left.\begin{aligned} \varphi &= \varphi_0^0; \quad q = \xi \Sigma_s \varphi \\ F(\vec{r}, u) &= \Sigma_0 f_0(u)\psi(\vec{r}) + f_0(u)\nabla D \nabla \psi, \quad \psi_0 = \psi. \end{aligned}\right\} \tag{B.12}$$

If M ≫ 1, the mean lethargy loss is small, and $F(\vec{r}, u)$ can be approximated by

$$F(\vec{r}, u) = (\Sigma_0 \psi + \bar{\mu}_0 \nabla D \nabla \psi)\,\delta(u). \tag{B.13}$$

Integrating (B.11) over $\underline{u}$ for $-\epsilon \leq u \leq +\epsilon$, we arrive at (since $q(\vec{r}, -\epsilon) = 0$ for $\epsilon > 0$)

$$q(\vec{r}, \epsilon) = \Sigma_0\psi + \vec{\mu}_0 \nabla D \nabla\psi + \int_{-\epsilon}^{+\epsilon} [\nabla D \nabla\psi - \Sigma_c\psi]\, du. \tag{B.14}$$

Let us now go to the limit $\epsilon \to 0$. The integral on the right side then vanishes, so that

$$q(\vec{r}, 0) = \Sigma_0\psi + \bar{\mu}\nabla D_0 \nabla\psi. \tag{B.15}$$

The second term on the right side of this equation is small, and we may ignore it without significant error. We then obtain

$$q(\vec{r}, 0) = \Sigma_0\psi. \tag{B.16}$$

As a result we arrive at the equations

$$\left.\begin{aligned} \nabla D_0 \nabla\psi - \Sigma_0\psi &= -Q(\vec{r}), \\ \nabla D \nabla\varphi - \Sigma_c\varphi &= \frac{\partial q}{\partial u}, \\ q(\vec{r}, 0) = \Sigma_0\psi; \quad D_0 &= \frac{1}{3\Sigma_0}. \end{aligned}\right\} \tag{B.17}$$

To this we add the thermal neutron diffusion equation

$$\nabla D_T \nabla\Phi - \Sigma_{cT}\Phi = -q.$$

Similarly, the adjoint reactor equations are

$$\left.\begin{aligned} \nabla D_0 \nabla\psi^* - \Sigma_0\psi^* &= -\Sigma_0\varphi^*(\vec{r}, 0), \\ \nabla D \nabla\varphi^* - \Sigma_c\varphi^* &= -\xi\Sigma_s \frac{\partial\varphi^*}{\partial u} - \Sigma_f Q^*, \\ \nabla D_T \nabla\Phi^* - \Sigma_{cT}\Phi^* &= -\Sigma_{fT} Q^*, \\ \text{where } Q^* = \nu_f\psi^*(\vec{r}), \quad & \varphi^*|_{u_T} = \Phi^*. \end{aligned}\right\} \tag{B.18}$$

These basic and adjoint reactor equations are easily reduced to multigroup sets of equations according to the methods described in Chapters VI and VII. In this appendix we are considering only procedure I of the multigroup representation for strong absorption of slowing-down neutrons (see Chapter VII). For this case we have

$$\nabla D^j \nabla\varphi^j - \Sigma^j\varphi^j = -f^j$$

(see Table I). Here

$$A^j = \xi\Sigma_s^j = \int_{u_{j-1}}^{u_j} \frac{1}{\xi\Sigma_s} \frac{p^j}{p^j(u)}\, du; \quad D^j = \frac{\displaystyle\int_{u_{j-1}}^{u_j} \frac{Dp^j}{\xi\Sigma_s} \frac{du}{p^j(u)}}{\displaystyle\int_{u_{j-1}}^{u_j} \frac{p^j}{\xi\Sigma_s} \cdot \frac{du}{p^j(u)}}.$$

The set of adjoint equations will be (Table II)

$$\nabla D^j \nabla\varphi^{*j-1} - \Sigma^j\varphi^{*j-1} = -f^{*j}$$

TABLE I

Effective Functions and Constants in the Multigroup Set of Reactor Equations for Strong Absorption of Slowing-Down Neutrons

j	φ^j	D^j	Σ^j	f^j
0	ψ	D_0	Σ_0	$Q(\vec{r})$
(1)	φ^1	D^1	$\frac{\xi\Sigma_s^1}{A^1}$	$\frac{1}{A^j} p^j \Sigma_0 \psi$
(2) — (m)	φ^j	D^j	$\frac{\xi\Sigma_s^j}{A^j}$	$\frac{1}{A^j} p^j \xi \Sigma_s^{j-1} \varphi^{j-1}$
($m+1$)	Φ	D_T	Σ_{cT}	$\xi\Sigma_s^m \varphi^m$

TABLE II

Effective Functions and Constants in the Multigroup Set of Adjoint Reactor **Equations** for Strong Absorption and Slowing-Down Neutrons

j	φ^{*j-1}	D^j	Σ^j	f^{*j}
($m+1$)	Φ^*	D_T	Σ_{cT}	$\Sigma_{fT} Q^*$
(m) — (1)	φ^{*j-1}	D^j	$\frac{1}{A^{*j}}$	$\frac{1}{A^{*j}} p^j \varphi^{*j} + \eta^j Q^*$
0	ψ	D_0	Σ_0	$\Sigma_0 \varphi^{*1}$

TABLE III

Effective Group Constants α^j and β^j for Calculating the Slowing-Down Sources With Strong Absorption of Slowing-Down Neutrons

j	α^j	β^j
0	α_1^1	0
(1) — ($m-1$)	$\alpha_1^{j+1} + \alpha_2^j$	β^j
m	α_2^m	β^m
($m+1$)	Σ_{fT}	0

Here

$$A^{*j} = \int_{u_{j-1}}^{u_j} \frac{p^j(u)}{\xi\Sigma_s}\,du; \quad D^j = \frac{\int_{u_{j-1}}^{u_j} \frac{D}{\xi\Sigma_s} p^j(u)\,du}{\int_{u_{j-1}}^{u_j} \frac{1}{\xi\Sigma_s} p^j(u)\,du},$$

$$\eta^j = \int_{u_{j-1}}^{u_j} \frac{\Sigma_f}{\xi\Sigma_s} e^{-\int_{u_{j-1}}^{u} \frac{\Sigma_c}{\xi\Sigma_s}du'}\,du.$$

The function $Q(\vec{r})$ is calculated according to

$$Q(\vec{r}) = \nu_f \sum_{j=0}^{m+1} (\alpha^j\varphi^j - \beta^j f^j),$$

(the coefficients α^j and β^j are given in Table III). Here

$$\alpha_1^j = \xi\Sigma_s^{j-1} \int_{u_{j-1}}^{u_j} \frac{\Sigma_f}{\xi\Sigma_s} p^j(u)\,du; \quad \alpha_2^j = \Sigma_j \int_{u_{j-1}}^{u_j} \frac{\Sigma_f}{\xi\Sigma_s} A^j(u)\,du;$$

$$\beta^j = \int_{u_{j-1}}^{u_j} \frac{\Sigma_f}{\xi\Sigma_s} \cdot A^j(u)\,du.$$

We note in conclusion that the generalization of this method to the case in which neutrons may be captured on the first collision presents no difficulty.

C. MATRICES AND MATRIX OPERATIONS

The array of numbers

$$A = \begin{Vmatrix} a_{11} & a_{12} & \dots & a_{1n} \\ a_{21} & a_{22} & \dots & a_{2n} \\ \cdot & \cdot & \cdot & \cdot \\ a_{n1} & a_{n2} & \dots & a_{nn} \end{Vmatrix}$$

is called a square matrix.

We shall consider some definitions and simple operations of matrix calculus.

Multiplication of the matrix by a number

$$\alpha A = \| \alpha a_{ik} \|.$$

Addition and subtraction of matrices

$$A \pm B = \| a_{ik} \pm b_{ik} \|.$$

Multiplication of matrices

$$AB = \left\| \sum_{j=1}^{n} a_{ij} b_{jk} \right\|.$$

The unit matrix E

$$E = \| \delta_{ik} \|,$$

where

$$\delta_{ik} \doteq \begin{cases} 1 & (i = k), \\ 0 & (i \neq k), \end{cases}$$

and for which

$$AE = EA = A.$$

The properties of matrix operations

$$\begin{aligned} A + (B + C) = (A + B) + C, &\quad A(BC) = (AB)C, \\ A + B = B + A, &\quad \alpha(AB) = (\alpha A)B = A(\alpha B), \\ (\alpha + \beta)A = \alpha A + \beta A, &\quad (A + B)C = AC + BC, \\ \alpha(A + B) = \alpha A + \alpha B, &\quad C(A + B) = CA + CB, \end{aligned}$$

where **A**, **B** and **C** are matrices, and α and β are numbers.

The inverse matrix $\mathbf{A}^{-1}$

$$A^{-1} = \frac{1}{|A|} C,$$

where $|\mathbf{A}|$ is the nonzero determinant of the elements of $\mathbf{A}$, and $\mathbf{C} = \|A_{ki}\|$ is the matrix whose elements are the minors A_{ki} of the elements a_{ki} in the determinant of $\mathbf{A}$.

Among the properties of the inverse are

$$A^{-1}A = AA^{-1} = E,$$
$$(AB)^{-1} = B^{-1}A^{-1}.$$

A matrix operating on a vector

Let $\mathbf{A} = \|a_{ik}\|$ be a matrix and $\mathbf{b} = |b_i|$ be a vector, where

$$|b_i| = \begin{vmatrix} b_1 \\ b_2 \\ \cdot \\ \cdot \\ \cdot \\ b_n \end{vmatrix}.$$

Then the product of a matrix and vector also gives a vector, which is defined by

$$A\mathbf{b} = \left| \sum_{j=1}^{n} a_{ij}b_j \right|.$$

Matrix calculus is described in more detail elsewhere [21, 83].

In conclusion we give some very important practical methods used in calculating the inverse of a matrix and the product of a matrix and a vector.

A Method for Increasing the Accuracy of the Inverse Matrix Elements

Due to rounding-off, the inverse matrix elements may be calculated with insufficient accuracy. It then becomes necessary to perform additional calculations to correct the values first obtained. This is usually done by a successive approximation method due to G. Hotteling (see [21]). Essentially, the method is the following.

Let $\mathbf{A}$ be a matrix whose elements are known to a given accuracy, and let $\mathbf{A}^{-1}$ be the inverse matrix which we wish to find with the given accuracy. We make use of the fact that

$$AA^{-1} = E.$$

When $\mathbf{A}^{-1}$ is calculated with insufficient accuracy, we have

$$AA_0^{-1} = E + F,$$

where $\mathbf{A}_0^{-1}$ is the approximate value of the inverse matrix, and $\mathbf{F}$ is the matrix of errors.

We now construct the following iteration process:

$$A_m^{-1} = A_{m-1}^{-1}(E + F_{m-1}), \quad F_m = -E + AA_m^{-1},$$

where $\underline{m}$ is the order of approximation.

There exist certain definite criteria for the convergence of this process [83]. In most cases of practical interest, the second or third approximation gives a sufficiently accurate result.

A Method for Increasing the Accuracy of the Result of Operating on a Vector by a Matrix

In solving equations by the method of spherical harmonics (for spherical and cylindrical geometry) accuracy is often lost in calculating the components of the vector obtained when a matrix operates on another vector.* Let us assume that the elements of $\mathbf{A}$ and the components of $\mathbf{b}$ are known with sufficient accuracy. We wish, with a given accuracy, to find the vector

---

*In particular, accuracy is lost in using Equation (66.2) for points with low values of $\underline{k}$.

$$\mathbf{c} = \mathbf{A}\mathbf{b}.$$

Let $\mathbf{c}_0$ be the approximate value of the desired vector. Then

$$\mathbf{c} = \mathbf{c}_0 + \mathbf{F},$$

where $\mathbf{F}$ is the error. Since $\mathbf{c}$ is unknown, so is $\mathbf{F}$.

To find the error vector we solve the following inverse problem: we consider $\mathbf{c}_0$ known and use the equation

$$\mathbf{A}\mathbf{b}_0 = \mathbf{c}_0$$

to find the vector

$$\mathbf{b}_0 = \mathbf{A}^{-1}\mathbf{c}_0.$$

Then

$$F_0 = A(b - b_0).$$

We thus have

$$c_1 = c_0 + A(b - b_0), \qquad b_0 = A^{-1} c_0.$$

Repeating this process, we obtain the general formula

$$c_m = c_{m-1} + A(b - b_{m-1}), \qquad b_m = A^{-1} c_m.$$

Without attempting to formulate this method more rigorously, we note only that it usually converges very rapidly and can as a rule be used in the second or third approximation to obtain the results with sufficient accuracy.

D. SUBCRITICAL REACTOR CALCULATIONS

In performing physical experiments on nuclear reactors, it is often necessary to use stationary subcritical assemblies, for which $K_{eff} < 1$. To maintain a steady state in such a reactor, one introduces additional external neutron sources. It is clear that this changes the mathematical problem of describing the stationary neutron spectrum in the reactor. Thus, the homogeneous eigenvalue problem becomes the inhomogeneous problem

$$\begin{gathered}\nabla D\nabla\varphi-\Sigma_c\varphi=\frac{\partial\xi\Sigma_s\varphi}{\partial u}-\chi(u)\left(\int\limits_{-\infty}^{u_T}\Sigma_f\varphi\,du+\Sigma_{fT}\Phi\right)-\eta(u)\,q(\vec{r}),\\ \nabla D_T\nabla\Phi-\Sigma_{cT}\Phi=-\xi\Sigma_s\varphi(\vec{r},u_T),\\ \varphi|_{S_e}=\Phi|_{S_e}=0,\end{gathered} \tag{D.1}$$

where $\eta(u)$ is the spectrum of neutrons emitted by sources distributed through the reactor according to $q(\vec{r})$.

Neutrons are emitted by the source and are elastically scattered in the reactor and slowed down, part of them absorbed by uranium nuclei and causing fission. This fission gives rise to neutrons of a new generation whose spectrum is given by $\chi(u)$. As opposed to the neutron distribution in a critical spectrum, the distribution of these neutrons is described not by the first eigenfunction of the reactor, but by a function which is a superposition of the fission integrals from point sources distributed through the reactor according to $q(\vec{r})$.

After this generation of neutrons is slowed down, the spectrum will be more similar to the first eigenfunction. In the further cycles of the chain reaction the spectrum asymptotically approaches the eigenfunction of the critical reactor. It should, however, be borne in mind that if $K_{eff} < 1$, the total number of neutrons in each generation will decrease due to leakage and capture. It is clear that the closer the reactor is to critical, the greater the number of cycles in it. It may thus occur that the overwhelming majority of neutrons produced originally by the sources may be distributed according to the reactor eigenfunction. If the reactor is highly subcritical, the number of neutrons distributed according to this eigenfunction will be small.

Since in a stationary subcritical reactor there exist at all times neutrons of all possible generations, the total neutron spectrum $\varphi(\vec{r}, u)$ is obtained by summing over all these generations.

We thus arrive at the qualitative conclusion that if the reactor is almost critical, its neutron spectrum will hardly differ from the first eigenfunction of the problem. In a highly subcritical reactor this spectrum will be very similar to that from point sources distributed over the reactor volume according to $q(\vec{r})$.

At the same time, we arrive at another important mathematical conclusion. If the reactor is close to critical, the total neutron spectrum can be obtained only after many iterations, since we must account for neutrons of many generations. If, however, the reactor is far from critical, the number of necessary iterations is relatively small.

When we consider the fact that physical experiments are usually performed on reactors close to critical, it is seen that the solution of (D.1) by means of successive approximations is not reasonable. In this appendix we will describe a direct method good for a large class of practically important problems.

We write the basic reactor equation (D.1) in the form of the multigroup diffusion equations

$$\nabla D^j \nabla \varphi^j - \Sigma^j \varphi^j = -W^j \varphi^{j-1} - S^j \sum_{l=1}^{m+1} c^l \varphi^l - \varkappa^j q(\vec{r}) \tag{D.2}$$

$$(j = 1, 2, \ldots m+1),$$

where

$$\Sigma^j = \frac{\xi \Sigma_s^j}{A^j}, \quad W^j = p^j \frac{\xi \Sigma_s^{j-1}}{A^j}, \quad S^j = \frac{\nu_f}{1-\varepsilon \nu_f} \frac{\chi^j}{A^j}, \quad \varkappa^j = \frac{\eta^j}{A^j}$$

$$\eta^j = \int_{u_{j-1}}^{u_j} \eta \frac{p^j}{p^j(u)} du, \quad D^{m+1} = D_T, \quad \Sigma^{m+1} = \Sigma_T, \quad W^{m+1} = \xi \Sigma_s^{j-1}, \quad S^{m+1} = 0,$$

and the c^l are coefficients from a quadrature formula.

We have restricted our considerations to the simplest procedure I of the group representation (see Section 27).

We write Equations (D.2) in the matrix form*

$$\nabla \boldsymbol{D} \nabla \boldsymbol{\varphi} - \boldsymbol{\Sigma} \boldsymbol{\varphi} = -\boldsymbol{\varkappa} q(\vec{r}), \tag{D.3}$$

$$\boldsymbol{\varphi}|_{S_e} = 0,$$

where $\boldsymbol{\varphi}$ and $\boldsymbol{\kappa}$ are vectors with components $\{\varphi^j\}$ and $\{\kappa^j\}$, and $\mathbf{D}$ and $\mathbf{\Sigma}$ are matrices; $\mathbf{D} = \| \delta_{lj} D^j \|$, $\Sigma = \| \Sigma_{lj} \|$,

$$\Sigma_{j,l} = \begin{cases} -c^l S^j & (l > j) \\ -c^j S^j + \Sigma^j & (l = j) \\ -c^l S^j - \delta_{l+1,\,j} W^j & (l < j) \end{cases}$$

and

$$\delta_{l,\,j} = \begin{cases} 0, & (l \neq j) \\ 1, & (l = j). \end{cases}$$

For a one-dimensional plane, cylindrical, or spherical geometry, Equation (D.3) becomes

$$\frac{1}{r^\alpha} \frac{d}{dr} r^\alpha \boldsymbol{D} \frac{d\boldsymbol{\varphi}}{dr} - \boldsymbol{\Sigma} \boldsymbol{\varphi} = -\boldsymbol{\varkappa} q(r), \tag{D.3'}$$

where α is a parameter given by:

$$\alpha = \begin{cases} 0 \text{ for the plane geometry} \\ 1 \text{ for the cylindrical geometry} \\ 2 \text{ for the spherical geometry.} \end{cases}$$

We now represent (D.3') in finite-difference form. Again restricting our considerations to procedure I of the methods described in Chapter VIII, and making use of the rules of matrix calculus, we arrive at

$$\boldsymbol{a}_k \boldsymbol{\varphi}_{k+1} - \boldsymbol{b}_k \boldsymbol{\varphi}_k + \boldsymbol{c}_k \boldsymbol{\varphi}_{k-1} = -\boldsymbol{F}_k, \tag{D.4}$$

where

$$\boldsymbol{a}_k = \frac{\left(1 + \alpha \frac{\Delta r_{k+1/2}}{2r_k}\right) \boldsymbol{D}_{k+1/2}}{\Delta r_{k+1/2}}, \quad \boldsymbol{c}_k = \frac{\left(1 - \alpha \frac{\Delta r_{k-1/2}}{2r_k}\right) \boldsymbol{D}_{k-1/2}}{\Delta r_{k-1/2}},$$

$$\boldsymbol{b}_k = \boldsymbol{a}_k + \boldsymbol{c}_k + (\boldsymbol{\Sigma} \Delta r)_k, \quad \boldsymbol{F}_k = (\boldsymbol{\varkappa} \Delta r q)_k.$$

* The first to indicate the possibility of representing the set of reactor equations in matrix form and then factoring this set were E.S. Kuznetsov and I.G. Krutikova.

Finally, we write (D.4) in the form

$$\varphi_{k+1} - B_k\varphi_k + C_k\varphi_{k-1} = -f_k, \tag{D.5}$$

where

$$B_k = a_k^{-1} b_k, \quad C_k = a_k^{-1} c_k, \quad f_k = a_k^{-1} F_k.$$

Let us now obtain the necessary boundary conditions. We first consider the center r = 0 of the reactor. We multiply (D.3') by r^α and integrate over $0 \le r \le \Delta r/2$, arriving at

$$I_{1/2} - I_0 - \Sigma \int_0^{\frac{\Delta r}{2}} \varphi r^\alpha \, dr = -\varkappa \int_0^{\frac{\Delta r}{2}} q r^\alpha \, dr, \tag{D.6}$$

where

$$I_{1/2} = r^\alpha D \frac{d\varphi}{dr}\bigg|_{r=\frac{\Delta r}{2}}.$$

Bearing in mind that

$$I_0 = 0,$$

and making use of the approximate relations

$$\int_0^{\frac{\Delta r}{2}} \varphi r^\alpha \, dr = \frac{1}{1+\alpha}\left(\frac{\Delta r}{2}\right)^{1+\alpha} \varphi_0$$

$$\int_0^{\frac{\Delta r}{2}} q r^\alpha \, dr = \frac{1}{1+\alpha}\left(\frac{\Delta r}{2}\right)^{1+\alpha} q_0,$$

$$I_{1/2} = \frac{\Delta r^{\alpha-1}}{2^\alpha} D(\varphi_1 - \varphi_0),$$

we arrive at

$$\varphi_1 - B_0\varphi_0 = -f_0, \tag{D.7}$$

where

$$\left.\begin{aligned} B_0 &= E + \frac{\Delta r^2}{2(1+\alpha)} D^{-1}\Sigma, \\ f_0 &= \frac{\Delta r^2}{2(1+\alpha)} D^{-1}\varkappa q_0. \end{aligned}\right\} \tag{D.7'}$$

Together with the boundary condition

$$\varphi_n = 0, \tag{D.8}$$

and the extrapolated reactor boundary, Equations (D.5), (D.7), and (D.7') specify the problem completely.

This set of equations is solved by matrix factorization:

$$\left.\begin{aligned} \beta_{k+1} &= C_{k+1}(B_k - \beta_k)^{-1}, \\ z_{k+1} &= \beta_{k+1}(z_k + f_k), \\ \varphi_k &= C_{k+1}^{-1}(\beta_{k+1}\varphi_{k+1} + z_{k+1}) \end{aligned}\right\} \tag{D.9}$$

with the condition that

$$\beta_1 = C_1 B_0^{-1}, \quad z_1 = \beta_1 f_0, \quad \varphi_n = 0. \tag{D.10}$$

In conclusion, let us consider the important special case in which the external sources are concentrated at the center of the reactor. We will assume that q(r) is of the form

$$q(r) = \begin{cases} q_0 & \left(0 \leqslant r \leqslant \frac{\Delta r}{2}\right) \\ 0 & \left(r > \frac{\Delta r}{2}\right). \end{cases} \tag{D.11}$$

It is clear that a source so distributed is not a point source, but we may use it in finite-difference calculations.

With (D.11) we arrive at the problem given by

$$\left.\begin{aligned} \varphi_{k+1} - B_k\varphi_k + C_k\varphi_{k-1} &= 0, \\ \varphi_1 - B_0\varphi_0 &= -f_0, \\ \varphi_n &= 0. \end{aligned}\right\} \tag{D.12}$$

This can be solved by the method of difference factorization described in (D.9) and (D.10), with the quantities $f_k (k \neq 0)$ set equal to zero.

The problem in which there is a neutron flux incident on the external reactor boundary is solved similarly.

E. NEUTRON THERMALIZATION*

In deriving the basic reactor equations in Chapter I, we assumed that the thermal neutron spectrum hardly differs from an equilibrium spectrum in an infinite homogeneous medium, and that the space distribution of the thermal neutrons can therefore be satisfactorily described by a single kinetic Boltzmann equation. Essentially, we have assumed that it is possible to separate the space and energy variables in performing the solution for the thermal neutrons. Such an approximation, however, is found to be a very rough approximation of reality, especially when dealing with reactors with many different zones which have different scattering and absorbing properties and different crystalline and molecular compositions.

One must therefore use a more accurate multigroup model for neutron thermalization. In this appendix we give a general method for reducing the equations for the thermal neutron spectrum to a multigroup set, and describe effective numerical methods for solving these equations within the framework of the diffusion approximation.

Let us thus consider the kinetic equation for the thermal neutrons with velocity $0 \leq v \leq v_C$. In the stationary case, as was shown in Chapter I, this equation is

$$\vec{\Omega}\nabla\varphi + \frac{\gamma(\vec{r}, v) + \dot{w}(\vec{r}, v)}{v}\varphi(\vec{r}, v) =$$

$$= \int_0^{v_C} \frac{dv'}{v'} \int d\Omega' \varphi(\vec{r}, \vec{\Omega}', v') g(\vec{\Omega}, \vec{\Omega}', v, v') + S(\vec{r}, \vec{\Omega}, v), \tag{E.1}$$

where $S(\vec{r}, \vec{\Omega}, v)$ is the fission source function.

Going to the diffusion approximation, we shall attempt to solve this equation in the form

$$\varphi(\vec{r}, \vec{\Omega}, v) = \frac{1}{4\pi}[\varphi_0(\vec{r}, v) + 3\vec{\Omega}\vec{\varphi}_1(\vec{r}, v)].$$

Using the same kind of an expansion for $S(\vec{r}, \vec{\Omega}, v)$, we arrive at

$$\left.\begin{aligned} \nabla\vec{\varphi}_1(\vec{r}, v) &= -\frac{\gamma + w}{v}\varphi_0(\vec{r}, v) + \int_0^{C} \frac{g_0(v, v')}{v'}\varphi_0(\vec{r}, v')\, dv' + S_0(\vec{r}, v), \\ \frac{1}{3}\nabla\varphi_0(\vec{r}, v) &= -\frac{\gamma + w}{v}\vec{\varphi}_1(\vec{r}, v) + \int_0^{v_C} \frac{g_1(v, v')}{v'}\vec{\varphi}_1(\vec{r}, v')\, dv' + \vec{S}_1(\vec{r}, v), \end{aligned}\right\} \tag{E.2}$$

where g_0 and g_1 are given by

$$g_0(v, v') = \int g(\vec{\Omega}, \vec{\Omega}'; v, v')\, d\vec{\Omega}, \quad g_1(v, v') = \int g(\vec{\Omega}, \vec{\Omega}'; v, v')(\vec{\Omega}, \vec{\Omega}')\, d\vec{\Omega}.$$

*This appendix has been written at the author's request by V.V. Smelov.

We break up the region $0 \le v \le v_C$ into a finite number of regions $v_l < v < v_{l+1}$ ($l = 1, 2, \ldots, n$; $v_1 = 0$, $v_{n+1} = v_C$), in each of which we attempt to solve (E.2) in the form

$$\varphi_0(\vec{r}, v) = \frac{M(v)}{M_l} \Phi_0^{(l)}(\vec{r}), \quad \vec{\varphi}_1(\vec{r}, v) = \frac{M(v)}{M_l} \vec{\Phi}_1^{(l)}(\vec{r}). \tag{E.3}$$

Here $M(v)$ is a function which is piecewise constant in space with discontinuities on the boundaries of the zones, within each of which it gives the equilibrium energy spectrum of the neutron flux in a homogeneous infinite medium filled with the substance of the zone under consideration, and

$$M_l = \int_{v_l}^{v_{l+1}} M(v)\, dv.$$

Equations (E.2) can now be rewritten in the form

$$\nabla \vec{\varphi}_1(\vec{r}, v) = -\frac{\gamma + w}{v} \varphi_0(\vec{r}, v) + \sum_{l=1}^{n} \frac{\Phi_0^{(l)}(\vec{r})}{M_l} \int_{v_l}^{v_{l+1}} M(v') \frac{g_0(v, v')}{v'} dv' + S_0(\vec{r}, v);$$

$$\frac{1}{3} \nabla \varphi_0(\vec{r}, v) = -\frac{\gamma + w}{v} \vec{\varphi}_1(\vec{r}, v) + \sum_{l=1}^{n} \frac{\vec{\Phi}_1^{(l)}(\vec{r})}{M_l} \int_{v_l}^{v_{l+1}} M(v') \frac{g_1(v, v')}{v'} dv' + \vec{S}_1(\vec{r}, v). \tag{E.2'}$$

Using (E.3), we integrate over $v_j < v < v_{j+1}$, obtaining

$$\nabla \vec{\Phi}_1^{(j)}(\vec{r}) = -\frac{\Phi_0^{(j)}(\vec{r})}{M_j} \int_{v_j}^{v_{j+1}} (\gamma + w) M(v) \frac{dv}{v} + \sum_{l=1}^{n} \frac{\Phi_0^{(l)}(\vec{r})}{M_l} \int_{v_j}^{v_{j+1}} dv \int_{v_l}^{v_{l+1}} M(v') g_0(v, v') \frac{dv'}{v'} + \int_{v_j}^{v_{j+1}} S_0(\vec{r}, v)\, dv,$$

$$\frac{1}{3} \nabla \Phi_0^{(j)}(\vec{r}) = -\frac{\vec{\Phi}_1^{(j)}(\vec{r})}{M_j} \int_{v_j}^{v_{j+1}} (\gamma + w) M(v) \frac{dv}{v} + \sum_{l=1}^{n} \frac{\vec{\Phi}_1^{(l)}(\vec{r})}{M_l} \int_{v_j}^{v_{j+1}} dv \int_{v_l}^{v_{l+1}} M(v') g_1(v, v') \frac{dv'}{v'} + \int_{v_j}^{v_{j+1}} \vec{S}_1(\vec{r}, v)\, dv.$$

This set of equations can be written

$$\left.\begin{aligned} \nabla \vec{\Phi}_1^{(j)}(\vec{r}) + \sum_{l=1}^{n} A_{jl} \Phi_0^{(l)}(\vec{r}) &= S_0^{(j)}(\vec{r}), \\ \frac{1}{3} \nabla \Phi_0^{(j)}(\vec{r}) + \sum_{l=1}^{n} B_{jl} \vec{\Phi}_1^{(l)}(\vec{r}) &= \vec{S}_1^{(j)}(\vec{r}), \end{aligned}\right\} \quad (j = 1, 2, \ldots n) \tag{E.4}$$

where

$$A_{jl} = \delta_{jl} \cdot \frac{1}{M_j} \int_{v_j}^{v_{j+1}} (\gamma + w) M(v) \frac{dv}{v} - \frac{1}{M_l} \int_{v_j}^{v_{j+1}} dv \int_{v_l}^{v_{l+1}} M(v') g_0(v, v') \frac{dv'}{v'}$$

$$B_{jl}=\delta_{jl}\cdot\frac{1}{M_j}\int\limits_{v_j}^{v_{j+1}}(\gamma+w)\,M(v)\,\frac{dv}{v}-\frac{1}{M_l}\int\limits_{v_j}^{v_{j+1}}dv\int\limits_{v_l}^{v_{l+1}}M(v')\,g_1(v,v')\,\frac{dv'}{v'},$$

$$S_0^{(j)}(\vec{r})=\int\limits_{v_j}^{v_{j+1}}S(\vec{r},v)\,dv,\quad \vec{S}_1^{(j)}(\vec{r})=\int\limits_{v_j}^{v_{j+1}}\vec{S}_1(\vec{r},v)\,dv,$$

$$\delta_{jl}=\begin{cases}1 & (j=l),\\ 0 & (j\neq l).\end{cases}$$

Let us now find a convenient control formula. Summing the A_{jl} coefficients over both indices and using the relation

$$w(v,\vec{r})=\int\limits_0^{v_C}g_0(v',v)\,dv',$$

we obtain

$$\sum_{j,\,l=1}^{n}A_{jl}=\sum_{j=1}^{n}\frac{1}{M_j}\int\limits_{v_j}^{v_{j+1}}\gamma M(v)\,\frac{dv}{v}\,. \tag{E.5}$$

If this equation is not fulfilled with sufficient accuracy, the A_{jl} coefficients should be corrected (by multiplying, for instance, the double integrals in the expressions for the A_{jl} by a correction factor c) to obtain the necessary accuracy in (E.5).

What remains is to solve Equations (E.4). In most cases of interest, the method of successive approximations is not effective for this solution, since, in addition to neutrons moving downward through energy groups, as is the case for slowing down at high energies, they move also upwards. It then becomes necessary to solve the problem directly.

We do this by using the already well-developed method of matrix factorization. We thus write (E.4) in the matrix form

$$\left.\begin{aligned}\nabla\boldsymbol{\psi}+\boldsymbol{\Sigma}_0\boldsymbol{\Phi}&=\boldsymbol{f}_0,\\ \frac{1}{3}\nabla\boldsymbol{\Phi}+\boldsymbol{\Sigma}_1\boldsymbol{\psi}&=\boldsymbol{f}_1,\end{aligned}\right\} \tag{E.6}$$

where $\boldsymbol{\Phi}$, $\boldsymbol{\psi}$, $\boldsymbol{f}_0$ and $\boldsymbol{f}_1$ are vectors with components $\{\Phi_0^j\}$, $\{\vec{\Phi}_1^j\}$, $\{S_0^j\}$ and $\{\vec{S}_1^j\}$, respectively, and where $\boldsymbol{\Sigma}_0$ and $\boldsymbol{\Sigma}_1$ are matrices given by

$$\boldsymbol{\Sigma}_0=\|A_{jl}\|,\quad \boldsymbol{\Sigma}_1=\|B_{jl}\|\,.$$

To (E.6) we add the boundary condition on the external extrapolated reactor surface

$$\boldsymbol{\Phi}=0 \text{ on } S_e. \tag{E.7}$$

It is easily seen that the quantities M_j introduced in (E.3) make it possible to look for continuous solutions $\boldsymbol{\Phi}(\vec{r})$, $\boldsymbol{\psi}(\vec{r})$ at all points of the reactor.

In the one-dimensional plane, cylindrical and spherical geometries, (E.6) becomes

$$\left.\begin{aligned}\frac{1}{r^{\alpha}}\frac{d}{dr}r^{\alpha}\boldsymbol{\psi}+\boldsymbol{\Sigma}_0\boldsymbol{\Phi}&=\boldsymbol{f}_0,\\ \frac{1}{3}\frac{d}{dr}\boldsymbol{\Phi}+\boldsymbol{\Sigma}_1\boldsymbol{\psi}&=\boldsymbol{f}_1,\end{aligned}\right\} \tag{E.8}$$

where

$$\alpha = \begin{cases} 0 \text{ for the plane geometry} \\ 1 \text{ for the cylindrical geometry} \\ 2 \text{ for the spherical geometry.} \end{cases}$$

Replacing these differential equations by finite-difference equations, as described in Section 78, we arrive at

$$\Phi_{k+1} - B_k\Phi_k + C_k\Phi_{k+1} = -F_k, \tag{E.9}$$

where

$$B_k = a_k^{-1} b_k, \quad C_k = a_k^{-1} c_k, \quad F_k = a_k^{-1} \varphi_k \quad (k = 1, 2, \ldots, n),$$

$$B_0 = E + \frac{3\Delta r^2}{2(1+\alpha)} \Sigma_0 \Sigma_1, \quad F_0 = \frac{3\Delta r^2}{2(1+\alpha)} \Sigma_1 f_0^{*},$$

$$a_k = \frac{1}{3}\left(\frac{r_{k+1/2}}{r_k}\right)^{\alpha} (\Sigma_1 \Delta r)^{-1}_{k+1/2}, \quad c_k = \frac{1}{3}\left(\frac{r_{k-1/2}}{r_k}\right)^{\alpha} (\Sigma_1 \Delta r)^{-1}_{k-1/2},$$

$$b_k = a_k + c_k + (\Sigma_0 \Delta r)_k,$$

$$\varphi_k = (f_0 \Delta r)_k - 3a_k (f_1 \Delta r)_{k+1/2} + 3c_k (f_1 \Delta r)_{k-1/2}.$$

We now proceed to the solution by the method of matrix factorization, which is described in detail for a similar class of mathematical problems in Appendix D. We present here only the final result, namely

$$\left.\begin{aligned} \beta_{k+1} &= C_{k+1}(B_k - \beta_k)^{-1}, \\ z_{k+1} &= \beta_{k+1}(z_k + F_k), \\ \Phi_k &= C_{k+1}^{-1}(\beta_{k+1}\Phi_{k+1} + z_{k+1}), \end{aligned}\right\} \tag{E.10}$$

with the condition that

$$\left.\begin{aligned} \beta_1 &= C_1 B_0^{-1}, \\ z_1 &= \beta_1 F_0, \\ \Phi_n &= 0. \end{aligned}\right\} \tag{E.11}$$

In conclusion we mention a particularly important application of the methods described in this appendix, namely, calculation of the thermal neutron spectrum in the cell of a heterogeneous reactor. In this case instead of (E.7), the condition which we must add to the matrix equations (E.6) is

$$\psi = 0 \text{ on S}. \tag{E.12}$$

where S is the surface of the Wigner-Seitz cell. The finite-difference representation of this boundary condition is most convenient to state in terms of a fictitious boundary point lying one-half of an interpolation interval outside S. Then for the one-dimensional geometry we have

$$\Phi_{n-1} = \Phi_n, \tag{E.13}$$

which follows from the second of (E.6) if we recall that on the boundary of the cell $f_1 = 0$. We now use (E.13) to obtain an explicit expression for Φ_n. This is done by using the last of (E.10) and writing it (with $k = n - 1$) in the form

$$\Phi_{n-1} = C_n^{-1}(\beta_n \Phi_n + z_n). \tag{E.14}$$

Eliminating Φ_{n-1} from (E.13) and (E.14), we arrive at

---

* Values for B_0 and F_0 in the case of a plane problem, have meaning only when the problem is symmetrical.

$$\Phi_n = (C_n - \beta_n)^{-1} z_n. \tag{E.15}$$

Thus we again obtain the problem of Equations (E.10) and (E.11), except that we must replace the second of Conditions (E.11) by (E.15).

None of the above mathematical operations depend on any assumptions about the concrete mechanism by which the neutrons interact with matter in the reactor, and they are therefore of a quite general character.

F. TABLES OF SPECIAL FUNCTIONS

TABLE

The Doppler Effect in Resonance Capture of Neutrons by a Homogeneous Medium

$$\eta(\xi, h) = \frac{2\sqrt{1+h}}{\pi h} \int_0^\infty \frac{\psi(x, \xi)}{\psi(x, \xi) + \frac{1}{h}} \, dx,$$

$$\text{where } \psi(x, \xi) = \frac{\xi}{2\sqrt{\pi}} \int_{-\infty}^{\infty} \frac{e^{-\frac{\xi^2}{4}(x-y)^2}}{1+y^2} \, dy.$$

ξ \ h	0,8	1,0	5,0	10,0	30,0	50,0	100	300	500	1000
0,04	1,314	1,379	2,187	2,731	3,338	3,447	3,346	2,784	2,464	2,052
0,05	1,311	1,373	2,134	2,573	3,068	3,106	2,943	2,401	2,121	1,779
0,075	1,296	1,354	2,017	2,342	2,585	2,532	2,320	1,858	1,655	1,424
0,10	1,280	1,334	1,917	2,160	2,264	2,173	1,963	1,581	1,427	1,261
0,15	1,258	1,305	1,760	1,899	1,868	1,764	1,584	1,317	1,223	1,128
0,20	1,236	1,278	1,639	1,718	1,635	1,538	1,389	1,192	1,129	1,071
0,30	1,200	1,233	1,471	1,490	1,376	1,305	1,205	1,091	1,059	1,030
0,40	1,172	1,197	1,362	1,350	1,253	1,195	1,125	1,051	1,032	1,016
0,50	1,148	1,179	1,287	1,268	1,180	1,135	1,083	1,033	1,020	1,010
0,70	1,113	1,128	1,189	1,168	1,102	1,071	1,043	1,016	1,009	1,004
1,00	1,080	1,090	1,117	1,095	1,054	1,037	1,021	1,007	1,004	1,002
2,00	1,032	1,034	1,036	1,028	1,013	1,009	1,004	1,000	1,000	1,000

TABLE II

Self-Shielding of a Resonance Level as a Function of the Total Cross Section (Σ^j) at the Maximum of the Resonance, and the Thickness (d); $\beta = \Sigma^j d$

$$f(\beta) = \int_0^1 e^{-\frac{\beta}{2\mu}} \left[I_0\left(\frac{\beta}{2\mu}\right) + I_1\left(\frac{\beta}{2\mu}\right)\right] d\mu \quad \text{for a plane layer}$$

$$f(\beta) = \frac{4}{\pi}\int_0^{\pi/2} \cos^2\psi d\psi \int_0^{\pi/2} e^{-\frac{\beta}{2}\frac{\cos\psi}{\sin\vartheta}} \left[I_0\left(\frac{\beta}{2}\frac{\cos\psi}{\sin\vartheta}\right) + I_1\left(\frac{\beta}{2}\frac{\cos\psi}{\sin\vartheta}\right)\right] \sin\vartheta d\vartheta \quad \text{for a cylindrical block}$$

β	f (β)		β	f (β)	
	plane layer	cylindrical block		plane layer	cylindrical block
0	1	1	3,00	0,4089	0,5637
0,25	0,8266	0,9274	4,00	0,3598	0,5044
0,50	0,7336	0,8683	5,00	0,3247	0,4596
0,75	0,6672	0,8182	6,00	0,2979	0,4248
1,00	0,6155	0,7749	7,00	0,2768	0,3965
1,50	0,5385	0,7038	8,00	0,2596	0,3732
2,00	0,4823	0,6470	9,00	0,2450	0,3534
2,50	0,4417	0,6018	10,00	0,2327	0,3366

TABLE III

The Factor Accounting for the Probability That a Neutron May Pass Through a Block in the Direction $\vec{\Omega}$ Without Colliding: $\beta = \Sigma_{sn} l$ (where Σ_{sn} is the Potential Scattering Cross Section in the Block, and l is the Neutron Path Length in the Block)

$$\Phi(\beta) = [1 - \operatorname{erf}(\sqrt{\beta})](1 + 2\beta) + \frac{2\sqrt{\beta}}{\sqrt{\pi}} e^{-\beta} - 2\beta.$$

β	Φ (β)	β	Φ (β)
0	0	2,00	0,9884
0,05	0,4129	2,50	0,9944
0,10	0,5369	3,00	0,9983
0,15	0,6169	4,00	0,9993
0,20	0,6751	5,00	0,9997
0,25	0,7200	6,00	1,000
0,50	0,8487	7,00	1,000
0,75	0,9094	8,00	1,000
1,00	0,9430	9,00	1,000
1,50	0,9753	10,00	1,000

TABLE IV

The Factor Accounting for the Probability That a Neutron May Pass Through a Block Without Colliding: $\beta = \Sigma_{sn} d$ (Where Σ_{sn} is the Cross Section for Potential Scattering in the Block, and d is the Block Thickness)

$$F(\beta) = 2 \int_0^1 \Phi\left(\frac{\beta}{\mu}\right) \mu d\mu \quad \text{for a plane layer}$$

$$F(\beta) = \frac{4}{\pi} \int_0^{\pi/2} \cos\psi d\psi \int_0^{\pi/2} \Phi\left(\beta \frac{\cos\psi}{\sin\vartheta}\right) \sin^2\vartheta d\vartheta \quad \text{for a cylindrical block}$$

β	$F(\beta)$		β	$F(\beta)$	
	plane layer	cylindrical block		plane layer	cylindrical block
0	0	0	2,00	0,9958	0,971
0,05	0,5035	0,3991	2,50	0,9981	0,981
0,10	0,6348	0,5187	3,00	0,9991	0,987
0,15	0,7128	0,5958	4,00	0,9997	0,993
0,20	0,7673	0,6521	5,00	0,9999	0,997
0,25	0,8066	0,6957	6,00	1,00	0,998
0,50	0,9103	0,8224	7,00	1,00	1,00
0,75	0,9518	0,8842	8,00	1,00	1,00
1,00	0,9726	0,9192	9,00	1,00	1,00
1,50	0,9894	0,958	10,00	1,00	1,00

TABLE V

Effective Resonance Integral for a Cylindrical Block of Natural Uranium

$$J(x) \cdot 10^{24} = 5 + 9.5 \frac{1}{\sqrt{x}} \quad \text{(Gurevich-Pomeranchuk theory)}$$

$$J(x) \cdot 10^{24} = 9.25 + 24.7 \frac{4}{18.8} \frac{1}{x} \quad \text{(Wigner theory)}$$

$$J(x) \cdot 10^{24} = 9.25 + 24.7 \frac{4}{18.8} \frac{1}{x} F(0.394\,x) \quad \text{(Orlov theory)}$$

x	$J(x) \cdot 10^{24}$		
	Gurevich-Pomeranchuk	Wigner	Orlov
0.25	24,00	30,27	20,09
0.375	20,51	23,26	17,56
0.500	18,43	19,76	16,07
0.625	17,02	17,66	15,08
0.750	15,97	16,26	14,32
1.00	14,50	14,50	13,33
1,25	13,50	13,45	12,70
1,50	12,76	12,75	12,22
2,00	11,72	11,88	11,59
2,50	11,01	11,35	11,18
3,00	10,48	11,00	10,89
4,50	9,478	10,42	10,38
6,00	8,878	10,13	10,11
7,50	8,469	9,950	9,941

Here we have set $\Sigma_{sn}^{U} = 0.394$ cm^{-1}, $\tilde{p} = 18.8$ g/cm^3 (where $\tilde{p}$ is the U^{238} density).

TABLE VI

The Factor Accounting Both for the Probability That a Neutron May Pass Through a Block Without Colliding and the Mutual Shielding of Uranium Blocks: $\alpha = \Sigma_s^0(D - d)$; $\beta = \Sigma_{sn} d$ (Where Σ_s^0 is the Scattering Cross Section in the Moderator; Σ_{sn} is the Potential Scattering Cross Section in the Block; D is the Distance Between Blocks, and d is the Thickness of a Block)

$$F(\alpha, \beta) = \frac{2\sqrt{\beta}}{\pi} \int_0^1 \sqrt{\mu}\, d\mu \int_{-\infty}^{\infty} \frac{1 - e^{-\left(\frac{1}{y^2} + \frac{\beta}{\mu}\right)}}{1 + \frac{\beta y^2}{\mu}} \cdot \frac{1 - e^{-\frac{\alpha}{\mu}}}{1 - e^{-\frac{\alpha}{\mu} - \left(\frac{1}{y^2} + \frac{\beta}{\mu}\right)}}\, dy.$$

β \ α	0	0,05	0,15	0,25	0,75	1,50	2,50	6,00	10,0
0,01	0	0,05832	0,1184	0,1525	0,2149	0,2359	0,2420	0,2439	0,2439
0,05	0	0,07935	0,1794	0,2472	0,4236	0,5154	0,5499	0,5626	0,5628
0,10	0	0,08545	0,2024	0,2826	0,4922	0,6046	0,6487	0,6661	0,6664
0,15	0	0,08779	0,2136	0,3016	0,5309	0,6526	0,7000	0,7189	0,7190
0,20	0	0,08904	0,2203	0,3134	0,5590	0,6886	0,7388	0,7584	0,7587
0,25	0	0,08972	0,2242	0,3216	0,5809	0,7176	0,7702	0,7909	0,7912
0,50	0	0,09108	0,2331	0,3404	0,6417	0,8062	0,8697	0,8944	0,8947
0,75	0	0,09145	0,2360	0,3466	0,6663	0,8460	0,9162	0,9441	0,9445
1,00	0	0,09162	0,2369	0,3492	0,6774	0,8651	0,9397	0,9693	0,9697
1,50	0	0,09170	0,2377	0,3510	0,6856	0,8800	0,9582	0,9897	0,9901
2,00	0	0,09171	0,2381	0,3514	0,6879	0,8844	0,9639	0,9960	0,9965
6,00	0	0,09173	0,2381	0,3516	0,6890	0,8867	0,9669	0,9994	1,000
10,0	0	0,09173	0,2381	0,3516	0,6890	0,8867	0,9669	0,9994	1,000

TABLE VII

The Dependence of $\Phi(\beta)$ on the Block Thickness and the Parameter ξ Related to the Neutron Energy in the Neighborhood of Resonance

$$\beta = \frac{d}{\xi^2}, \quad \xi = \frac{2}{\Gamma_j}(E - E^j)$$

$$\Phi(\beta) = 2E_3(\beta) \quad \text{for a plane layer}$$

$$\Phi(\beta) = \frac{4}{\pi}\int_0^{\pi/2} \sin^2\vartheta d\vartheta \int_0^1 e^{-\frac{\beta}{\sin\vartheta}\mu} \frac{\mu}{\sqrt{1-\mu^2}} d\mu \quad \text{for a cylindrical block}$$

β	Φ (β)		β	Φ (β)	
	plane layer	cylindrical block		plane layer	cylindrical block
0	1	1	1,50	0,1135	0,2720
0,10	0,8326	0,9058	2,00	0,06026	0,1856
0,20	0,7038	0,8228	2,50	0,03260	0,1310
0,30	0,6000	0,7488	3,00	0,01786	0,0952
0,40	0,5146	0,6826	4,00	0,00552	0,0544
0,50	0,4430	0,6233	5,00	0,00176	0,0329
0,60	0,3832	0,5703	6,00	0,00057	0,0225
0,70	0,3322	0,5219	7,00	0,00019	0,0154
0,80	0,2886	0,4787	8,00	0,00006	0,0080
0,90	0,2514	0,4392	9,00	0,00002	0,0040
1,00	0,2194	0,4060	10,00	0,00001	0

TABLE VIII

Mutual Shielding Factor for Resonances: $\eta = 1/(1 + L\Sigma_S^0)$ **(**Σ_S^0 **is the Scattering Cross Section in the Moderator, and L is the Mean Free Path of the Neutron in the Moderator for Collision With a Block)**

$$R(\eta) = \frac{1}{\sqrt{\pi}} \frac{\eta}{\left\langle \frac{\overline{l'}}{\sqrt{\overline{l'}}} \right\rangle} \int_0^\infty \frac{1-\Phi\left(\frac{1}{y^2}\right)}{1-\eta\Phi\left(\frac{1}{y^2}\right)} dy.$$

η	$R(\eta)$ plane layer	$R(\eta)$ cylindrical block	η	$R(\eta)$ plane layer	$R(\eta)$ cylindrical block
0	1	1	0,7	0,5235	0,5610
0,1	0,9435	0,9512	0,8	0,4244	0,4604
0,2	0,8835	0,8995	0,9	0,2981	0,3277
0,3	0,8212	0,8441	0,95	0,2104	0,2328
0,4	0,7552	0,7832	0,975	0,1486	0,1646
0,5	0,6853	0,7185	1,00	0	0
0,6	0,6086	0,6451			

TABLE IX

Self-Shielding Coefficients B(β**);** $\beta = \Sigma d$ **(**Σ **is the Total Cross Section in the Block, and** <u>**d**</u> **is the Block Thickness)**

$$B(\beta) = \frac{\frac{1}{2} - E_3(\beta)}{\beta} \quad \text{for a plane layer}$$

$$B(\beta) = \frac{4}{\pi} \int_0^{\pi/2} \cos\psi d\psi \int_0^{\pi/2} \sin^2\vartheta \frac{1-e^{-\beta\frac{\cos\psi}{\sin\vartheta}}}{\beta} d\vartheta \quad \text{for a cylindrical block}$$

β	$B(\beta)$ plane layer	$B(\beta)$ cylindrical block	β	$B(\beta)$ plane layer	$B(\beta)$ cylindrical block
0	1	1	3,00	0,1638	0,3016
0,25	0,7013	0,8572	4,00	0,1243	0,2365
0,50	0,5568	0,7525	5,00	0,0998	0,1932
0,75	0,4603	0,6665	6,00	0,0833	0,1628
1,00	0,3903	0,5960	7.00	0,0714	0,1405
1,50	0,2955	0,4867	8,00	0,0625	0,1234
2,00	0,2349	0,4071	9,00	0,0555	0,1100
2,50	0,1935	0,3480	10,00	0,0500	0,0992

TABLE X

The Function T(α, β)

$$T(\alpha;\ \beta)=\int_0^{\infty} e^{-\xi}\ln\sqrt{1+\alpha\xi+\beta\xi^2}\,d\xi.$$

α \ β	0	0,2	0,4	0,6	0,8	1,0
0	0	0,2498	0,4000	0,5149	0,6101	0,6919
0,1	0,0918	0,3088	0,4481	0,5565	0,6470	0,7252
0,2	0,1710	0,3635	0,4934	0,5960	0,6823	0,7573
0,4	0,3045	0,4626	0,5771	0,6697	0,7488	0,8182
0,6	0,4159	0,5506	0,6529	0,7375	0,8104	0,8750
0,8	0,5123	0,6300	0,7225	0,8002	0,8680	0,9284
1,00	0,5977	0,7024	0,7867	0,8587	0,9221	0,9789

TABLE XI

Φ(a) as a Function of the Dimensionless Block Radius a

$$\Phi(a)=\frac{4}{\pi}\int_0^{\pi/2}\cos\psi d\psi\int_0^{\pi/2}\sin^2\vartheta d\vartheta T\left(\frac{2\sin\vartheta\cos\psi}{a};\ \frac{\sin^2\vartheta}{a^2}\right).$$

a	$\Phi(a)$	a	$\Phi(a)$
0	∞	0,60	0,6965
0,025	2,6436	0,70	0,6271
0,050	2,3050	0,80	0,5711
0,10	1,8353	0,90	0,5243
0,20	1,3339	1,0	0,4847
0,30	1,0697	5,0	0,1240
0,40	0,9026	10,0	0,0642
0,50	0,7844	∞	0

TABLE XII

q^0_∞(a) as a Function of the Dimensionless Block Radius a

$$q^0_\infty(a) = \left(\frac{4}{3a} - \Phi(a)\right)^{-1}.$$

a	$q^0_\infty(a)$	a	$q^0_\infty(a)$
0	0	0,60	0,6554
0,025	0,01973	0,70	0,7827
0,050	0,04105	0,80	0,9127
0,10	0,08697	0,90	1,0447
0,20	0,1875	1,00	1,1784
0,30	0,2963	5,00	7,0092
0,40	0,4114	10,00	14,456
0,50	0,5313	∞	∞

TABLE XIII

The Neutron Transmission Function A (β) in the Block; β = Σd (Σ is the Total Cross Section in the Block, and d is the Block Thickness)

$$A(\beta) = 1 - 2E_3(\beta) \text{ for a plane layer}$$

$$A(\beta) = \frac{4}{\pi}\int_0^{\pi/2} \cos\psi\, d\psi \int_0^{\pi/2} \sin^2\vartheta \left(1 - e^{-\beta\frac{\cos\psi}{\sin\vartheta}}\right) d\vartheta \text{ for a cylindrical block}$$

β	A (β)		β	A (β)	
	plane layer	cylindrical block		plane layer	cylindrical block
0	0	0	3,00	0,9828	0,9120
0,25	0,3506	0,2160	4,00	0,9944	0,9536
0,50	0,5568	0,3793	5,00	0,9980	0,9737
0,75	0,6904	0,5039	6,00	0,9996	0,9846
1,00	0,7806	0,6008	7,00	1,0000	0,9914
1,50	0,8865	0,7359	8,00	1,0000	0,9951
2,00	0,9396	0,8207	9,00	1,0000	0,9979
2,50	0,9675	0,8770	10,00	1,0000	1,0000

TABLE XIV

F(β) for a Plane Layer; β = Σd (Σ is the Total Cross Section, and d is the Thickness of the Absorbing Layer)

$$F(\beta) = 1 - 3E_4(\beta).$$

β	F (β)	β	F (β)
0	0	1,5	0,8620
0,1	0,1369	2,00	0,9249
0,2	0,2518	2,50	0,9587
0,3	0,3493	3,00	0,9770
0,4	0,4327	4,00	0,9927
0,5	0,5044	5,00	0,9977
0,6	0,5662	6,00	0,9992
0,7	0,6196	7,00	0,9997
0,8	0,6661	8,00	0,9999
0,9	0,7066	9,00	1,0000
1,00	0,7418	10,00	1,0000

TABLE XV

F(a, B) for a Cylindrical Block; a is the Dimensionless Block Radius, B = Σd (Σ is the Total Cross Section, and d is the Block Diameter)

$$F(a, \beta)=\frac{4}{\Phi(a)}\int_0^{\pi/2}\cos\psi\, d\psi\int_0^{\pi/2}\sin^2\vartheta\, T\left(\frac{2\sin\vartheta\cos\psi}{a};\ \frac{\sin^2\vartheta}{a^2}\right)\left(1-e^{-\beta\frac{\cos\psi}{\sin\vartheta}}\right)d\vartheta.$$

a \ β	0	0,2	0,4	0,6	0,8	1,0	1,5	2,0	3,0	4,0	5,0	6,0	7,0	8,0
0	0	0,1766	0,3164	0,4289	0,5204	0,5953	0,7300	0,8142	0,9034	0,9434	0,9627	0,9734	0,9800	0,9852
0,025	0	0,1723	0,3101	0,4219	0,5131	0,5879	0,7230	0,8080	0,8988	0,9398	0,9600	0,9715	0,9785	0,9835
0,050	0	0,1710	0,3084	0,4200	0,5112	0,5862	0,7216	0,8067	0,8980	0,9387	0,9594	0,9707	0,9779	0,9824
0,100	0	0,1711	0,3090	0,4208	0,5089	0,5874	0,7235	0,8085	0,8995	0,9405	0,9608	0,9720	0,9789	0,9835
0,200	0	0,1715	0,3097	0,4223	0,5144	0,5900	0,7266	0,8123	0,9032	0,9437	0,9632	0,9737	0,9809	0,9852
0,300	0	0,1718	0,3104	0,4232	0,5157	0,5915	0,7280	0,8141	0,9051	0,9454	0,9645	0,9750	0,9819	0,9857
0,400	0	0,1720	0,3109	0,4240	0,5166	0,5925	0,7292	0,8153	0,9062	0,9463	0,9656	0,9758	0,9825	0,9862
0,500	0	0,1722	0,3114	0,4247	0,5172	0,5933	0,7304	0,8163	0,9071	0,9471	0,9664	0,9766	0,9830	0,9866
0,600	0	0,1725	0,3118	0,4253	0,5179	0,5941	0,7314	0,8174	0,9080	0,9478	0,9672	0,9775	0,9834	0,9870
0,700	0	0,1726	0,3122	0,4259	0,5185	0,5947	0,7322	0,8183	0,9088	0,9485	0,9678	0,9782	0,9836	0,9873
0,800	0	0,1728	0,3125	0,4260	0,5191	0,5953	0,7330	0,8192	0,9095	0,9491	0,9684	0,9787	0,9838	0,9876
0,900	0	0,1729	0,3127	0,4263	0,5197	0,5959	0,7336	0,8199	0,9102	0,9498	0,9687	0,9790	0,9840	0,9878
1,000	0	0,1731	0,3130	0,4264	0,5202	0,5965	0,7343	0,8205	0,9108	0,9503	0,9690	0,9793	0,9842	0,9880
5,000	0	0,1766	0,3192	0,4353	0,5300	0,6078	0,7464	0,8328	0,9215	0,9597	0,9766	0,9851	0,9896	0,9924
10,000	0	0,1778	0,3215	0,4382	0,5334	0,6115	0,7507	0,8370	0,9259	0,9630	0,9795	0,9874	0,9915	0,9939
∞	0	0,1809	0,3254	0,4430	0,5388	0,6170	0,7567	0,8425	0,9306	0,9668	0,9826	0,9899	0,9936	0,9955

TABLE XVI

Comparison of $\Phi(a)$ and $\ln\frac{1}{a}$ (a is the Dimensionless Radius of the Cylindrical Block)

$$\Phi(a)=\frac{4}{\pi}\int_0^{\pi/2}\cos\psi d\psi\int_0^{\pi/2}\sin^2\vartheta d\vartheta\, T\left(\frac{2\sin\vartheta\cos\psi}{a};\ \frac{\sin^2\vartheta}{a^2}\right).$$

a	$\Phi(a)$	$\ln\frac{1}{a}$	$\Phi(a)/\ln\frac{1}{a}$
0	∞	∞	0,7853
0,025	2,6436	3,6889	0,7166
0,050	2,3050	2,9957	0,7694
0,10	1,8353	2,3026	0,7971
0,20	1,3339	1,6094	0,8288
0,30	1,0697	1,2040	0,8884
0,40	0,9026	0,9163	0,9850
0,50	0,7844	0,6931	1,1318
0,60	0,6965	0,5108	1,3635
0,70	0,6271	0,3567	1,7581
0,80	0,5711	0,2231	2,5598
0,90	0,5243	0,1054	4,9743
1,00	0,4847	0	$\pm\infty$
5,00	0,1240	$-1,6094$	$-0,07705$
10,00	0,0642	$-2,3026$	$-0,02788$
∞	0	$-\infty$	0

LITERATURE CITED

[1] A.I. Alikhanov, V.V. Vladimirskii, et al., "Experimental heavy-water physical reactor," Coll: Reactor Design and Theory (1955).

[2] E.L. Ince, Ordinary Differential Equations [Russian translation] (DNTVU, 1939).

[3] A.I. Akhiezer and I.Ia. Pomeranchuk, Some Problems in Nuclear Theory (GITTL, 1950).

[4] G.A. Bat', "Calculations for a heterogeneous reactor with few blocks," Coll: Reactor Design and Theory (1955).

[5] G.A. Bat' and D.F. Zaretskii, "Review of effective boundary conditions," J. Atomic Energy (USSR) (1958) (in press).

[6] D.I. Blokhintsev, N.A. Dollezhal' and A.K. Krasin, "Reactor for an electric power plant of the Acad. Sci. USSR," J. Atomic Energy (USSR) 1 (1956).

[7] D.I. Blokhintsev, M.E. Minashin and Iu.A. Sergeev, "Physical and thermal reactor calculations for an electric power plant of the Acad. Sci. USSR," J. Atomic Energy (USSR) 1 (1956).

[8] D.I. Blokhintsev and N.A. Nikolaev, "The first atomic electric power plant of the USSR and paths of development of atomic energy," Coll: Reactor Design and Theory (1955).

[9] D.I. Blokhintsev, "The fundamentals of the physics and engineering of atomic energy" (Report at meeting of the Academy of Sciences, Ukrainian SSR), Ukr. Fiz. Zhurn. 1, 3, 209 (1956).

[10] N.N. Bogoliubov, Problems of Dynamical Theory in Statistical Mechanics (Gostekhizdat, 1946).

[11] V.S. Vladimirov, "Concerning a certain integro-differential equation," Izv. AN SSSR, Ser. Matem. 21 (1957).

[12] V.S. Vladimirov, "On the integro-differential equation of particle transport," Izv. AN SSSR, Ser. Matem. 21 (1957).

[13] V.S. Vladimirov, "On the use of the Monte Carlo method to find the lowest eigenvalue and eigenfunction belonging to it for a linear integral equation," Teoriia Veroiatnostei i ee Primenenie 1, 1 (1956).

[14] V.S. Vladimirov, "An approximate solution of a certain boundary value problem in second-order differential equations," Prikl. Matem. i Mekh. XIX, 3 (1955).

[15] V.S. Vladimirov, "Numerical solution of the kinetic equation for a sphere," Coll: Calculational Mathematics,No. 3 (1955).

[16] A.D. Galanin, "Calculation of the thermal utilization coefficient in a heterogeneous reactor," Coll: Reactor Design and Theory (1955).

[17] A.D. Galanin, "The use of the effective boundary condition method to calculate the critical size of a reactor," Coll: Reactor Design and Theory (1955).

[18] A.D. Galanin, "Critical size of a heterogeneous reactor with a small number of blocks," Coll: Reactor Design and Theory (1955).

[19] A.D. Galanin, "Neutron absorption and multiplication during slowing-down in the two-group theory," Coll: Reactor Design and Theory (1955).

[20] A.D. Galanin, "The theory of thermal-neutron nuclear reactors," Supplements No. 2 and 3 to J. Atomic Energy (USSR) (1958).

[21] T. R. Gantmaher, Matrix Theory [Russian translation] (GITTL, 1953).

[22] A.O. Gel'fand, The Calculus of Finite Differences (GITTL, 1952).

[23] S. Glasstone and M. Edlund, The Elements of Nuclear Reactor Theory [Russian translation] (IL, 1954).

[24] I.V. Gordeev, V.V. Orlov and T.Kh. Sedel'nikov, "The temperature dependence of the effective resonance absorption integral," J. Atomic Energy (USSR) 9 (1957).

[25] C. Goodman, The Science and Engineering of Nuclear Power, vols. I and II [Russian translation] (IL, 1948, 1950).

[26] I.I. Gurevich and I.Ia. Pomeranchuk, "The theory of resonance absorption in heterogeneous systems," Coll: Reactor Design and Theory (1955).

[27] M.B. Egiazarov, V.S. Dikarev and V.G. Madeev, "Measurement of neutron resonance absorption in a uranium-graphite lattice," (Academy of Sciences of the USSR Conference on the Peaceful Uses of Atomic Energy, Div. of Physicomathematical Sciences, July 1-5, 1955).

[28] A.I. Zhukov, O.V. Lokutsievskii, A.A. Samarskii and V.S. Riaben'kii, Report to the Third All-Union Mathematical Congress (1956).

[29] D.F. Zaretskii and D.D. Odintsov, "Effective boundary conditions for 'grey bodies'," Coll: Reactor Design and Theory (1955).

[30] B.L. Ioffe and L.B. Okun', "Burnup of fuel in nuclear reactors," J. Atomic Energy (USSR) 4 (1956).

[31] L.V. Kantorovich and V.I. Krylov, Approximation Methods in Higher Analysis (GITTL, 1950).

[32] I. Kaplan and D. Chernik, "Uranium-graphite lattices. The Brookhaven reactor," Coll: Experimental Reactors and Reactor Physics (1956). [In Russian]

[33] E. Kogen, "Review of neutron thermalization theory," Coll: Experimental Reactors and Reactor Physics (1956). [In Russian].

[34] L. Collatz, Numerical Methods for Solving Differential Equations [Russian translation] (IL, 1953).

[35] K.Ia. Kondrat'ev, The Sun's Radiant Energy (Gidrometeoizdat, 1954).

[36] A.K. Krasin and B.G. Dubovskii, "Beryllium reactor physics," J. Atomic Energy (USSR) 4 (1956).

[37] M.G. Krein, "The theory of self-adjoint extensions of semi-bounded Hermitian operators, with applications," Matematicheskii sb. 21 (63), 363 (1947).

[38] E.S. Kuznetsov, "Radiative equilibrium of a gas envelope surrounding an absolute black body," Izv. AN SSSR, Ser. Geofiz. 3 (1951).

[39] E.S. Kuznetsov, "A general method for constructing approximate equations of radiant energy transfer," Izv. AN SSSR, Ser. Geofiz. 4 (1951).

[40] E.S. Kuznetsov and B.V. Ovchinskii, "The results of a numerical solution of the integral equations for the theory of light scattered by the atmosphere," Tr. Geofiz. Inst. AN SSSR 4 (131) (1949).

[41] E.S. Kuznetsov, "The theory of nonhorizontal visibility," Izv. AN SSSR, Ser. Geofiz. 5 (1943).

[42] R. Courant and D. Hilbert, Methods of Mathematical Physics, vol. I [Russian translation] (GITTL, 1951).

[43] R. Courant, K. Friedrichs and H. Lewy, "On the difference equations of mathematical physics," Uspekhi Matem. Nauk VIII (1940).

[44] I.V. Kurchatov, Speech to the 1956 Session of the Supreme Soviet, USSR.

[45] I.V. Kurchatov, "Some questions concerning the development of atomic energy in the USSR (a lecture read on April 25, 1956 at the British Center for Scientific Research at Harwell), J. Atomic Energy (USSR) 3 (1956).

[46] O.A. Ladyzhenskaia, The Mixed Problem for the Hyperbolic Equation (GITTL, 1953).

[47] O.V. Lokutsievskii, "Numerical methods for solving partial differential equations" (All-Union Conference on Functional Analysis and Its Application), Uspekhi Matem. Nauk XI, 3 (69), 224 (1956).

[48] L.D. Landau, N.N. Meiman and I.M. Khalatnikov, "Numerical integration of partial differential equations by net methods," Proceedings of the 3rd All-Union Mathematical Congress, II (1956).

[49] L.A. Liusternik, "A remark on the numerical solution of Laplace boundary value problems and the calculation of eigenvalues by net methods," MIAN SSSR XX (1947).

[50] G.I. Marchuk, "On the question of the multigroup method for nuclear reactor calculations," Supplement No. 1 to J. Atomic Energy (USSR) (1958).

[51] G.I. Marchuk, "Finite-difference diffusion equations," Supplement No. 1 to J. Atomic Energy (USSR) (1958).

[52] G.I. Marchuk, "The multigroup method for calculating a reactor for an electric power plant," J. Atomic Energy (USSR) 2 (1956).

[53] G.I. Marchuk, "On approximate methods for nuclear reactor calculations" (Collection of Speeches Given at the Academy of Sciences USSR Meeting of 1955).

[54] Nuclear Reactors (Publication of the US AEC) [Russian translation] (IL, 1950) part 1.

[55] N.N. Meiman, "Some applications of finite difference methods to differential equations," Proceedings of the Third All-Union Mathematical Congress, I (1956) p. 60.

[56] A.E. Milne, Numerical Analysis [Russian translation] (IL, 1951).

[57] A.E. Milne, Numerical Solution of Differential Equations [Russian translation] (IL, 1955).

[58] S.G. Mikhlin, Direct Methods in Mathematical Physics (Gostekhizdat, 1951).

[59] M.V. Nikolaeva, "On Southwell's relaxation method," Trudy MIAN SSSR XXVIII (1949).

[60] D. Okrent et al., "Review of theoretical and experimental fundamentals of fast-neutron reactor physics," Coll: Experimental Reactors and Reactor Physics (1956).

[61] V.V. Orlov, "Mutual screening of layers of a resonance neutron absorber," J. Atomic Energy (USSR) (1958) (in press).

[62] V.V. Orlov and V.P. Kocherin, "Neutron slowing-down length," J. Atomic Energy (USSR) (1958) (in press).

[63] V.V. Orlov, T.V. Goloshvili and A.I. Vaskin, "Resonance absorption of neutrons in a block," J. Atomic Energy (USSR) (1958) (in press).

[64] R. Peierls, The Quantum Theory of Solids [Russian translation] (IL, 1956).

[65] Iu.V. Petrov, "Resonance absorption for a dense distribution of small blocks," J. Atomic Energy (USSR) 4 (1957).

[66] I.G. Petrovskii, Lectures on Partial Differential Equations (Gostekhizdat, 1953).

[67] V.Ia. Pupko, "Relation between critical charge and critical volume for various types of reactors," Supplement No. 1 to J. Atomic Energy (USSR) (1958).

[68] I.M. Ryzhik and I.S. Gradshtein, Tables of Integrals, Sums, Series, and Products (GITTL, 1951).

[69] V.S. Riaben'kii and A.F. Filippov, On the Stability of Difference Equations (GITTL, 1956).

[70] A.A. Samarskii, Speech to the All-Union Congress on Functional Analysis and Its Applications (Moscow, 1956).

[71] V.K. Saul'ev, "Note on the problem of cylindrical reactor calculations," J. Atomic Energy (USSR) 7 (1957).

[72] V.K. Saul'ev, "On a method for the numerical integration of diffusion equations," Doklady Akad. Nauk SSSR 115, 6 (1957).

[73] V.V. Smelov, "On the question of neutron thermalization," J. Atomic Energy (USSR) 10 (1957).

[74] V.I. Smirnov, A Course in Higher Mathematics, vol. IV (GITTL, 1951).

[75] S.L. Sobolev, "Some contemporary problems in calculational mathematics," Proceedings of the Third All-Union Mathematical Congress, II (1956).

[76] S.L. Sobolev, The Equations of Mathematical Physics (Gostekhizdat, 1954).

[77] S.L. Sobolev, L.A. Liusternik and L.V. Kantorovich, "Functional analysis in calculational mathematics," Proceedings of the 3rd All-Union Mathematical Congress, II (1956).

[78] A.N. Tikhonov and A.A. Samarskii, Equations of Mathematical Physics (GITTL, 1953).

[79] A.N. Tikhonov and A.A. Samarskii, "On difference methods for equations with discontinuous coefficients," Doklady Akad. Nauk SSSR 108, 3 (1956).

[80] K.D. Tolstov, F.L. Shapiro and I.V. Shtranikh, Mean Neutron Velocities in Various Media (Academy of Sciences USSR Conference on the Peaceful Uses of Atomic Energy, Div. Physicomathematical Sciences, July 1-5, 1955).

[81] L.N. Usachev, Fast-Neutron Reactor Calculation Methods (dissertation) (1954).

[82] L.N. Usachev, "The equation for neutron importance, reactor kinetics, and perturbation theory," Coll: Reactor Design and Theory (1955).

[83] V.N. Faddeeva, Calculational Methods in Linear Algebra (GITTL, 1950).

[84] S.M. Feinberg, "Heterogeneous methods for reactor calculations (Review of results and comparison with experiment)," Coll: Reactor Design and Theory (1955).

[85] S.M. Feinberg, "Some theoretical problems in a uranium-water lattice," Report to the Academy of Sciences USSR (1956).

[86] V.S. Fursov, Work of the Academy of Sciences USSR on Uranium-Graphite Reactors (Academy of Sciences USSR Conference on the Peaceful Uses of Atomic Energy, plenary session, July 1-5, 1955).

[87] S. Chandrasekhar, Radiation Transfer [Russian translation] (IL, 1953).

[88] D. Chernik, "The theory of uranium-water lattices," Coll: Experimental Reactors and Reactor Physics (1956).

[89] Ia.V. Shevelev, "Neutron diffusion in a plane uranium-water lattice," J. Atomic Energy (USSR) II, 3 (1957).

[90] D.V. Shirkov, "The synthetic-kernel method for neutron diffusion problems in a hydrogenous medium," Supplement No. 1 to J. Atomic Energy (USSR) (1958).

[91] D.V. Shirkov, "The synthetic-kernel method for neutron diffusion problems in nonhydrogenous moderators," Supplement No. 1 to J. Atomic Energy (USSR) (1958).

[92] D. Hughes, Pile Neutron Research [Russian translation] (IL, 1954).

[93] E. Amaldi and E. Fermi, "On the absorption and the diffusion of slow neutrons," Phys. Rev. 50, 899 (1936).

[94] H.A. Bethe, L. Tonks and H. Hurwitz, "Neutron penetration and slowing down at intermediate distances through medium and heavy nuclei," Phys. Rev. 80, No. 1, 11 (1950).

[95] G. O'Brien, M. Hyman and S. Kaplan, "Study of numerical solution of partial differential equations," J. of Math. and Phys. XXIX, No. 4 (1951).

[96] I. Le Caine, "Application of a variational method to Milne's problem," Phys. Rev. 72, No. 7, 564 (1947).

[97] G.V. Dardel and R. Persson, "Determination of neutron resonance parameters from measurements of the absorption integral: application to the main resonance of uranium – 238," Nature 170, No. 4339, 1117 (1952).

[98] B. Davison and J.B. Sykes, Neutron Transport Theory (Oxford, 1957).

[99] J. Douglas, "On the relation between stability and convergence in the numerical solution of linear, parabolic and hyperbolic differential equations," J. Soc. Indust. Appl. Math. 4 (1956).

[100] R. Eddy, "Stability in the numerical solution of initial value problems in partial differential equations," Naval Ordnance Laboratory Memorandum 10 (1949).

[101] R. Erlich and H. Hurwitz, "Multigroup methods for neutron diffusion problems," Nucleonics 12, No. 2, 23 (1954).

[102] E. Fermi, Ricerca Scient. 2, No. 1-2, 13 (1936).

[103] E. Fort and S. Frankel, "Stability conditions in the numerical treatment of parabolic differential equations," MTAC 7, No. 43, 135 (1953).

[104] S. Frankel, "Convergence rates of iterative treatments of partial differential equations," MTAC 4, (1950).

[105] K. Fuchs, "Perturbation theory in neutron multiplication problems," Proc. Phys. Soc. Sect. A, 62, 791 (1949).

[106] F. Hildebrand, "On the convergence of numerical solutions of the heat-flow equations," J. of Math. and Phys. XXXI (1952).

[107] G. Holte, "On the space-energy distribution of slowed-down neutrons," Ark. Fysik 2, 6, 523 (1950).

[108] H. Hurwitz and R. Erlich, Comparison of Theory and Experiment for Intermediate Assemblies (Report No. 608 to the International Conference on the Peaceful Uses of Atomic Energy, 1955).

[109] H. Hurwitz, M.S. Nelkin and G.J. Habetler, "Neutron thermalization," Nuclear Science and Engineering 1, 280 (1956).

[110] H. Hurwitz and P.F. Zweifel, "Slowing down of neutrons by hydrogenous moderators," J. of Appl. Phys. 26, No. 8, 923 (1955).

[111] M. Human, "On the noniterative numerical solution of boundary value problems," Appl. Sci. Res. B. 2 (1952).

[112] F. John, "On integration of parabolic equations by difference methods," Comm. Pure Appl. Math. 5 (1952).

[113] O. Kellog, "On the existence of closure of sets of characteristic functions," Math. Ann. 86, 14 (1922).

[114] V. Kourganoff and J.W. Busbridge, Basic Methods in Transfer Problems (Oxford, 1952).

[115] H. Lewy, "On the convergence of solutions of difference equations," Coll: Studies and Essays Presented to R. Courant on his 60th Birthday (New York, 1948).

[116] H. Liebmann, "Die angenährten Ermittlung harmonischer Funktionen und komformer Abbildung," Sitzber. Bayr. Acad. Wiss. Math. Phys. kl. (1918).

[117] M.E. Mandl and J. Howlett, A Method for Calculating the Critical Mass of an Intermediate Reactor (Report No. 430 to the International Conference on Peaceful Uses of Atomic Energy, 1955).

[118] R.E. Marchak, H. Brooks and H. Hurwitz, "Introduction to the theory of diffusion and slowing down of neutrons," Nucleonics 4, No. 5, 6 (1949); 5, No. 1, 2 (1949).

[119] R.E. Marchak, "A note on the spherical harmonic method applied to the Milne problem for a sphere," Phys. Rev. 71, No. 7, 443 (1947).

[120] R.E. Marchak, "Theory of the slowing down of neutrons by elastic collisions with atomic nuclei," Rev. Mod. Phys. 19, No. 3, 185 (1947).

[121] R.E. Marchak, "Variational method for asymptotic neutron densities," Phys. Rev. 71, No. 10, 688 (1947).

[122] C. Mark, "Neutron density near a plane surface," Phys. Rev. 72, No. 7, 558 (1947).

[123] E. Melkonian, W.W. Havens Jr. and L.J. Rainwater, "Slow neutron velocity spectrometer studies. V, Re, Ta, Ru, Cr, Ca," Phys. Rev. 92, No. 6, 702 (1953).

[124] R. Mertens, "Sur la diffusion multiple de particules chargés," C. r. Acad. Sci. 236, No. 18, 1753 (1953)

[125] R. Mertens, "Sur la resolution en n-ieme approximation des problemes de diffusion multiple," C. r. Acad. Sci. 237, No. 25, 1644 (1953).

[126] R. Mertens, "Sur le probleme de Milne pour une loi de diffusion simple de Rayleigh," C. r. Acad. Sci. 238, No. 1, 53 (1954).

[127] W. Meyer, "Methode zur angenährten Lösung von Eigenwertproblem mit Anwendungen auf Schwingungsprobleme," Ann. Physik 8, 297 (1931).

[128] R. Peierls, "Critical conditions in neutron multiplication," Proc. Camb. Phil. Soc. 35, 610 (1939).

[129] G. Placzek, "On the theory of the slowing down of neutrons in heavy substances," Phys. Rev. 69, No. 9 and 10, 423 (1946).

[130] G. Placzek and W. Seidel, "Milne's problem in transport theory," Phys. Rev. 72, No. 7, 550 (1947).

[131] G. Placzek and G. Volkoff, Canad. J. Research Sect. A, 25 (1947).

[132] F. Rellich, "Störungstheorie der Spektralzerlegung," Math. Ann. 113, No. 4 (1936).

[133] J. Sheldon, "On the numerical solution of elliptic difference equations," MTAC IX, No. 51, 101 (1955).

[134] B. Shortley, C. Weller et al., "Numerical solution of axisymmetrical problems, with applications to electrostatics and torsion," J. Appl. Physics 18, No. 1, 116 (1947).

[135] G. Soodak and K. Campbell, Elementary Pile Theory (New York, 1950).

[136] R. Southwell, Relaxation Methods in Engineering Science (Oxford Univ. Press, 1943).

[137] B. Spinrad, "Anisotropic diffusion lengths in diffusion theory," J. Appl. Phys. 26, No. 5, 548 (1955).

[138] A. Thompson, "Numerical computation of neutron distribution and critical size," J. Appl. Phys. 22, No. 10, 1223 (1951).

[139] M. Verde and G.C. Wick, "Some stationary distributions of neutrons in an infinite medium," Phys. Rev. 71, No. 12, 852 (1947).

[140] M. Wang and E. Guth, "On the theory of multiple scattering, particularly of charged particles," Phys. Rev. 84, No. 6, 1092 (1951).

[141] A. Weinberg and M. Aivin, "Current status of nuclear reactor theory," Am. J. Phys. 20, No. 7, 401 (1952).

[142] G.C. Wick, "On the space distribution of slow neutrons," Phys. Rev. 75, No. 5, 738 (1949).

[143] G.C. Wick, Z. Physik 121 (1943).

[144] N. Wiener and E. Hopf, "Über eine Klasse singulärer Integralgleichungen," Berliner Ber. Math. Phys. Kl. 696 (1931).

[145] E.P. Wigner, "Theoretical physics in the Chicago Metallurgical Laboratory," J. Appl. Phys. 17, No. 11, 857 (1946).

[146] E.J. Wilkins, R.L. Hellens and P.F. Zweifel, Status of Experimental and Theoretical Information on Neutron Slowing-Down Distributions in Hydrogenous Media (Report No. 597 to the International Conference on Peaceful Uses of Atomic Energy, 1955).